Multi-Sensor Systems and Data Fusion in Remote Sensing

Multi-Sensor Systems and Data Fusion in Remote Sensing

Editors

Piotr Kaniewski
Mateusz Pasternak
Stefano Mattoccia

MDPI • Basel • Beijing • Wuhan • Barcelona • Belgrade • Manchester • Tokyo • Cluj • Tianjin

Editors
Piotr Kaniewski
Military University of Technology
Poland

Mateusz Pasternak
Military University of Technology
Poland

Stefano Mattoccia
University of Bologna
Italy

Editorial Office
MDPI
St. Alban-Anlage 66
4052 Basel, Switzerland

This is a reprint of articles from the Special Issue published online in the open access journal *Remote Sensing* (ISSN 2072-4292) (available at: https://www.mdpi.com/journal/remotesensing/special_issues/Multisensors_rs).

For citation purposes, cite each article independently as indicated on the article page online and as indicated below:

LastName, A.A.; LastName, B.B.; LastName, C.C. Article Title. *Journal Name* **Year**, *Volume Number*, Page Range.

ISBN 978-3-0365-6798-3 (Hbk)
ISBN 978-3-0365-6799-0 (PDF)

Cover image courtesy of Mateusz Pasternak

Contents

About the Editors

Piotr Kaniewski

Prof. Dr. Piotr Kaniewski studied Radiotechnical Systems of Aircraft at The Military University of Technology in Warsaw. He graduated and received his MSc in 1994, PhD in 1998, and was habilitated in 2011. He worked as an Engineer, Assistant, Assistant Professor, and currently works as an Associate Professor at The Faculty of Electronics at The Military University of Technology, where, since 2012, he is the Director of the Institute of Radioelectronics. His current research is focused on navigation systems dedicated for special purposes, such as supporting synthetic aperture radars (MOCO), supporting ground penetrating radars (GPR) with accurate scanning trajectory information, distributed navigation algorithms for UAV swarms, and navigation systems for GNSS denied environments, especially using SLAM on UAVs and UWB ranging modules for indoor navigation.

Mateusz Pasternak

Prof. Dr. Mateusz Pasternak received his Master of Science degree in Electronics from the Department of Electronics of the Military University of Technology (MUT) in 1989. After that, he joined the Acoustoelectronic Faculty at the Institute of Applied Physics MUT as a candidate for a doctor's degree. Initially, he started his research activity from the comprehensive study of surface acoustic wave (SAW) propagation and scattering properties that was completed with the design and manufacture of a variety of advanced SAW devices such as special high-frequency SAW filters and convolvers for radar applications. He received his Ph. D. degree in July 1995. Since 2007, his field of interest has been extended with ultrawideband radars technology, especially ground-penetrating radars (GPR). In the frame of these efforts, four GPR systems were designed and manufactured. Simultaneously, he continued the research on SAW device theory and applications focusing on SAW chemical sensors. As a result of this work, the next-generation sensor series were developed. In 2022, became a Professor. Currently, he is employed in the Institute of Radioelectronics, MUT. He has published about 100 scientific papers and 4 books.

Stefano Mattoccia

Stefano Mattoccia is currently an Associate Professor at the Department of Computer Science and Engineering of the University of Bologna. His research activity concerns computer vision, mainly focusing on depth perception and related tasks. In these fields, he co-authored more than 130 scientific publications. He is a Senior IEEE member.

Preface to "Multi-Sensor Systems and Data Fusion in Remote Sensing"

This book collects papers belonging to the Special Issue of MDPI Remote Sensing entitled "Multi-Sensor Systems and Data Fusion in Remote Sensing". The currently observed technological progress, including the emergence of new sensors, development of sophisticated platforms for those sensors, and advances in signal and data processing, creates an opportunity for scientists and engineers to develop new and more capable integrated multi-sensor systems. Extended frequency bands, improved resolutions, and data rates of the new sensors as well as the common use of distributed sensors increase the influx of data in contemporary multi-sensor systems. At the same time, the users' expectations with respect to the size of the observed area or volume, data resolution, accuracy, speed of operation, and functionality of remote sensing systems are still increasing. These facts pose new challenges for the data fusion algorithms that must often employ the newest achievements from the areas of big data mining, statistical estimation, artificial intelligence, etc. The presented material provides a fresh insight into the newest developments in the fields of multi-sensor systems and data fusion and will be of interest to the entire remote sensing community.

Piotr Kaniewski, Mateusz Pasternak, and Stefano Mattoccia
Editors

 MDPI

Article

Hexagonal Grid-Based Framework for Mobile Robot Navigation

Piotr Duszak [1,*], Barbara Siemiątkowska [1] and Rafał Więckowski [2]

[1] Institute of Automatic Control and Robotics, Warsaw University of Technology, 02-525 Warsaw, Poland; barbara.siemiatkowska@pw.edu.pl
[2] Łukasiewicz Research Network—Industrial Research Institute for Automation and Measurements PIAP, 02-486 Warsaw, Poland; rafal.wieckowski@piap.lukasiewicz.gov.pl
* Correspondence: piotr.duszak@pw.edu.pl

Abstract: The paper addresses the problem of mobile robots' navigation using a hexagonal lattice. We carried out experiments in which we used a vehicle equipped with a set of sensors. Based on the data, a traversable map was created. The experimental results proved that hexagonal maps of an environment can be easily built based on sensor readings. The path planning method has many advantages: the situation in which obstacles surround the position of the robot or the target is easily detected, and we can influence the properties of the path, e.g., the distance from obstacles or the type of surface can be taken into account. A path can be smoothed more easily than with a rectangular grid.

Keywords: mapping; data fusion; path planning; hexagonal grid

Citation: Duszak, P.; Siemiątkowska, B.; Więckowski, R. Hexagonal Grid-Based Framework for Mobile Robot Navigation. *Remote Sens.* **2021**, *13*, 4216. https://doi.org/10.3390/rs13214216

Academic Editor: Piotr Kaniewski

Received: 6 August 2021
Accepted: 18 October 2021
Published: 21 October 2021

1. Introduction

Over the last few decades, we have observed a rapid development of mobile robotics. Robots such as autonomous vacuum cleaners and lawnmowers are used as standard equipment. The market for autonomous cars and transport within industrial plants is developing.

In 2020, DARPA launched a program called Racer (Robotic Autonomy in Complex Environments with Resiliency) (https://www.darpa.mil/news-events/2020-10-07 accessed on 15 October 2021).

The RACER program aims to develop universal solutions designed to work with various UGV platforms in challenging terrain, taking into account the terrain conditions, at least in terms of adjusting the UGV speed to the environmental conditions. The program aims to develop an autonomy system that will not limit the UGV platform while driving off-road. In other words, if the UGV platform has been designed to drive in rough terrain up to 70 km/h, the autonomy system is to be able to safely navigate the vehicle up to this speed. The program's primary objective is to develop an algorithm that adjusts the local path of the UGV in real time to maximalize the speed of the UGV while driving. Figure 1 presents an example of local path planning with traversability estimation. The algorithm adjusts the path of the UGV to avoid the snowdrift on the front of the platform.

Research on this type of autonomy system at the Łukasiewicz Research Network—Industrial Research Institute for Automation and Measurements PIAP has been conducted since the launch of the project in 2018 called ATENA—Autonomous system for terrain UGV platforms with the following leader function, implemented in the field of scientific research and development work for the defense and security of the state financed by the National Center for Research and Development in Poland under the program Future technologies for defense—a competition of young scientists. The off-road autonomy system was developed and adapted to work with an off-road car. As part of the project, two technology demonstrators were created, functional in following the leader based on the operation of the vision systems. Łukasiewicz—PIAP ATENA system demonstrator has been tested under real conditions in various weather conditions and is fully operative in the following leader function in rough terrain with obstacle avoidance up to 30 km/h. The main advantage of this system is the fact that the vehicle that is following the leader creates the traversability estimation in real time and calculates its own path to follow the leader, in other words, the following vehicle, not just repeating the leader path; it calculates its own adjustment to the set off-road ability. The developed system is equipped with a module for autonomous navigation in an unknown environment. We paid special attention to solving the issue of map construction and planning a collision-free path.

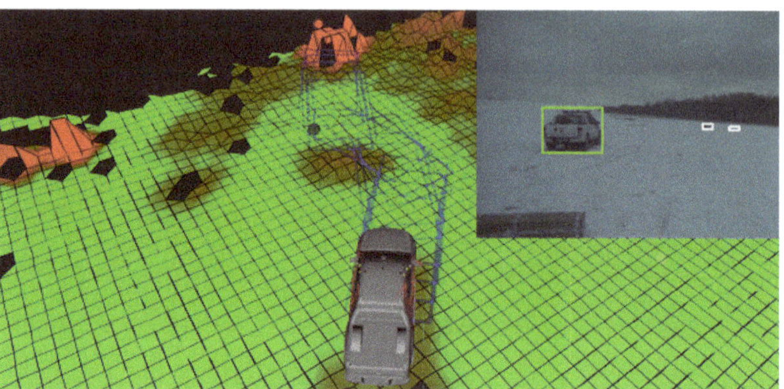

Figure 1. Visualization of path planning by a UGV in rough terrain in winter conditions. On the right corner is the image from the roof camera on the UGV.

Figure 2 presents the Łukasiewicz—PIAP ATENA system demonstrators on which the research was performed. The demonstration of the ATENA system contains the off-road car with the Łukasiewicz—PIAP drive-by-wire system and Łukasiewicz—PIAP autonomy controller. The drive-by-wire system allows control of the steering wheel, adjusts the velocity of the vehicle, and controls the brakes, without losing the ability to control the vehicle by a human.

Figure 2. Two Łukasiewicz—PIAP ATENA system demonstrators on the test terrain in winter conditions.

This article, which is an extended version of our conference paper [1], presents a method of path planning and building a map of the environment using hexagonal grids. We applied our approach to 3D data obtained using a set of sensors mounted on the ATENA system demonstrator. The data were collected on the premises of the Łukasiewicz Research Network—Industrial Research Institute for Automation and Measurements PIAP. We demonstrate that our algorithm allows us to build accurate models of the environment. The area of the collected data is a rectangle with a circumference of approximately 400 m (Figure 3). The created map is useful for path planning especially in unstructured and rough terrain when the ATENA system demonstrator must search for the lost leader in near historical localization of it or the case of a human leader when the traveling path must be absolutely different than the leader path. The vehicle must calculate its own path not based on the leader path, but based on leader localization.

Figure 3. The view of the terrain used in this article.

This paper is organized as follows: After the Introduction and the section discussing related work in Section 3, we briefly describe the mapping module. In Section 4, we present a collision-free path planning algorithm. Section 5 contains experimental results illustrating the advantages of our approach. The article concludes with a summary and bibliography.

2. Related Work

An essential part of mobile robots is the navigation technique [2]. In [2], it was underlined that the the system has to balance among accuracy, efficiency, and robustness. The navigation consists of three main modules: map building, localization, and collision-free path planning. A map of the environment is necessary for a mobile robot to perform its tasks. When mapping, a robot has to deal with different kinds of noise. Errors are divided into systematic, resulting from defects in the equipment, and nonsystematic, resulting from the conditions of use. Systematic odometry error is caused by a discretized sampling of wheel increments and wheel slippage. Nonsystematic errors result from terrain roughness and wheel slippage. Measurement noise problems also occur with sensors such as laser rangefinders, cameras, etc. Rapid and accurate techniques of data collection, calibration, and processing are required to improve the accuracy [2]. In the literature, the map building problem is described as a chicken-and-egg problem. The task is to build an accurate map of the environment based on the robot's position and determine the robot's pose based on the created map and the sensors' indication. The reduced accuracy of the map and robot pose have a negative influence on the execution of the path planning task. There are two kinds of navigation systems: reactive navigation and map-based navigation. In the first method, the mobile robot has no map and acts based on the senors' indication [3]. In the case of the map-based navigation, the robot is able to sense, plan, and act. It plans [3] an obstacle-free path to a predetermined destination. Usually, it is assumed that the planned path is optimal, e.g., shortest and fastest. The method consists of four steps:

- The robot observes the environment using sensors;
- Noise is removed;
- The robot determines its pose, and the map is updated;
- A collision-free path is planned.

The first three steps are performed in a loop. The path is replanned if previously undetected obstacles appear on the robot's path.

The choice of the path planning method is closely related to environment representation. The maps described in the literature are divided into two main groups: metric maps and nonmetric maps (topological and semantic). Metric maps are represented as a grid of cells [4] or as a set of features [5].

The grid-based map initially proposed in [4] is one of the primary methods of an environment representation. In this approach, an environment is divided into square areas, and an occupancy value is attached to the corresponding grid cell. Usually, Bayesian theory is used in order to update the occupancy value based on the sensors' reading. Grid-based representation requires an enormous amount of memory, but is able to represent the uniformed objects. The experimental results presented in [6] showed that the improved grid map allows the robot to plan a collision-free path and navigate safely in a static and dynamic environment. A square-based grid map is a popular approach in the ROS system using the Universal Grid Map Library [7]. This library was used in the ATENA system demonstrator with a traversability estimation cost map. The square-based grid map is widely use in path planning based on the Hybrid A* algorithm and uses estimated terrain traversability to find the path that optimizes both traversability and distance for the UGV [8]. A similar approach was tested in winter conditions in the ATENA system demonstrator; the results of the test were sufficient, but it is possible to optimize this approach with a hexagonal grid map. The square cells are also used in rough terrain with high vegetation density when the cells contain information about the "go" and "no go" information [9]. The grid map is also popular because of its ability to represent terrain by the 2.5D grid-type elevation map. Each cell holds a value of the height of the region. This approach can be used even in rough terrain [10]. A hexagonal grid map can be used in mapping static and dynamic environments. The shape of the cell is not important for dynamic approaches, but the advantages of the shape are preserved [11].

Feature-based maps are compact, so they do not require much memory. It is assumed that the features are predefined, so the structure of the environment has to be known in advance.

Square cells used in grid-based representation have some disadvantages. The distance from the center of a cell to the center of a diagonally adjacent cell is greater than the distance to the center of cells with which it shares an edge. Neighboring cells do not always share edges: diagonal cells contact only at a point. Curved shapes are not well represented on a rectangular lattice. In biological vision systems (for example, the human retina), photoreceptors are typically arranged in a hexagonal lattice. It has been shown that hexagonal grids have numerous advantages [12]. First of all, the distance between a given cell and its immediate neighbors is the same along any of the six main directions; curved structures are represented more accurately than by rectangular pixels. A smaller number of hexagonal pixels is required to represent the map. This allows reducing the computation time and required storage space. In [13,14], it was shown that for a given resolution capability of the sensors, hexagon sampling gives a smaller quantization error. In the paper [15], it was shown that that hexagonal grid map representation was better than the quadrangular grid map representation for cooperative robot exploration, but the problem of collision-free path planning in a real environment was not considered in this article.

In the case of a square grid, we can easily calculate the coordinates of the cell of the array corresponding to the point (x,y). For hexagonal meshes, the calculation is more complex. The lack of an effective method of representing hexagonal meshes and a simple transformation algorithm from Cartesian coordinates to hexagonal coordinates precluded their usage. In [16], the array set addressing (ASA) method was described. The approach is based on representing the hexagonal grid as two rectangular arrays. Figure 4 presents the method of hexagonal grid representation. Different arrays are represented using different colors (pink and blue). The arrays are distinguished using a single binary coordinate. The complete address of a cell in a hexagonal grid is uniquely represented by three coordinates:

$$(a, r, c) \in \{0, 1\} \times \mathbb{Z} \times \mathbb{Z} \tag{1}$$

where a—binary index of an array, r—row index, c—column number, and $\mathbb{Z}$—positive integers. The transformation from hexagonal (a, r, c) representation to Cartesian is defined by the formula:

$$\begin{aligned} x &= d \cdot \tfrac{a}{2} + c \\ y &= d\sqrt{3} \cdot (\tfrac{a}{2} + r) \end{aligned} \tag{2}$$

where d is the distance between the center of gravity and a vertex of the hexagon.

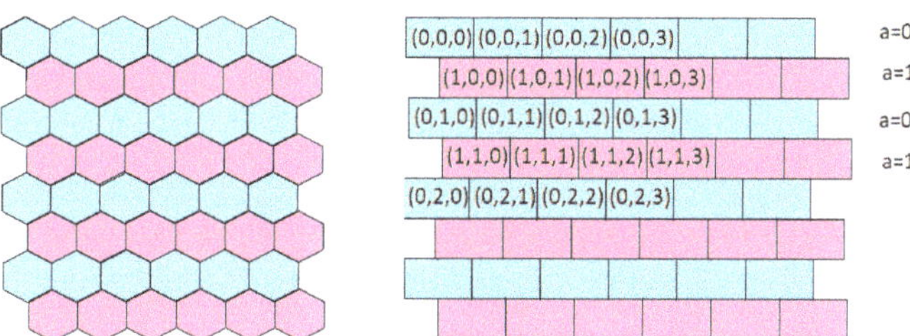

Figure 4. Array set addressing method.

An efficient method of transformation from Cartesian coordinates to hexagonal representation was presented in [16].

Figure 5 presents the addresses of the nearest neighbors of the cell (a, r, c) In this article, the hexagonal grid-based representation of the robot environment is presented. It is shown that hexagonal maps of an environment can be easily built based on sensors readings and is useful in collision-free path planning and mobile robot localization.

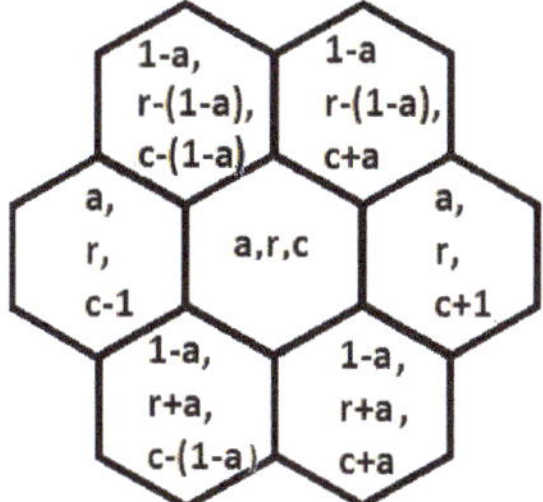

Figure 5. The addresses of the nearest neighbors of the cell (a, r, c).

3. Mapping

3.1. Sensors

The ATENA vehicle is equipped with a set of sensors located on the roof of the off-road car. The sensor set consists of three LiDAR sensors—Velodyne VLP-16 (sixteen layers, 100 m range), five Basler acA1920-48gc (50 fps at 2.3 MP resolution) cameras, and the Xsens MTi-G-710 IMU sensor.

LiDAR sensors are located on the roof of the vehicle in such a way as to create a synergy effect using all three sensors (Figure 6). The main central sensor is tilted by a dozen or so percent so that the cloud of points covers the shape of the road just in front of the ATENA vehicle. Two additional LiDAR sensors are placed on the roof divergently on the sides to form vertical lines of points, which enables better detection of horizontal objects such as barriers, fences, etc. The IMU sensor is placed in the geometric center of the vehicle.

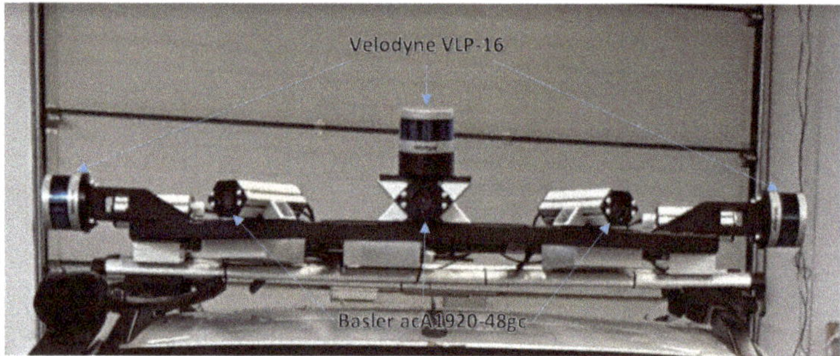

Figure 6. Sensors on the roof of the vehicle with the ATENA system.

The present research data were collected using the ATENA core operating with the ROS package. One of the core functionalities of the ATENA program is the fusion of data from Velodyne LiDARs, information from the Xsens MTi-G-710 IMU sensor, and the odometry data from the CAN network of the vehicle. The system builds a 3D world model around the vehicle and creates a map as it moves. For the ATENA system, the model is built in a closed area limited to a square of 50 m × 50 m. To prepare a map of the surroundings, this limitation was turned off, which allowed building a full map of the surroundings around the building.

This work used the resulting point cloud (Figure 7). The point cloud was created by fusing data from the three LiDAR sensors and data from odometry (measurement on the wheels by the onboard system of a vehicle and information from the IMU sensor).

Figure 7. Point cloud built by the Łukasiewicz—PIAP ATENA demonstrator.

3.2. Map Building

A point cloud can be useful for environment visualization, object detection or feature determination, and semantic segmentation [17,18], but it is not suitable for path planning. Therefore, there is a need to process it into a map. In our approach, a grid-based traversability [19] map was chosen. This is a very popular representation of the environment, and it is also very convenient for path planning. It can be used for algorithms such as A* [20], the diffusion method [21], and reinforcement learning path planning [22].

Two types of lattices were used: square and hexagonal. The former was used as a reference, and the latter is a new approach. There are reasons to believe that the latter is better. As shown in [1], straight lines, circles, and polynomials are statistically better represented on a hexagonal grid than on a square grid. This is important because straight lines are very common on maps, as buildings are composed of such lines. A robot's path, due to the need to avoid obstacles, can resemble a polynomial graph.

In the study described in [1], the quality of the representation of different curves (lines, circles, and polynomials) for square and hexagonal grids was compared. For lines, this was performed as follows. First, the parameters of the straight line were randomly generated. Then, it was drawn on both types of grids (with the same resolution). After this, the average distance between each cell belonging to the representation and the original straight line was calculated. This experiment was repeated 10,000 times, and the average error for all these lines was calculated. The quality of circles and polynomials was checked analogously. Summary results are presented in Table 1. As one can see, for each type of curve, the hexagonal grid represented it better.

Table 1. Mean error of the representation of different curves on the square and hexagonal lattice.

Curve Type	Mean Error for Square Grid	Mean Error for Hexagonal Grid
line	0.136	0.110
circle	0.138	0.110
polynomial	0.138	0.110

To create a grid-based map from a point cloud, we need discretization. Therefore, when creating a map on a square lattice, each point was assigned to a cell as follows:

$$\begin{aligned} x_s &= \lfloor x/w \rfloor \\ y_s &= \lfloor y/w \rfloor \end{aligned} \tag{3}$$

where:

x, y—original coordinates in the point cloud;

x_s, y_s—discrete coordinates in the square grid map;

w—width of one cell.

Obviously, several points may be assigned to one cell. The height of this cell is calculated based on the average, according to the formula:

$$z_s = \frac{1}{n}\Sigma_{i=1}^{n} z_i \tag{4}$$

where:

z_s—height of the cell in the square grid map;

n—number of points assigned to the cell;

z_i—original z coordinates for all points, which were assigned to a given cell.

The results can be seen in Figure 8. The scale near the map determines the elevation.

The first step necessary to build a map on a hexagonal grid is to convert the Cartesian coordinates to the coordinates used for hexagons. When creating a map, the so-called cube coordinate system [23,24] (this can be seen in Figure 9) is much more useful than the ASA. Therefore, the former was used when generating the map, and the latter was used when planning the path (a simple way to switch between these coordinate systems is given at the end of this subsection). The transformation between the Cartesian and cube systems is given by the following formulas:

$$\begin{aligned} x_c &= x \\ y_c &= \tfrac{\sqrt{3}}{2}y - \tfrac{1}{2}x \\ \zeta_c &= -y_c - x_c \end{aligned} \tag{5}$$

where:

x, y—original Cartesian coordinates in the point cloud;

x_c, y_c, ζ_c—continuous cube-hexagonal coordinates.

The next step is to discretize the variables. The following algorithm was used for this purpose [23]. First, auxiliary variables are calculated analogously to the square grid:

$$\begin{aligned} \hat{x}_h &= \lfloor x_c/w \rfloor \\ \hat{y}_h &= \lfloor y_c/w \rfloor \\ \hat{\zeta}_h &= \lfloor \zeta_c/w \rfloor \end{aligned} \tag{6}$$

where $\hat{x}_h, \hat{y}_h, \hat{\zeta}_h$—discretized hexagonal coordinates. If the variables satisfy the following condition $\hat{x}_h + \hat{y}_h + \hat{\zeta}_h = 0$, they become the final discrete coordinates on the hexagonal grid. Otherwise, it is necessary to correct the coordinates based on the following Algorithm 1 ($\{\bullet\}$ denotes the fractional part).

Algorithm 1 Discretization of hexagonal coordinates.

```
if x̂_h + ŷ_h + ζ̂_h = 0 then
    x_h ← x̂_h
    y_h ← ŷ_h
    ζ_h ← ζ̂_h
else
    if {x_c/w} ≥ {y_c/w} and {x_c/w} ≥ {ζ_c/w} then
        x_h ← −ŷ_h − ẑ_h
        y_h ← ŷ_h
        ζ_h ← ζ̂_h
    else if {y_c/w} ≥ {x_c/w} and {y_c/w} ≥ {ζ_c/w} then
        x_h ← x̂_h
        y_h ← −x̂_h − ζ̂_h
        ζ_h ← ζ̂_h
    else
        x_h ← x̂_h
        y_h ← ŷ_h
        ζ_h ← −x̂_c − ŷ_h
    end if
end if
```

The height is calculated analogously as before. The results can be seen in Figure 10. In Figure 11, a comparison between the square and hexagonal map is shown. The image shows a close-up of the same southern section of the map, so that the differences between the lattices can be seen more clearly.

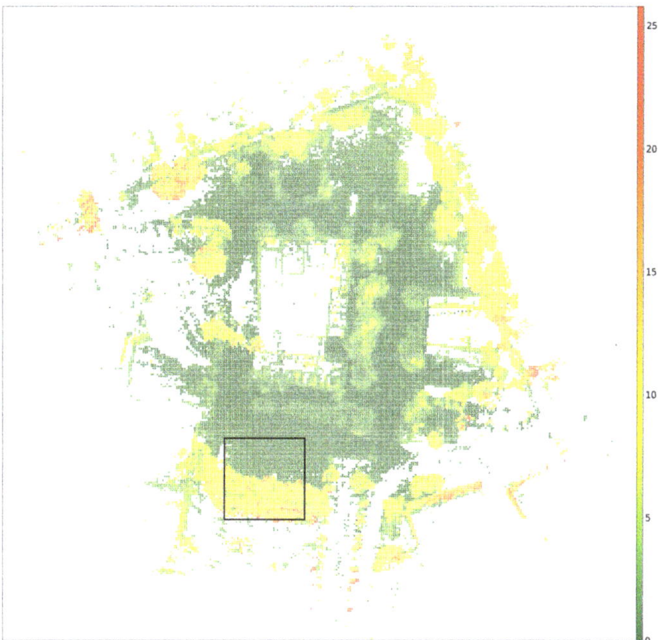

Figure 8. The 2.5D map on a square grid. The color indicates at what height an obstacle has been detected within the cell. The scale is placed on the right side of the figure. The black rectangle marks the section shown in the enlarged version in Figure 11.

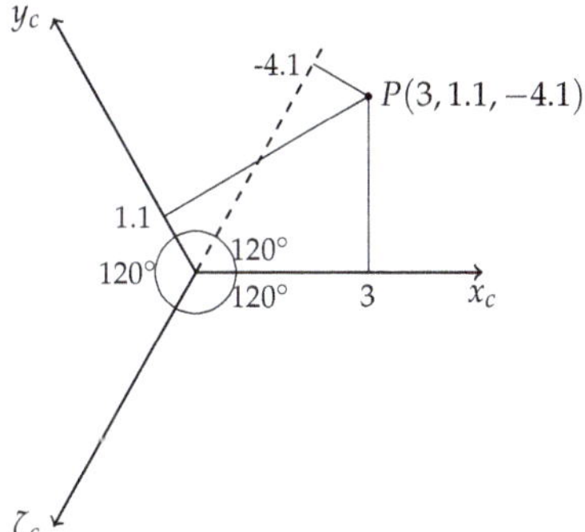

Figure 9. Cube coordinate system with an example point.

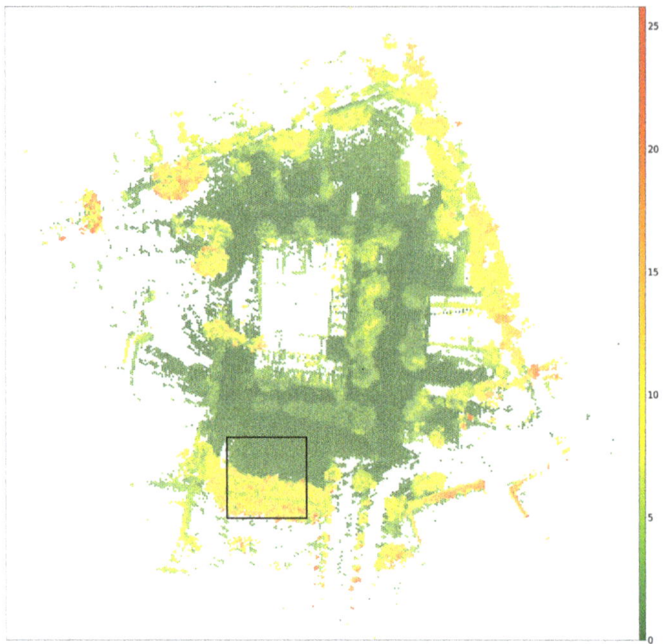

Figure 10. The 2.5D map on a hexagonal grid. The color indicates at what height an obstacle has been detected within the cell. The scale is placed on the right side of the figure. The black rectangle marks the section shown in the enlarged version in Figure 11.

(**a**) Hexagonal map

Figure 11. *Cont.*

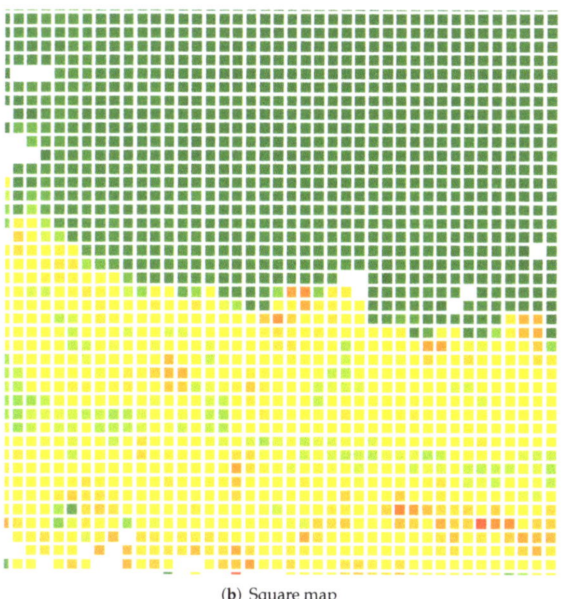

(b) Square map

Figure 11. Close-up comparison of the grids.

The last step is a conversion between the cube coordinates and ASA according to the formula:

$$\begin{aligned} a &= x_h \mod 2 \\ r &= \lfloor x_h/2 \rfloor \\ c &= -\zeta_h - \lfloor (x_h+1)/2 \rfloor \end{aligned} \quad (7)$$

4. Path Planning

In our conference paper [1], we showed that the hexagonal map is useful during path planning. Figure 12 presents collision-free paths computed based on hexagonal and square grids.

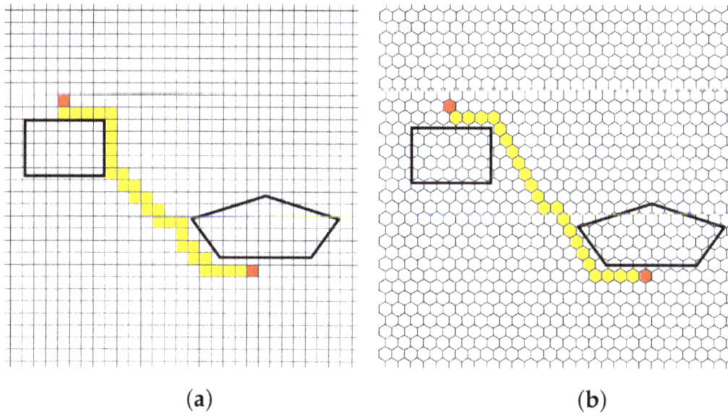

(a) (b)

Figure 12. Path planning using square grids (a) and hexagonal grids (b). Black lines represent fragments of obstacles; yellow cells represent the planned path [1].

In this article, we suggest using the diffusion method for path planning [21]. The method is not effective for large areas with few obstacles. For small areas with a large number of obstacles, the path generation time is shorter than for other methods (potential field, RRT). An additional advantage of the approach is that we can easily take into account the cost of driving over different surfaces. Therefore, we decided to use the diffusion method. In this approach, collision-free path planning is performed on a hexagonal grid. Obstacle-free cells represent the possible robot positions (states). Two states are distinguished: the robot position (c_R) and the goal position (c_G).

In the first step, a diffusion map is initialized. A big value is assigned to the cell, which represents the goal position (c_G), and the values 0.0 are attached to other cells.

In classical path planning systems, we divide cells into two classes: free from obstacles and occupied. For 2.5D maps, class membership is determined by thresholding. A cell is occupied if the observed height exceeds a certain threshold, and free of obstacles otherwise. In classical systems, we look for the shortest path, but we look for the path with the shortest travel time in many cases. The travel time (robot speed) can be related to the type of ground, distance to obstacles, etc. To solve this problem, we introduced an additional parameter cf. In the current version of the system, the value of this parameter is zero for free cells and infinity (very large integer) for occupied cells. In future works, we want to build a surface recognition system and then the parameter will be fully used. In the case of a perfectly smooth surface (asphalt), the value of this parameter equals 0.0. In the case of uneven terrain, the value is increased proportionally to the increase in the cost (time) of movement.

For each unoccupied cell (c_{aij}), the value (v_{aij}) is calculated according to the formula:

$$v_{aij} = max_{c_{ekl} \in N_{aij}}(v_{ekl} - cf_{ekl} - dist(c_{ekl}, c_{aij})) \tag{8}$$

where:

cf_{ekl}—the value of the cost function assigned to the cell c_{ekl}, $0 \leq c_{ekl} \leq \infty$;

N_{aij}—neighborhood of the cell c_{aij}.

This process continues until stability is established.

During the next step, the list of cells is generated. The first cell represents the robot position. The next one is indicated by the neighbor of (c_R) with a maximum value of v_{akl}. The process continues until the cell with the maximum function value v is reached.

The method has several advantages:

- The situation in which the position of the robot or the target is surrounded by obstacles is easily detected;
- By specifying the values of the cf function, we can influence the properties of the path—e.g., the distance from obstacles or the type of surface can be taken into account;
- A square cell has four neighbors, and a hexagonal cell has six adjacent cells;
- A path can be smoothed more easily than with a rectangular grid.

5. Experimental Results

Experiments were conducted in a real static environment. The path planning method has been tested for a hexagonal map of the environment based on the robot's sensor data. Figures 13–16 show paths generated for several different situations.

R denotes the location of the robot. G denotes the location of the target. Red color indicates cells occupied by obstacles; green color indicates cells free from obstacles; white color indicates cells whose state is unknown; black cells represent the path. Figure 13 shows a situation in which the robot avoids moving through unknown areas (cf = 100), but the distance to obstacles has not been taken into account. It is clear that the path runs dangerously close to the obstacles.

In Figure 14, an additional cost was added to the cells directly adjacent to obstacles.

In Figure 15, the neighborhood radius was extended by three cells.

For different values of the radius of interaction of the obstacles, we obtain different paths. The path shown in Figure 15 is longer than the path in Figure 13, but it is safer.

Figure 16 shows a situation where the cost function is the same for free and unknown cells and the radius of interaction of obstacles is three cells. The path is completely different from that shown in the previous figures.

We performed three series of experiments to compare the lengths of the path generated by the diffusion method in the case of hexagonal and rectangular grids. In each series, 1000 start and destination points were generated. Paths were planned using the methods mentioned above. The generated lists of cells were converted into lists of segments, and then, the lengths of the paths were calculated.

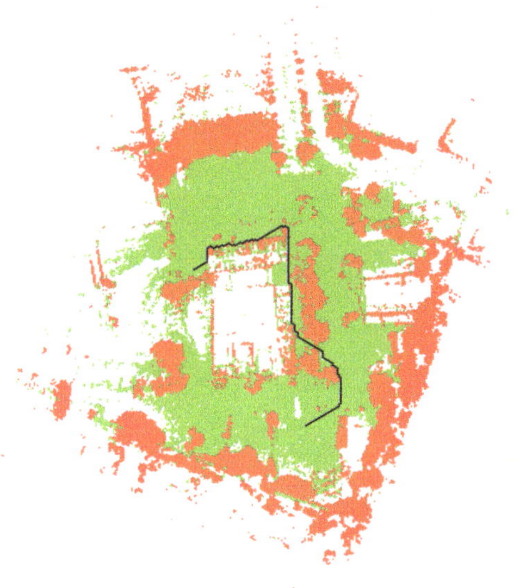

Figure 13. Collision free path—classic approach, green for obstacle-free cells, red for obstacle-occupied cells, and black for the planned path.

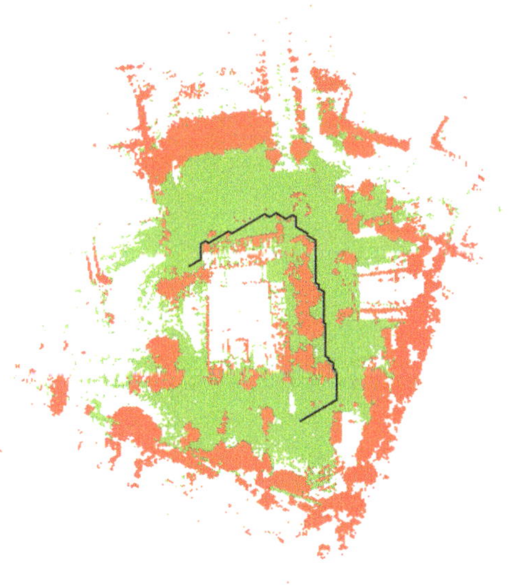

Figure 14. Collision free path—distance to the obstacles is taken into account (radius of neighborhood = 1), green for obstacle-free cells, red for obstacle-occupied cells, and black for the planned path.

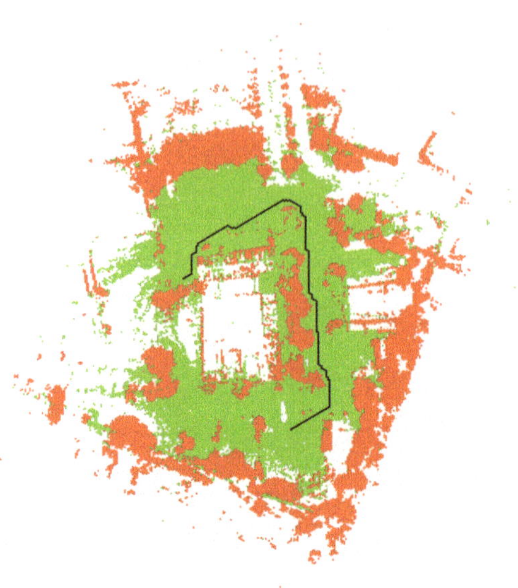

Figure 15. Collision free path—distance to the obstacles is taken into account (radius of neighborhood = 3), green for obstacle-free cells, red for obstacle-occupied cells, and black for the planned path.

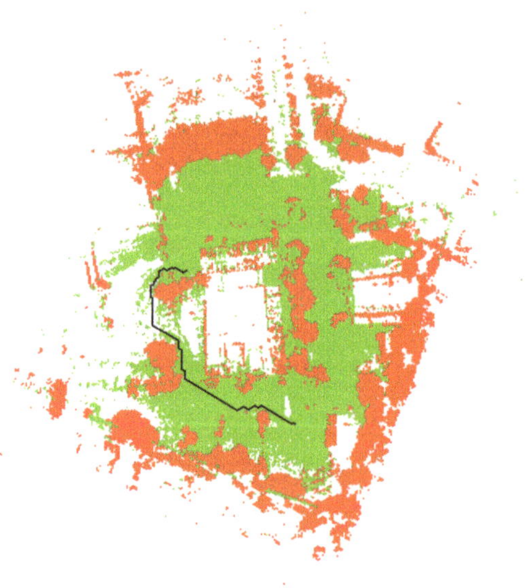

Figure 16. Collision free path—the cost function is the same for free and unknown cells, green for obstacle-free cells, red for obstacle-occupied cells, and black for the planned path.

The value of parameter dd is computed as follows:

$$dd^i = \frac{(dr^i - dh^i)}{dh^i} \quad (9)$$

where: i—the number of experiments; dr—the length of a path generated using square grids; dh—the length of a path generated using hexagonal grids.

In grid-based path planning methods, we need to expand obstacles by a given number of cells. This operation is equivalent to the dilation used in machine vision. We performed each series of experiments for a different neighborhood value (from r = 0 to r = 2). The results of the experiments are presented in Table 2. Figure 17 shows histograms of the parameter dd and corresponding Gaussian KDE distributions.

It can be seen that when we do not expand the obstacles, the gain from using the hexagonal grids is about 3%, while it exceeds 10% when we expand the obstacles. The shortened path length for the hexagonal grid is due to the fact that the hexagonal grid represents the shape of the obstacles better than the rectangular grid. For r = 2, the graph is not symmetrical with respect to the mean value, and in 82% of the cases, the path planned with the hexagonal grid is shorter than with the rectangular grid.

We did not notice a significant effect of map form (rectangular grid, hexagonal grid) on path planning time. The most time-consuming stage was diffusion. Regardless of the grid type, the number of cells occupied by obstacles affected the diffusion process. The diffusion process took more than one second (computer, map) for an empty environment. Fortunately, the absence of barriers is easy to detect, and we can plan the path using ordinary geometrical methods. In a maze-type environment with many obstacles, the diffusion time did not exceed 0.5 s (PC, Windows 10, i5-1035G1 CPU 1.00 GHz, 1.19 GHz, RAM 16 GB, map: 640 × 320 cells; a cell represents 1 m^2 in area). An essential advantage of the hexagonal mesh is that it approximates the shape of objects much better (Section 3). As a result, in many situations, a collision-free path is not found when using a rectangular mesh, but is found when using hexagonal one. As a result, the generated path transforms to the interpretable data by the Łukasiewicz—PIAP drive-by-wire system, which works on the universal automotive-grade controller for complex mobile working machines. The low-level program was designed to control the steering wheel by the original power steering of the off-road car. The velocity of the vehicle is controlled by the algorithm by using the electronic throttle and added electric ABS pomp. The universal controller is responsible for adjusting the velocity of the vehicle and the steering angle of the wheels to the set value by the CAN frame sent by the Łukasiewicz—PIAP autonomy controller.

Table 2. Mean values of parameter dd, for different neighborhood values.

Neighborhood Value	Mean dd
0	0.03
1	0.12
2	0.15

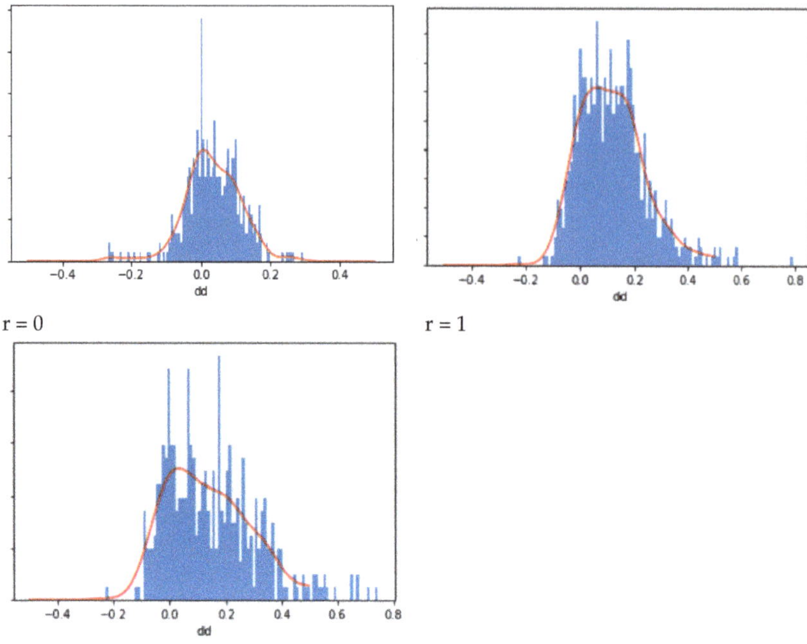

Figure 17. Histograms of parameter dd; the red line represents the estimated probability density function.

6. Conclusions and Future Works

In this paper, we presented the possibility of using hexagonal grids in outdoor navigation. Experimental results showed the advantages of a hexagonal grid over a square grid. In our further work, we will use the diffusion map to determine safe vehicle speeds. Furthermore, we plan to determine robot localization based on matching hexagonal maps from two places.

The method could be used in the autonomy systems for the outdoor navigation in the UGV control systems in convoys, reconnaissance, surveillance missions, etc. This type of mapping can bring the goals of speedup and improved local path planning in real time on autonomous systems to reach the level of not limiting the UGV ability on rough terrains. We also plan to apply elements of the presented algorithm to the tasks described in the papers [25,26].

Author Contributions: Methodology and writing: B.S., P.D. and R.W.; software for path planning, B.S.; software for mapping, P.D.; low-level control, R.W. All authors have read and agreed to the published version of the manuscript.

Funding: This research received no external funding.

Institutional Review Board Statement: Not applicable.

Informed Consent Statement: Not applicable.

Data Availability Statement: The source code and data used to support the findings of this study are available from the corresponding author upon request.

Conflicts of Interest: The authors declare no conflict of interest.

References

1. Duszak, P.; Siemiątkowska, B. The application of hexagonal grids in mobile robot Navigation. In Proceedings of the Conference Mechatronics, Recent Advances Towards Industry, Advances in Intelligent Systems and Computing, Kunming, China, 22–24 May 2020; Springer: Berlin/Heidelberg, Germany, 2020; Volume 1044, pp. 198–205, ISBN 978-3-030-29992-7. [CrossRef]
2. Reinoso, O.; Payá, L. Special Issue on Mobile Robots Navigation. *Appl. Sci.* **2020**, *10*, 1317. [CrossRef]
3. Siegwart, R.; Nourbakhsh, I.R. *Introduction to Autonomous Mobile Robots*; Bradford Company: Cambridge, MA, USA, 2004.
4. Elfes, A. Using occupancy grids for mobile robot perception and navigation. *Computer* **1989**, *22*, 46–57. [CrossRef]
5. Rodriguez-Losada, D.; Matia, F.; Galan, R. Building geometric feature based maps for indoor service robots. *Robot. Auton. Syst.* **2006**, *54*, 558. [CrossRef]
6. Zhang, Y.; Tian, G.; Shao, X.; Cheng, J. Effective Safety Strategy for Mobile Robots Based on Laser-Visual Fusion in Home Environments. *IEEE Trans. Syst. Man Cybern. Syst.* **2021**, *99*, 1–13.
7. Fankhauser, P.; Hutter, M. A Universal Grid Map Library: Implementation and Use Case for Rough Terrain Navigation. In *Robot Operating System (ROS). Studies in Computational Intelligence*; Koubaa, A., Eds.; Springer: Cham, Switzerland, 2016; Volume 625. [CrossRef]
8. Thoresen, M.; Nielsen, N.H.; Mathiassen, K.; Pettersen, K.Y. Path Planning for UGVs Based on Traversability Hybrid A. *IEEE Robot. Autom. Lett.* **2021**, *6*, 1216–1223. [CrossRef]
9. Ahtiainen, J.; Stoyanov, T.; Saarinen, J. Normal Distributions Transform Traversability Maps: LiDAR-Only Approach for Traversability Mapping in Outdoor Environments. *J. Field Robot.* **2017**, *34*, 600–621. [CrossRef]
10. Belter, D.; Skrzypczyński, P. Rough terrain mapping and classification for foothold selection in a walking robot. *J. Field Robot.* **2011**, *28*, 497–528. [CrossRef]
11. Thrun, S. *Robotic Mapping: A Survey*; 2002; pp. 1–35. Available online: http://citeseerx.ist.psu.edu/viewdoc/download?doi=10.1.1.319.3077&rep=rep1&type=pdf (accessed on 15 October 2021).
12. Middleton, L.; Sivaswamy, L. *Hexagon Image Processing—A Practical Approach*; Springer: London, UK, 2005.
13. Yabushita, A.; Ogawa, K. Image reconstruction with a hexagonal grid. In Proceedings of the Nuclear Science Symposium Conference Record, Norfolk, VA, USA, 10–16 November 2002.
14. Jeevan, K.; Krishnakumar, S. An Image Steganography Method Using Pseudo Hexagonal Image. *Int. J. Pure Appl. Math.* **2018**, *118*, 2729–2735.
15. Quijano, H.J.; Garrido, L. Improving Cooperative Robot Exploration Using an Hexagonal World Representation. In Proceedings of the Conference: Electronics, Robotics and Automotive Mechanics Conference, Cuernavaca, Morelos, Mexico, 25–28 September 2007.
16. Rummelt, N. Array Set Addressing: Enabling Efficient Hexagonally Sampled Image Processing. Ph.D. Thesis, Unversity of Florida, Gainesville, FL, USA, 2010.
17. Wang, Y.; Zheng, N.; Bian, Z. A Closed-Form Solution to Planar Feature-Based Registration of LiDAR Point Clouds. *ISPRS Int. J. Geo-Inf.* **2021**, *10*, 435. [CrossRef]
18. Zhu, R.; Guo, Z.; Zhang, X. Forest 3D Reconstruction and Individual Tree Parameter Extraction Combining Close-Range Photo Enhancement and Feature Matching. *Remote Sens.* **2021**, *13*, 1633. [CrossRef]
19. Martínez, J.L.; Morales, J.; Sánchez, M.; Morán, M.; Reina, A.J.; Fernández-Lozano, J.J. Reactive Navigation on Natural Environments by Continuous Classification of Ground Traversability. *Sensors* **2020**, *20*, 6423. [CrossRef] [PubMed]
20. Duchoň, F.; Huňady, D.; Dekan, M.; Babinec, A. Optimal Navigation for Mobile Robot in Known Environment. *Appl. Mech. Mater.* **2013** *282*, 33–38. [CrossRef]
21. Siemiątkowska, B.; Dubrawski, A. Cellular Neural Networks for Navigation of a Mobile Robot. *Lect. Notes Comput. Sci.* **1998**, *1424*, 147–154.
22. Lakshmanan, A.K.; Mohan, R.E.; Ramalingam, B.; Le, A.V.; Veerajagadeshwar, P.; Tiwari, K.; Ilyas, M. Complete coverage path planning using reinforcement learning for Tetromino based cleaning and maintenance robot. *Autom. Constr.* **2020**, *112*, 103078. [CrossRef]
23. Her, I. Geometric transformations on the hexagonal grid. *IEEE Trans. Image Process.* **1995**, *4*, 1213–1222. [CrossRef] [PubMed]
24. Snyder, W.E.; Qi, H.; Sander, W.A. Coordinate system for hexagonal pixels. In *Proceedings of Medical Imaging 1999: Image Processing*; The International Socicty for Optical Engineering: Bellingham, WA, USA, 2004.
25. Stecz, W.; Gromada, K. UAV Mission Planning with SAR Application. *Sensors* **2020**, *20*, 1080. [CrossRef] [PubMed]
26. Siemiatkowska, B.; Stecz, W. A Framework for Planning and Execution of Drone Swarm Missions in a Hostile Environment. *Sensors* **2021**, *12*, 4150. [CrossRef] [PubMed]

remote sensing

MDPI

Article

Stratified Particle Filter Monocular SLAM

Pawel Slowak * and Piotr Kaniewski

Faculty of Electronics, Military University of Technology, ul. gen. S. Kaliskiego 2, 00-908 Warsaw, Poland; piotr.kaniewski@wat.edu.pl
* Correspondence: pawel.slowak@wat.edu.pl

Abstract: This paper presents a solution to the problem of simultaneous localization and mapping (SLAM), developed from a particle filter, utilizing a monocular camera as its main sensor. It implements a novel sample-weighting idea, based on the of sorting of particles into sets and separating those sets with an importance-factor offset. The grouping criteria for samples is the number of landmarks correctly matched by a given particle. This results in the stratification of samples and amplifies weighted differences. The proposed system is designed for a UAV, navigating outdoors, with a downward-pointed camera. To evaluate the proposed method, it is compared with different samples-weighting approaches, using simulated and real-world data. The conducted experiments show that the developed SLAM solution is more accurate and robust than other particle-filter methods, as it allows the employment of a smaller number of particles, lowering the overall computational complexity.

Keywords: SLAM; autonomous navigation; particle filter; monocular camera; IMU; UAV

Citation: Slowak, P.; Kaniewski, P. Stratified Particle Filter Monocular SLAM. *Remote Sens.* **2021**, *13*, 3233. https://doi.org/10.3390/rs13163233

Academic Editor: Weimin Huang

Received: 16 June 2021
Accepted: 10 August 2021
Published: 14 August 2021

1. Introduction

After over two decades of extensive work, the robotics research community has proposed a multitude of advanced simultaneous localization and mapping (SLAM) approaches [1,2]. Solving the SLAM problem robustly is a necessary element to achieve full autonomy in the field of robotics. Among the most important elements of the simultaneous localization and mapping procedure, one can list surroundings perception with data extraction and association, robot pose estimation, local map building and maintaining global map coherence, so that the so-called loop closure can be performed in a previously visited area. A vast number of different solutions to all SLAM components has been proposed using numerous state-of-the-art frameworks. To classify different approaches, it is useful to categorize SLAM systems in accordance with the type of sensor used, the class of mathematical back-end synthesizing the pose and map information and the method of map representation.

A wide variety of sensors have been employed in SLAM: monocular cameras, stereo cameras, laser range finders, sonars, global positioning system (GPS) receivers, inertial measurement units (IMUs), etc. Some of those sensors can be used only for position estimation, others only to gather information about the environment, while devices such as cameras can be utilized for both purposes. The type of the chosen sensor determines the character of data that will be processed during the SLAM procedure, which influences subsequent stages of a SLAM framework design task.

The second vital aspect of the simultaneous localization and mapping, according to which different SLAM approaches can be distinguished, is the mathematical apparatus that is employed during the pose and map estimation step. For the sake of brevity, we will only address the basic distinction of filtering versus batch optimization. Filter methods characterize the map and the current pose information as a probability density function (PDF) using a variant of a Kalman filter (e.g., an extended Kalman filter (EKF) [3,4] or an unscented Kalman filter (UKF) [5]) or a particle filter (PF) [6,7]), a detailed description of which is presented later. A common feature of all filter methods is the lack of explicit storage of information about previous system states, which is commonly referred to as online or recursive state estimation. This stochastic modeling approach is referred to as the Markov chain. On the other hand, the optimization SLAM algorithms, known also as batch- or grid-based methods, perform a global or semi-global error minimization procedure from a set of previous poses and measurements. Also the keyframe-based SLAM should be mentioned here, as it is the most commonly used [2] optimization method for a visual SLAM. Keyframes are a small subset of distinctive camera frames recorded along the sensor trajectory. Only those characteristic keyframes are processed in the pose-map error minimization, which is performed using global bundle adjustment (BA) [8].

In [9], the authors proved that the optimization approach of the global BA approach offers better performance than the recursive filtering when comparing accuracy, computational complexity and robustness in large scale applications. However, the authors did not include a particle filter method in their testing, justifying the omission by an assumption that it is wasteful to employ a particle filter for unimodal distributions estimation. However, categorizing the SLAM procedure, where an algorithm has to evaluate not only the robot or sensor pose but also the state of rarely static surroundings, as a strict unimodal distribution estimation can be considered an oversimplification. Particle filters provide a more robust, multimodal modeling approach. Moreover, PFs computational complexity, contrary to Kalman filters, does not scale cubically [10] in terms of the number of observed landmarks. It is linear—similar to batch SLAM algorithms. Therefore, we consider particle filters as a valid and reasonable solution to the concurrent localization and mapping problem.

Another useful means of the classification of SLAM methods is the map building approach, which can be done using two separate but interdependent criteria. To begin with, the map can be built either directly or indirectly. The first method is based on analyzing the surroundings using unprocessed sensor readings, while the latter identifies specific features in the environment according to a chosen extraction approach, which can be different geometric shapes (e.g., points [11], corners [12] and lines [13]) or more complex objects [14]. The way that the map elements are picked is directly associated with the resulting map type. A given SLAM procedure can produce either a dense map of the environment where every part of an observed area—for example every registered pixel—is associated with a distinct element of the map. This approach is commonly used in SLAM systems that utilize sensors which are capable of determining full three-dimensional measurements of the environment, like rangefinders [15] or RGBD cameras [16]. On the other hand, the registration of a sparse map is a process wherein the map is constructed around salient scene elements (characteristic points, regions or shapes), that can be correctly associated in subsequent sensor observations.

As different reviews show [1,17,18], although the development of SLAM systems offers a wide range of opportunities for modern autonomous systems, the simultaneous localization and mapping domain has yet to achieve complete success. Hence, the community needs to address various potential solutions to concurrent localization and mapping problems. Our manuscript presents a novel framework for an indirect, monocular SLAM, based on a Rao–Blackwellized particle filter. We incorporate a distinctive approach towards particle weighting, where the weights are stratified proportionally to the number of landmarks matched in a given camera frame. This approach offers better accuracy and robustness than standard resampling methods. Thus, it allows to lower the computational complexity of the SLAM algorithm by decreasing the number of particles needed for adequate performance. Furthermore, our algorithm has been examined using both simulation and real-world data, registered from a quadcopter. The presented approach was tested in various set-ups and all the results were consistent. The paper contains selected results of the simulations and experiments. The remainder of this article is outlined as follows: In Section 2, the related work in the field of particle filter SLAM is discussed; in Section 3, a detailed explanation of the proposed algorithm is presented; simulation and real experimental results are provided in Section 4 to validate the adopted approach. The conclusions of the paper are included in Section 5. This manuscript is a continuation of our previous works in the field of particle filter SLAM [19] but offers a new approach to the filtering algorithm.

2. Related Work

The earliest idea of a SLAM framework employing a Rao–Blackwellized particle filter was briefly discussed in [20]. The authors advocated the use of Bayesian inference in order to achieve autonomous localization and mapping capabilities. Undoubtedly, the most well-known solution to a Rao–Blackwellized particle filter SLAM was presented in [6], where lidar was used to build a grid map. Many later systems have been designed upon FastSLAM (both 1.0 and 2.0 [7]) frameworks—especially those utilizing rangefinders as main sensors [21–23].

Rapid progress in computer vision and computational capacity led to the proliferation of cameras in the field of robotics. Among the first systems that employed a particle filter and a camera to perform simultaneous localization and mapping, one should list [24]. The algorithm described in that manuscript processed edges, found with the Sobel mask, as landmarks, but did not benefit from the fact that different observations can be treated as probabilistically independent if one knows the camera position and orientation. As a result, 10,000 particles had to be used to estimate the state of the camera pose, together with only 15 landmarks. In [25], the authors implemented an algorithm that allowed one to sequentially approximate the full 6DoF posterior of a camera, together with up to eight, tracked 3DoF scene points. This method was successfully validated using 500 particles; however, a small set of previously known landmarks was required for it to work properly. Another system, developed by Sim et al. [26], is an early example of a Rao–Blackwellized particle filter framework that was constructed using a camera as the main sensor. With the usage of the SIFT algorithm [27], the authors' indoor mobile robot was able to successfully track more than 11,000 landmarks. Later, Eade and Drummond [15] constructed an algorithm capable of camera-tracking with as little as 50 particles and implemented the ingenious idea of inverse depth [28] as a third element of a 3DoF landmark-state vector. In [29], the authors extracted landmarks using speeded-up robust features (SURF) [30] and applied a global optimization algorithm to achieve optimal matching for its scene points.

While the earliest formulations of Rao–Blackwellized particle filters in a FastSLAM-like framework assumed that map elements are estimated using EKFs, other monocular SLAM approaches implemented alternative nonlinear Kalman filtering strategies, namely the UKF. In [31], the authors exploited the spatial structure of the environment and developed an algorithm that searched for and extracted locally planar objects as landmarks, whilst Lee [32] introduced a template prediction mechanism to compensate for camera motion. Both mentioned systems employed UKFs for landmark storage to overcome the issue concerning Jacobians approximations in EKFs.

The profound analysis presented in [9] marked a milestone in the SLAM systems domain, as more researchers have tended to shift towards keyframe simultaneous localization and mapping approaches over the last decade. Still, there have been numerous examples of efficient PF SLAM methods implemented since. One of them was proposed in [33] as a tool for pose-tracking for augmented reality. The algorithm implemented an idea to discard outliers indirectly—not during the data association-and-gating phase but after the particle-weighting procedure, as incorrect matches lower sample weights significantly, thus minimizing chances for a given particle to be resampled. The authors used both lines and points, extracted with a Harris corner detector [34] and Hough line transform, respectively. A different particle filter-based solution to the SLAM problem was outlined in [35] for an indoor aerial vehicle. Asserting that the robot was designed to navigate only inside manmade structures, that system exploited the abundance of straight lines in the camera images and facilitated human-like procedures to predict landmark depths. The ranging strategy assumed that the monocular camera altitude was known, and used this information to process relative poses of observed geometric structures in order to synthesize a simultaneous localization-and-mapping algorithm, similar to the FastSLAM approach.

As mentioned before, the monocular camera-observation model suffers greatly from a lack of depth information. Contrary to monocular-camera algorithms, RGBD-camera-based-SLAM approaches are able to directly initialize landmark depths using single-sensor reading. One paper [36] presents a remarkable stereo-camera-particle-filter-SLAM solution, where the authors proposed a smart procedure for outlier identification by landmark-position correlation analysis. Moreover, landmarks are efficiently detected and matched using the SURF algorithm. To tackle the unknown depth issue otherwise, one can place a pattern of known dimensions inside the camera field of view. In [37], the authors proposed the insertion of a chessboard inside the first few camera frames to allow an accurate depth estimation as well as a reduction of camera pose uncertainty for a monocular-camera-based SLAM system. To calculate the depth of subsequently observed landmarks, the described algorithm delayed their initialization until a triangulation procedure could be carried out.

In reference to SLAM being strictly a perception problem, one would intuitively seek its refinement in the modification of a given observation model. However, Zhou and Maskell proposed an improvement to the FastSLAM framework based on the motion model revision. In [38], the estimation of system dynamics is partitioned into two sub-models. The camera location was calculated with a particle filter, and its velocity with a Kalman filter. This idea allowed one to reduce the particle filter's dimensionality, as well as to achieve a better accuracy than one would with an analogous solution constructed upon the classic FastSLAM framework.

Further examples of particle-filter SLAMs also include non-pure-visual systems, where data from range finders are fused with images. Chen et al. [39] implemented a system wherein an urban search-and-rescue robot navigates using 2D lidar in a feature-based 3D map, constructed with a monocular camera. This allows for the obtaining of a real scale of the surroundings, as well as the maintenance of full, 6DoF motion and mapping capabilities. In [40], the authors propose a system wherein one robot performs the camera SLAM while others reuse the resulting map for simultaneous localization-and-map-scale estimation.

Among additional instances of PF-based systems in the field of robotics, recent works not related directly to the SLAM problem should be listed as well. The manuscript by Acevedo et al. [41] characterizes an algorithm which employs a particle filter to enable a group of networking robotic entities to search for a moving target. In [42], PF is used as a framework to solve 6DoF, visual pose-tracking, where Rao–Blackwellization is introduced by decoupling the translation and orientation data. In a paper by Di Yuan et al. [43], a PF-based system for redetection in the object-tracking approach for accurate localization in difficult conditions is described. In [44], the authors describe a self-localization technique that employs a PF, based on particle swarm optimization, that requires fewer particles to function correctly than comparable benchmark approaches.

3. Materials and Methods

Although the method that we propose in this paper can be easily adjusted to any monocular camera configuration, our SLAM framework was designed and validated under the assumption it will serve as a secondary navigational system for a surveillance UAV with its monocular camera pointed downward. Based on inertial-measurement-unit (IMU) measurements and subsequent camera frames, we were able to synthesize the registered data into a correct trajectory, together with a sparse map of the observed area.

3.1. Motion Model

To minimize the number of reference frames needed, our system directly estimates the pose (position and orientation) of a camera sensor, rather than the pose of a UAV itself. The state vector $\mathbf{x}_k$ during time step k consist of nine variables:

$$\mathbf{x}_k = \begin{bmatrix} x \\ y \\ z \\ v_x \\ v_y \\ v_z \\ \phi \\ \theta \\ \psi \end{bmatrix} \tag{1}$$

where x, y, z represent a localization in Cartesian coordinates, v_x, v_y, v_z are orthogonal components of a velocity vector and ϕ, θ and ψ are the roll, pitch and yaw orientation angles respectively. We choose the east-north-up (ENU) coordinate system as the global reference frame, where x is east, y is north and z is up. The source of camera orientation, relative to the ENU frame, is obtained from the onboard IMU.

While the camera itself has six degrees of freedom (6DoF), the bearing-only observation model causes monocular SLAM to be a 7DoF problem, where a map representation can be determined only up to a scale. To mitigate the issue of scale ambiguity and drift, which is an inherent problem of single-camera SLAM frameworks, we chose to select the onboard IMU as an additional sensor. It is used as the source of a stream of metric measurements that are control signals for trajectory prediction. The employment of an inertial measurement unit implicates the usage of a constant velocity (CV) motion model as the most appropriate. The discretized motion equation describing the camera movement relationships is given below:

$$\mathbf{x}_k = \mathbf{f}(\mathbf{x}_{k-1}, \mathbf{u}_k) + \mathbf{w}_k \tag{2}$$

where $\mathbf{x}_k$ is the predicted state vector containing estimates of the platform's kinematic parameters, $\mathbf{x}_{k-1}$ is the state vector estimated during the previous time step $k-1$, $\mathbf{f}$ is the state-transition nonlinear vector function, $\mathbf{u}_k$ represents IMU readings, and $\mathbf{w}_k$ is the additive Gaussian process noise which can be described by a zero-mean multivariate normal distribution $N(0, \mathbf{Q}_k)$ where $\mathbf{Q}_k$ denotes the process covariance matrix.

3.2. Sensor Model

As previously stated, a single camera is the main sensor utilized for map building in our SLAM system. Monocular cameras are common components of off-the-shelf UAVs. Among their advantages which can be considered the most beneficial in the field of robotics are their small size and low cost, as well as low power consumption. Moreover, the update rates and resolutions of these cameras are sufficient to track environmental changes during motion at velocities up to tens of meters per second. These features are the reason that monocular cameras are widely used for localization, structure from motion, mapping, SLAM etc. However, one can also identify consequential drawbacks of the single-camera usage. The main disadvantage, which has to be addressed when analyzing the monocular sensor model, is the lack of depth information. In other words, the perspective projection that transforms 3D real-world points into 2D camera pixels coordinates is lossy and causes a pose calculation problem for extracted real-world features. As landmark 3D positions cannot be straightforwardly computed, the approach to resolve this difficulty needs to be adopted. To address this issue, a given algorithm can either initialize a landmark immediately after its first observation (where the uncertainty of measured depth is set as significantly larger than for other coordinates), or delay the feature detection until observations from different perspectives provide conditions for temporal view stereoscopy analysis, to estimate information about the landmark's full, 3D pose. The first approach is called an undelayed initialization, while the latter is commonly referred to as delayed landmark initialization.

The adopted sensor model, together with the initialization strategy, starts with a single-camera frame registration. The camera's intrinsic parameters are known and can be denoted as $\mathbf{K}_{intr}$:

$$\mathbf{K}_{intr} = \begin{bmatrix} f_x & s & u_0 \\ 0 & f_y & v_0 \\ 0 & 0 & 1 \end{bmatrix} \tag{3}$$

where f_x and f_y represent focal lengths along the camera's axis and are equal for a pinhole camera model, u_0 and v_0 are principal point offset and s is the camera axis skew.

To extract distinctive scene points from images, we use the SURF algorithm [21]. The SURF detector outputs a set of pixel coordinates in a given image frame, where the determinant of a Hessian matrix reaches its maximum value. It is worth noting that SURF points are scale and rotation invariant. Exemplary extraction procedure results are presented in Figure 1.

Figure 1. Frame from a downward-looking camera with extracted SURF features.

To represent extracted scene points in the ENU coordinate frame, we use the concept of anchored modified-polar points, which can also be referred to as inverse-distance points (IDP) [45]. A single IDP point is defined by a six-element vector:

$$^{p}\mathbf{m}_k^i = \begin{bmatrix} x_0 \\ y_0 \\ z_0 \\ \varepsilon \\ \alpha \\ \rho \end{bmatrix} \tag{4}$$

where $^{p}\mathbf{m}_k^i$ is the state of the i-th landmark observed by the p-th particle at the time k. The first three elements of the state vector (x_0, y_0, z_0) are the ENU coordinates that encode the position of the particle from which the landmark was originally observed (the point $^{p}\mathbf{p}_0$ in Figure 2.). These coordinates are frequently referred to as the anchor point. Next, ε and

α are respectively the elevation and azimuth angles at which the observation was made, while $\rho = \frac{1}{d}$ is the inverse of the distance between the camera and the scene point.

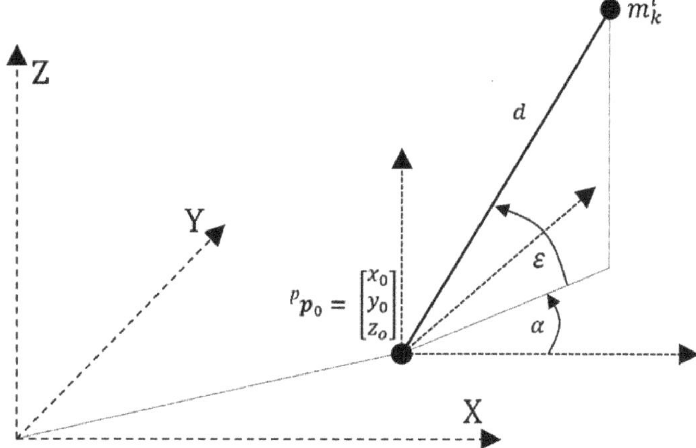

Figure 2. IDP landmark parametrization.

Figure 2 presents the idea of IDP landmark parametrization. The usage of anchored modified-polar points allows one to initialize landmarks in the map immediately. Still, the monocular-camera, salient-feature extraction step provides only two-dimensional measurements of three-dimensional objects' locations, where the distance d remains unknown. To resolve the issue of being unable to recover the true localization of the environment features, we choose the strategy of setting an initial inverse depth, of every registered landmark, as a preset value with reasonably large uncertainty. As the camera is pointed downwards, the starting depth value is either assumed equal to the UAV's altitude or is calculated using positions of nearby, previously seen scene points. The standard deviation of such observation is selected in a way so as to include infinite distance (inverse depth equal to 0) in the 3σ region. Using the undelayed initialization scheme allows one to comply with the adopted discrete-time Markov chain approach, and satisfies the property of a memoryless process, such that the currently processed state is sufficient to estimate the probability distribution of future states.

The first step of the initialization of a newly observed landmark is to transform its coordinates, expressed in pixels of the image plane, using the pinhole camera model. The inverse camera projection is performed in accordance to the equation below:

$$\begin{bmatrix} x_{cam} \\ y_{cam} \\ 1 \end{bmatrix} = \mathbf{K}_{intr}{}^{-1} \begin{bmatrix} u \\ v \\ 1 \end{bmatrix} \tag{5}$$

where u and v are pixel coordinates in a registered image frame while x_{cam} and y_{cam} are coordinates in the standard reference frame of the camera. Next, by transforming the resultant vector so that it is expressed in the global reference frame (ENU), the optical ray pointing from the camera center $^p\mathbf{p}_0$ to the extracted scene point is obtained:

$$\begin{bmatrix} x_{enu} \\ y_{enu} \\ z_{enu} \end{bmatrix} = {}^{enu}\mathbf{R}_{cam} \begin{bmatrix} x_{cam} \\ y_{cam} \\ 1 \end{bmatrix} \tag{6}$$

where $^{enu}\mathbf{R}_{cam}$ encodes the rotation from the camera reference frame to ENU coordinates. The usage of IDP parametrization implies the lack of need for the employment of 3D coordinates in a homogeneous form, as the obtained vector is subjected only to rotation.

The information that is conventionally contained in translation—when using 3D Euclidean points as landmark representation—is encoded in the anchor point ${}^{p}\mathbf{p}_0$.

Next, the vector's ENU coordinates are expressed using a modified-polar point convention:

$$\begin{bmatrix} \varepsilon \\ \alpha \\ \rho \end{bmatrix} = \begin{bmatrix} atan2\left(z_{enu}, \sqrt{x_{enu}{}^2 + y_{enu}{}^2}\right) \\ atan2(y_{enu}, x_{enu}) \\ \frac{1}{d} \end{bmatrix} \tag{7}$$

where ε and α are respectively the elevation and yaw angles and ρ is the inverse depth. The addition of the anchor point results in the acquiring of the complete IDP landmark parametrization:

$${}^{p}\mathbf{m}_k^i = \begin{bmatrix} {}^{p}\mathbf{p}_0 \\ \varepsilon \\ \alpha \\ \rho \end{bmatrix} = \begin{bmatrix} \begin{bmatrix} x_0 \\ y_0 \\ z_0 \end{bmatrix} \\ atan2\left(z_{enu}, \sqrt{x_{enu}{}^2 + y_{enu}{}^2}\right) \\ atan2(y_{enu}, x_{enu}) \\ \frac{1}{d} \end{bmatrix} \tag{8}$$

As landmarks are stored using separate EKFs, the initialization procedure has to comprise the calculation of the landmark covariance matrices as well. A covariance matrix ${}^{p}\mathbf{P}_k^i$ that describes the uncertainty of a transformation of the 2D point extracted from an image frame to its IDP representation is given by the following formula:

$${}^{p}\mathbf{P}_k^i = \left({}^{p}\mathbf{H}_k^i\right)^{-1} \mathbf{R}_k^p \left[\left({}^{p}\mathbf{H}_k^i\right)^{-1}\right]^T + \begin{bmatrix} 0 & 0 & 0 & 0 & 0 & 0 \\ 0 & 0 & 0 & 0 & 0 & 0 \\ 0 & 0 & 0 & 0 & 0 & 0 \\ 0 & 0 & 0 & 0 & 0 & 0 \\ 0 & 0 & 0 & 0 & 0 & 0 \\ 0 & 0 & 0 & 0 & 0 & \sigma_\rho^2 \end{bmatrix} \tag{9}$$

where $\mathbf{R}_k^p$ is a two-by-two sensor-noise matrix, describing the accuracy of the scene point localization in the image plane, and $\left({}^{p}\mathbf{H}_k^i\right)^{-1}$ follows the ordinary EKF notation and denotes a Jacobi matrix of the inverse observation function that describes the transformation from pixel coordinates to the ENU coordinate system. $\left({}^{p}\mathbf{H}_k^i\right)^{-1}$ is given by the following formula:

$$\left({}^{p}\mathbf{H}_k^i\right)^{-1} = \begin{bmatrix} 0 & 0 & 0 \\ 0 & 0 & 0 \\ 0 & 0 & 0 \\ 1 & 0 & 0 \\ 0 & 1 & 0 \\ 0 & 0 & 1 \end{bmatrix} {}^{p}\mathbf{J}_k^i {}^{enu}\mathbf{R}_{cam} \mathbf{K}_{intr}{}^{-1} \begin{bmatrix} 1 & 0 \\ 0 & 1 \\ 0 & 0 \end{bmatrix} \tag{10}$$

where ${}^{p}\mathbf{J}_k^i$ is the Jacobi matrix of a function that transforms the vector pointing from the ${}^{p}\mathbf{p}_0$ to the i-th landmark from 3D ENU Cartesian coordinates into modified spherical coordinate system representation and is equal to:

$${}^{p}\mathbf{J}_k^i = \begin{bmatrix} \frac{x_{enu} z_{enu} \left(x_{enu}{}^2 + y_{enu}{}^2 + z_{enu}{}^2\right)}{-\sqrt{x_{enu}{}^2 + y_{enu}{}^2}} & \frac{y_{enu} z_{enu} \left(x_{enu}{}^2 + y_{enu}{}^2 + z_{enu}{}^2\right)}{-\sqrt{x_{enu}{}^2 + y_{enu}{}^2}} & \frac{\sqrt{x_{enu}{}^2 + y_{enu}{}^2}}{x_{enu}{}^2 + y_{enu}{}^2 + z_{enu}{}^2} \\ \frac{-y_{enu}}{x_{enu}{}^2 + y_{enu}{}^2} & \frac{x_{enu}}{x_{enu}{}^2 + y_{enu}{}^2} & 0 \\ 0 & 0 & 0 \end{bmatrix} \tag{11}$$

During camera movement, in subsequent time steps, the sensor model is able to retrieve the bearings of extracted features from consecutive, multiple views. The SLAM algorithm gradually estimates real 3D poses of landmarks, using standard EKF update equations given below:

$$ {}^{p}\mathbf{y}_k^i = \mathbf{z}_k^{[i]} - h\left({}^{p}\mathbf{m}_k^{[i]}\right) \tag{12} $$

$$ {}^{p}\mathbf{S}_k^{\mathrm{i}} = {}^{p}\mathbf{H}_k^{\mathrm{i}}\,{}^{p}\mathbf{P}_{k-1}^{\mathrm{i}}\left({}^{p}\mathbf{H}_k^{\mathrm{i}}\right)^T + \mathbf{R}_k^p \tag{13} $$

$$ {}^{p}\mathbf{K}_k^i = {}^{p}\mathbf{P}_{k-1}^{\mathrm{i}}\left({}^{p}\mathbf{H}_k^{\mathrm{i}}\right)^T\left({}^{p}\mathbf{S}_k^{\mathrm{i}}\right)^{-1} \tag{14} $$

$$ {}^{p}\mathbf{m}_{k+1}^i = {}^{p}\mathbf{m}_k^{\mathrm{i}} + {}^{p}\mathbf{K}_k^i\,{}^{p}\mathbf{y}_k^i \tag{15} $$

$$ {}^{p}\mathbf{P}_k^{\mathrm{i}} = \left(\mathbf{I}_{6x6} - {}^{p}\mathbf{K}_k^i\,{}^{p}\mathbf{H}_k^{\mathrm{i}}\right){}^{p}\mathbf{P}_{k-1}^{\mathrm{i}} \tag{16} $$

where ${}^{p}\mathbf{y}_k^i$ is the measurement innovation (of the i-th landmark seen by the p-th particle), $\mathbf{z}_k^{[i]}$ is the measurement itself and h is the nonlinear vector function that describes the projection of a predicted scene point location from the 3D ENU coordinates to the image plane. It is the inverse of the transformation described in the Equation (5) through (11). Furthermore, ${}^{p}\mathbf{S}_k^{\mathrm{i}}$, ${}^{p}\mathbf{H}_k^{\mathrm{i}}$ and ${}^{p}\mathbf{K}_k^i$ are respectively the residual covariance, the Jacobian of the vector function h, and the Kalman gain—calculated for a given particle-landmark pair—during time step k.

3.3. Sequential Importance Resampling (SIR) Particle Filter

A particle filter is a mathematical tool capable of the accurate estimation of non-linear, non-Gaussian and multimodal distributions. Among different PF variants, the sequential importance resampling (SIR) approach is suited best for SLAM applications, and most of the previously mentioned works use it to address the simultaneous localization on the mapping problem. Our SLAM algorithm is built upon a SIR PF as well. The filtering operation is conducted by drawing a weighted set of samples that are generated in accordance with a predefined distribution from a set of particles from a previous time step. To address the issue of a particle filter degeneracy, a resampling procedure is run periodically to exclude samples with the lowest weights—i.e., less probable hypotheses.

To achieve a robust and efficient filtering procedure, the proposal's distribution of samples has to match the desired distribution as closely as possible. Therefore, particles representing poses of the camera are sampled at a frequency equal to the IMU data sample rate. This process is performed in accordance to the motion model described by Equation (2), and its PDF is assumed to be in the following form:

$$ p(\mathbf{x}_k|\mathbf{x}_{k-1}, \mathbf{u}_k) \tag{17} $$

Performing SLAM, rather than simple navigation, indicates the inclusion of a set of extracted landmarks in the filtering process. The straightforward approach of recovering the momentary camera pose and the map of its surroundings with a particle filter is given by the joint posterior:

$$ p\left(\mathbf{x}_k, \mathbf{m}_k^{[1:M]}\middle|\mathbf{x}_{k-1}, \mathbf{u}_k, \mathbf{z}_k\right) \tag{18} $$

where $m_k^{[1:M]}$ is a set of all M landmarks observed up to time k. The number of particles required to estimate the PDF accurately grows exponentially, as every newly added landmark increases the number of dimensions of the state space that has to be sampled. Therefore, for SLAM systems designed to navigate robustly and effectively through large areas, a change in this approach is necessary. To solve the issue of a rapid increase of the number of required particles, PF SLAM systems commonly benefit from the fact that the estimations of observed scene points can be treated as conditionally independent, if the robot trajectory is assumed to be known. The application of this relationship results in a special case of the Rao–Blackwellization (RB) of a particle filter and is based on marginalizing landmarks out of the state vector [20]. The standard RB PF is the mathematical basis of the adopted SLAM framework. The consequence of state-vector size reduction is a decrease in the number of samples needed to perform a SLAM routine accurately. The resulting joint posterior is:

$$p\left(x_k, m_k^{[1:M]} \middle| x_{k-1}, u_k, z_k\right) = p(x_k | x_{k-1}, u_k) p\left(m_k^{[1:M]} \middle| x_k, z_k\right) \tag{19}$$

and can be further factored out as:

$$\left(x_k, m_k^{[1:M]} \middle| x_{k-1}, u_k, z_k\right) = p(x_k | x_{k-1}, u_k) \prod_{i=1}^{M} \left(m_k^i \middle| x_k, z_k\right) \tag{20}$$

The implementation of this conceptual solution to Rao–Blackwellized SLAM is performed by the division of the estimation task among different filters. The main particle filter models the camera trajectory with a number of weighted samples, while landmark positions are estimated using EKFs, whose accuracy determine particle weights. Every particle has to maintain a separate EKF for every observed scene point. This implicates the computational complexity of $O(MN)$ for N-particle distribution, describing the pose of a camera and an M-landmark map.

3.4. Weighting Approach

The samples in a particle filter are weighted according to the likelihood functions that describe the accuracy of camera readings, given the predicted landmarks and sensor locations. The weight of a particle is inversely proportional to the innovation of observed landmarks $\mathbf{y}_k^{[1:L_k^p]}$ in a time step k which is measured in pixels, in the image frame:

$$w_k^p \sim w_{k-1}^p p\left(\mathbf{y}_k^{[1:L_k^p]} \middle| \mathbf{x}_k^{[p]}\right) \tag{21}$$

where L_k^p is the number of landmarks matched by the particle p at a time k. The weight of a given particle can be further calculated using the following formula:

$$w_k^p = w_{k-1}^p \left|2\pi \mathbf{S}_k^{[1:L_k^p]}\right|^{-1/2} exp\left[-\frac{1}{2}\left(\mathbf{y}_k^{[1:L_k^p]}\right)^T \left(\mathbf{S}_k^{[1:L_k^p]}\right)^{-1} \mathbf{y}_k^{[1:L_k^p]}\right] \tag{22}$$

where $\mathbf{S}_k^{[1:L_k^p]}$ is the innovation covariance matrix constructed for all landmarks that were matched with the previously seen scene points at a time step k. The particle cloud encoding exemplary estimates in a particle filter in the local body frame is shown in Figure 3.

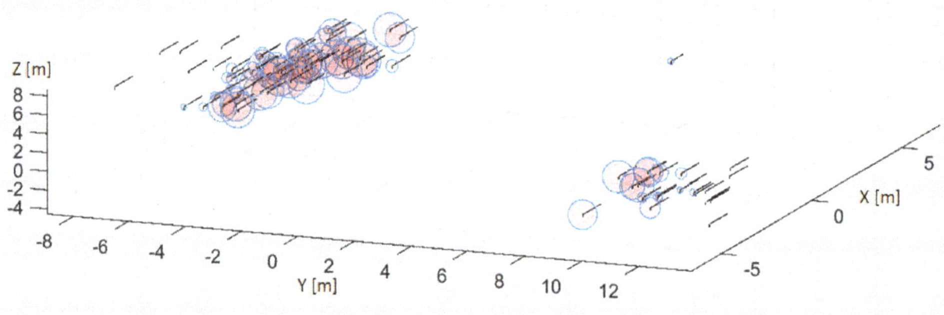

Figure 3. An example of bimodal PDF encoded using weighted particles.

For our algorithm to include particular landmarks in the sample-weight calculation, they have to be correctly associated with previously extracted scene points. The matching procedure consists of two steps. After the visual association of image features, which is based on a comparison of SURF descriptors (and was described in previous sections), we proceed to spatial gating, using the Mahalanobis distance to remove spatial outliers. This step introduces a potentially error-prone dependency. Namely, landmark matches have to be gated in accordance with measurement residuals $y_{k+1}^{[i]}$ for all the potential associations—for every particle. As the set of samples represents the complete PDF of a camera pose, the differences between predicted and observed landmark positions vary between particles, while the matching threshold remains constant. Consequently, the number of landmarks that pass the gating criterion may differ within the particle set. Hence, using the Mahalanobis distance as a straightforward solution to the outliers issue introduces distinctive ambiguities during the particle weighting and resampling procedures. While every observation is assumed to be independent and normally distributed, Equation (22) could be rewritten as:

$$w_k^p = w_{k-1}^p \prod_{i=1}^{L_k^p} \left|2\pi \mathbf{S}_k^{[i]}\right|^{-1/2} exp\left[-\frac{1}{2}\left(\mathbf{y}_k^{[i]}\right)^T \left(\mathbf{S}_k^{[i]}\right)^{-1} \mathbf{y}_k^{[i]}\right] \tag{23}$$

The magnitude of $\mathbf{y}_{k+1}^{[1:L_k^p]}$ depends on the deviation from the true trajectory of the camera for a given particle. Due to the nature of the computation and system parameters, every product factor of the resulting weight w_k^p is significantly smaller than one. Thus, every matched landmark $1 : L_k^p$ causes w_k^p to decrease. Consequently, although the number of matched landmarks is, in general, proportional to the accuracy of the trajectory estimation process, it is not unusual that, for two given particles whose poses resulted in matching a different number of landmarks, the one with more matches would have a lower importance factor. This is caused by the gating step, as the negative impact of the magnitudes of landmark-pose innovations on w_k^p is often lesser than the presence of additional product factors. A natural way to include the missed landmarks in the weight calculation would be to introduce a probability of incorrect association P_{ia} into the Equation (23):

$$w_k^p = w_{k-1}^p (1 - P_{ia})^{L_k^p} \prod_{i=1}^{L_k^p} \left|2\pi \mathbf{S}_k^{[i]}\right|^{-1/2} exp\left[-\frac{1}{2}\left(\mathbf{y}_k^{[i]}\right)^T \left(\mathbf{S}_k^{[i]}\right)^{-1} \mathbf{y}_k^{[i]}\right] {P_{ia}}^{L_{max} - L_k^p} \tag{24}$$

where L_{max} is the maximum number of matched landmarks by any of the particles during a given time step. The tested value of P_{ia} ranged between 0.2 and 0.001 [46]. However, simulations shown in the next section suggest that this idea is insufficient to solve the beforementioned problem—the weights of particles which matched fewer landmarks were often still larger, as the sole adjustments of the P_{ia} did not allow one to tune the weighting process accurately. To efficiently address this issue, we introduce the weight stratification scheme. The idea behind this approach is based on adopting the number of matched landmarks as a primary importance factor criterion. After initial weighting, samples are divided into subsets according to the number of correctly matched scene points and their weights are further adjusted. The adjustment is performed with the addition of offsets to particle weights in a way that separates each of the subsets, creating disjunctive strata of particles. The details of the stratification algorithm are described in the flowchart presented in Figure 4. The graphical interpretation of this procedure is illustrated in Figures 5 and 6. In this way, matching more landmarks by a given particle guarantees having a greater weight.

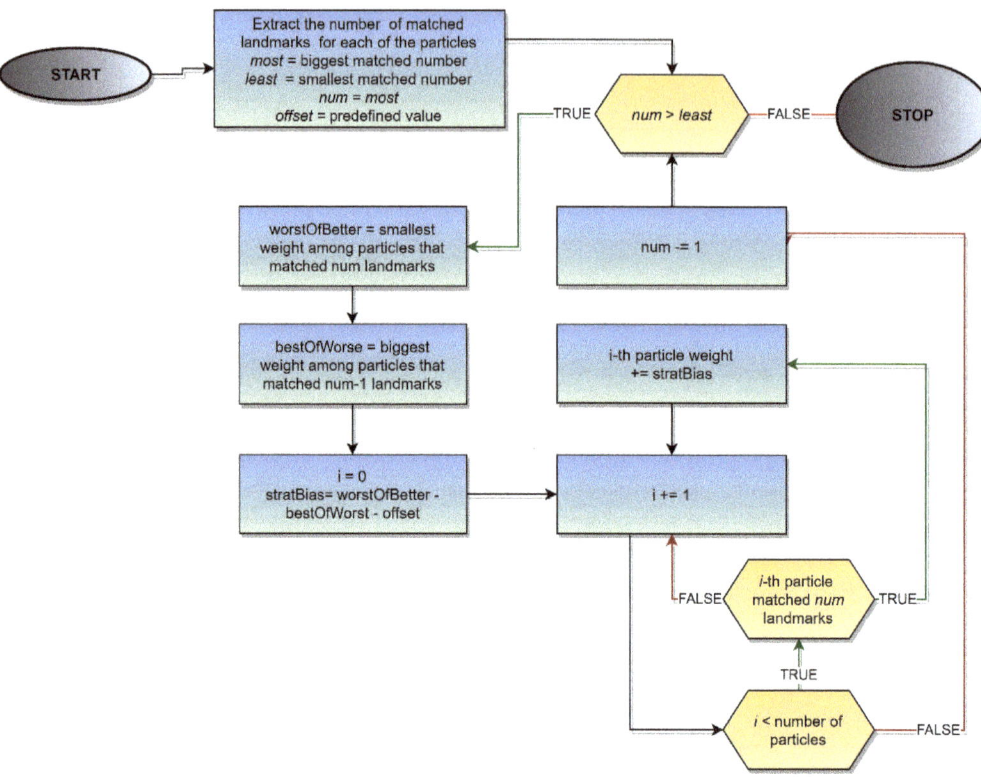

Figure 4. The flowchart of the weights stratification algorithm.

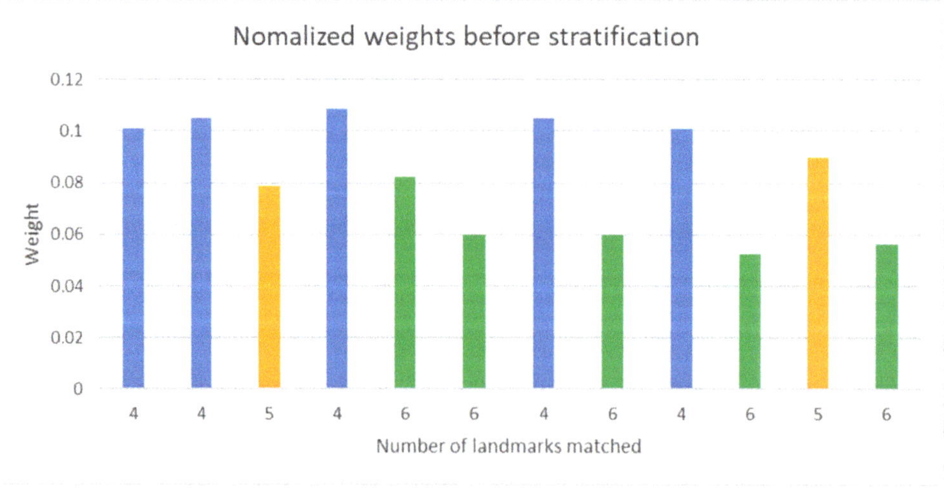

Figure 5. Weights before stratification.

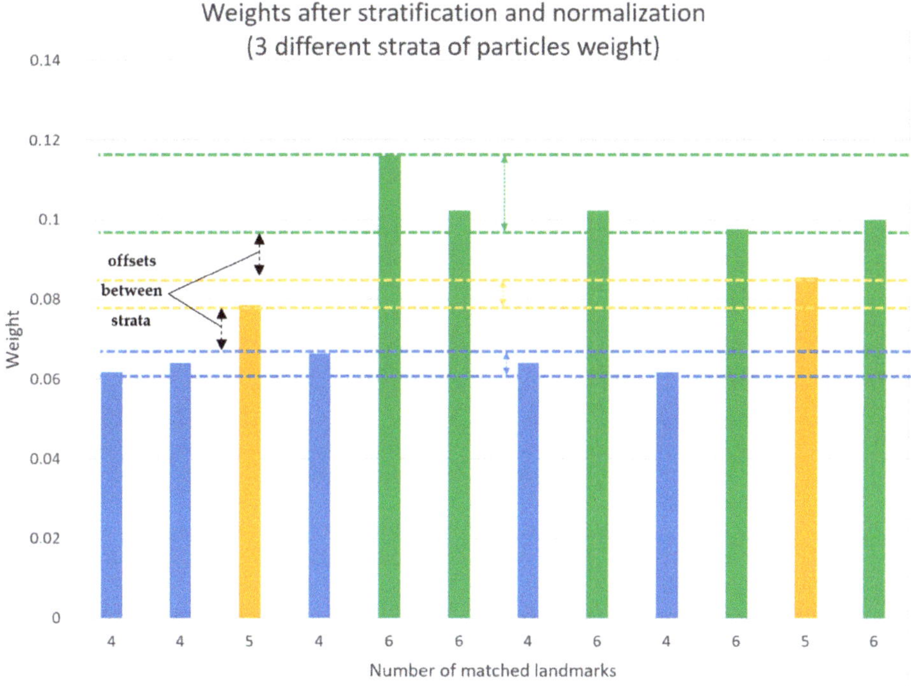

Figure 6. Weights after stratification.

The offset that separates the strata is equal to the following expression:

$$offset = (L_{\max} - L_k^s) \ln P_{ia} \tag{25}$$

where L_k^s is the number of landmarks matched by particles in a given strata s.

Applying the algorithm results in the stratification of sample subsets. Of note is the fact that the procedure is performed in such a manner that particle-weight ratios, in distinct subsets, are preserved. The weights adjustment outcome is presented in the figures below.

After the stratification and normalization, the system performs the resampling procedure, though only if the efficient number of particles N_{eff}, calculated using the expression below, is less than a quarter of the true number of particles.

$$N_{eff} = \frac{1}{\sum_{p=1}^{N} (w^p)^2} \tag{26}$$

This procedure allows one to replace samples of negligible weights with those representing the most probable state hypotheses only when significant disproportions in particle weights occur.

3.5. Landmark Management

To perform large scale simultaneous localization and mapping, a SLAM architecture has to address the issue of efficient landmark management. Particle-filter solutions are especially vulnerable to rapid increases in the number of scene points used for mapping, as every particle represents a unique map. Our SLAM system manages landmarks in a way that minimizes memory usage. To achieve it, we run two separate data sets. One of them stores landmark data that is shared among all particles, such as scene-point coordinates in an image frame, their SURF descriptors, and information on whether landmarks were identified in the currently processed frame and if they were newly observed. Furthermore, a unique ID is assigned to every scene point in the first set. The other database contains information that can be referred to as particle-dependent: landmark positions, covariance matrices, SURF-matching and Mahalanobis-gating results, the number of times landmarks were seen and updated, as well as the number of times landmarks were not observed if their pose indicated otherwise, last-observation and update times, the last angle of observation and the observation-likelihood values.

Furthermore, robust operation requires a flexible approach to landmark initialization and removal. The algorithm distinguishes four different states in which a landmark can be after the data-association procedure. First, a new landmark is extracted and initialized if the number of matched landmarks in the current frame is lower than the predefined threshold of a desired number of landmarks per frame. This prioritizes already-seen scene points over newly observed ones. In addition, scene points that were matched correctly are further examined in terms of angle of observation. If the angle between the current camera pose and the pose of the sensor during the last landmark observation is larger than a predefined threshold, which provides sufficient triangulation conditions, the landmark is updated, using EKF. Otherwise, the landmark is only marked as matched. This artificial limitation of landmark-update frequency is necessary, due to the properties of the EKF covariance-matrix-update equation. If a landmark is not observed, even though it is predicted to be inside the current sensor Field of View (FOV), it is marked as unmatched. After processing all observations, the variables that monitor the number of updates, correct matches and failed observations are updated for all particles and their associated scene points.

The last part of landmark management is landmark removal. The removal assessment is based on the algorithm described by the flowchart in Figure 7 and performed particle-wise.

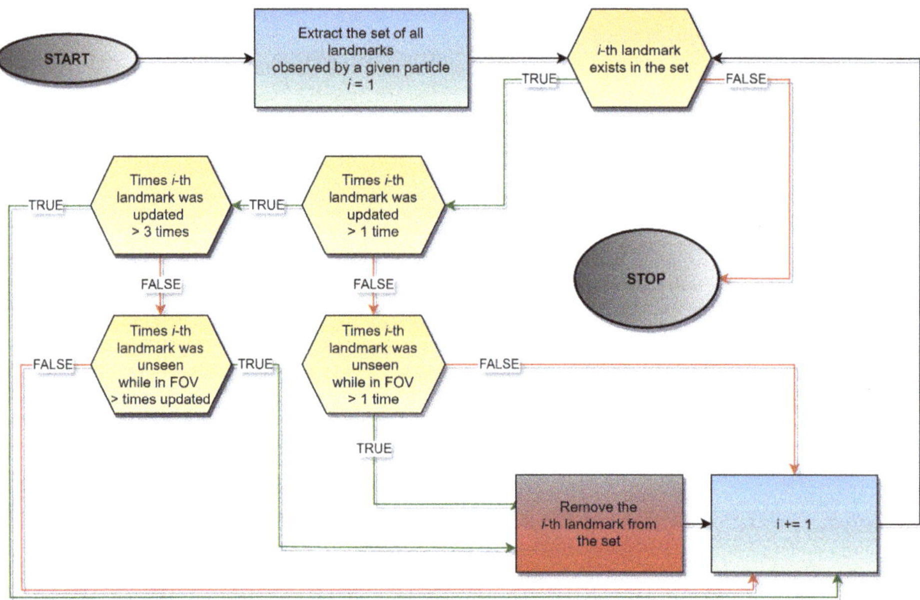

Figure 7. The flowchart of the landmark removal procedure.

Only after a given landmark is removed by all of the particles is its data in the dataset shared among deleted particles as well.

In terms of accuracy, it is more useful to increase the number of scene points, rather than the number of processed image frames per second [9]. Hence, we adjusted our system such that the frames-per-second rate is artificially lowered to less than five, so that the number of extracted, associated and initialized features during every frame could be maximized without increasing the computational burden.

Each loop of the presented particle filter algorithm starts after sensor-data acquisition. First, particle motion prediction is performed, which is characterized by a linear time complexity $O(N)$, where N is the number of particles. Next, the matching, gating and update is carried out for every newly registered landmark. Those procedures also have linear time complexity $O(M)$, where M is the number of landmarks. However, as every sample represents a unique map, landmarks processing has to be performed for every particle resulting in the $O(MN)$ complexity. The stratification and resampling steps have linear time complexity $O(N)$. Hence, the overall algorithm complexity can be reckoned as $O(MN)$.

4. Results

To evaluate the feasibility of our approach, we conducted a series of experiments, using a simulation environment, as well as analyzing real-world data collected by a UAV. Information gathered in both approaches was processed offline, with software developed in MATLAB. We aimed at comparing five gating and weighting approaches to the particle-filter SLAM problem:

1. Our novel weights stratification.
2. Adding a penalty for unmatched landmarks, in accordance to the Equation (24).
3. Weighting particles using only the lowest number of matched landmarks (so that all the samples are evaluated using an equal number of landmarks).
4. Using no gating, as in [33].
5. Gating without addressing the issue of difference in the number of matched landmarks between particles.

However, the significantly poorer performance of the last two methods led to their exclusion from the undertaken evaluation.

4.1. Simulation

Simulations were predominantly used to examine the overall cohesion of different solutions. First of all, this type of approach allowed the SLAM procedure to be performed in precisely defined conditions, e.g., exactly known UAV trajectory, camera orientation, sensor noise, etc. Availability of the reference trajectory is of particular value, as it enables the calculation of positioning errors.

The robotics simulator chosen to generate data for the evaluation of the performance of the compared methods was the Gazebo open-source software [47]. Gazebo was adopted, as it is the most-known robotic simulator, accepted as the de facto software platform for robotics [48–50]. The experiment configuration was built upon a set of two sensors following a predefined trajectory. The camera was pointed directly downwards and the IMU provided 3D accelerations and angular velocities. To achieve similarity between the real world and the simulation environment, we used an aerial photograph, taken from a UAV, and stretched it over the ground plane in Gazebo. The exemplary simulation setup is presented in Figure 8, where the sensor is marked with a red ellipse.

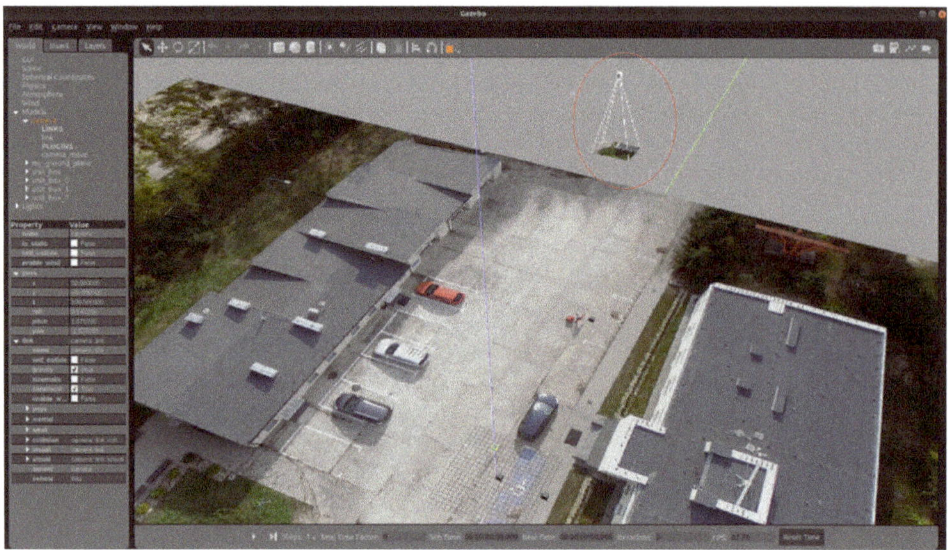

Figure 8. Simulation environment.

An exemplary map-building process (Figure 9) that emerged during one of our SLAM-algorithm runs is presented below. The landmarks are shown as either green circles (those scene points which were updated at least once) with a blue 1σ ellipsoid uncertainty region or as green stars (those scene points that were not yet updated). Ellipsoids in magenta denote landmarks being updated at the given time step. If the landmark is seen (correctly matched) in the currently processed frame, its edge is red, and if a landmark is newly observed, the marker is red.

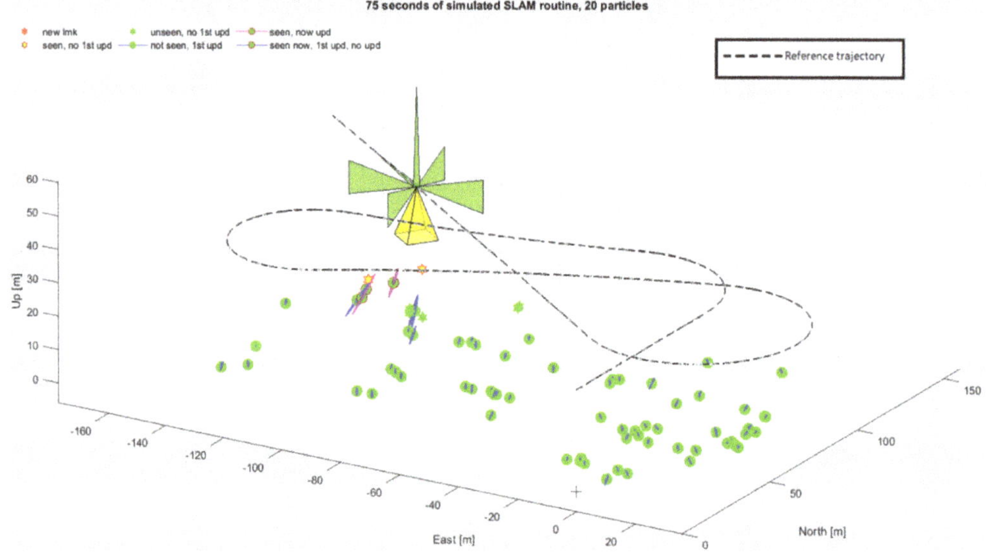

Figure 9. An exemplary map building process.

We tested the listed approaches, using a scenario in which the UAV was flying at an altitude of 30 m. The trajectory of the drone was chosen in a way that created four areas in which there was a possibility for a loop closure. The flight took about 75 s, during which the vehicle traveled almost 900 m. An example of a SLAM procedure's results, for simulations conducted using the above-discussed scenario, is presented in Figure 10, where the reference track is marked in black and all other colored lines represent the paths of single particles. The areas where a potential loop-closing procedure can be performed are marked with blue ellipses.

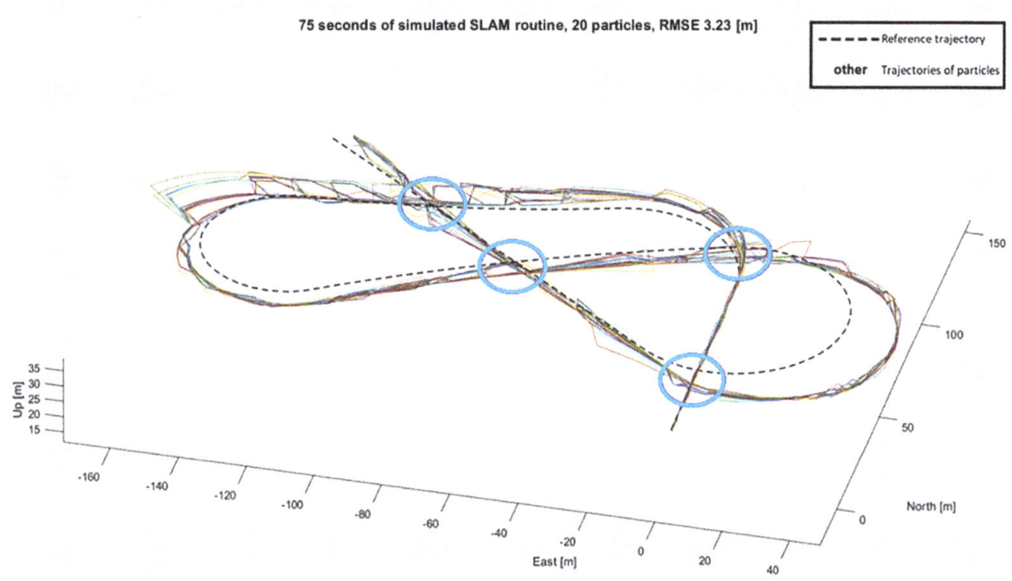

Figure 10. Simulated SLAM routine for 20 particles.

In the example above, the achieved root-mean-square (RMS) error of the trajectory, using only 20 particles, is less than 3.3 m.

To evaluate the performance of different approaches to gating and weighting procedures, we performed a set of simulation runs for each approach, after which we averaged the results. The data collected during the tests are presented in Table 1. The indexes of the weighting and gating approach match the order of the list in the first paragraph of this section.

Table 1. Comparison of simulation results.

Number of particles		20			40			80		
Weighting and gating approach		1	2	3	1	2	3	1	2	3
Percent of correct loop closures	loop 4	90	0	20	80	10	30	90	30	30
	loop 3	90	0	20	80	10	30	100	60	30
	loop 2	100	0	30	100	10	30	100	60	40
	loop 1	100	0	50	100	20	60	100	80	70
Number of resamplings		62.11	-	37.5	68.5	29	50.33	68.22	63.67	53
Root mean squared error [m]		20.34	-	34.23	13.545	29.95	28.68	7.80	17.33	15.97

The compared methods were analyzed in terms of overall accuracy—trajectory RMS error—as well as the ability to correctly close the loop. The loop closure was considered successful if there were landmark matches for scene points which had not been seen during the previous 5 s of observation. It was convenient to analyze this condition using a plot, an example of which can be seen in Figure 11, where all four successful loop closures are visible.

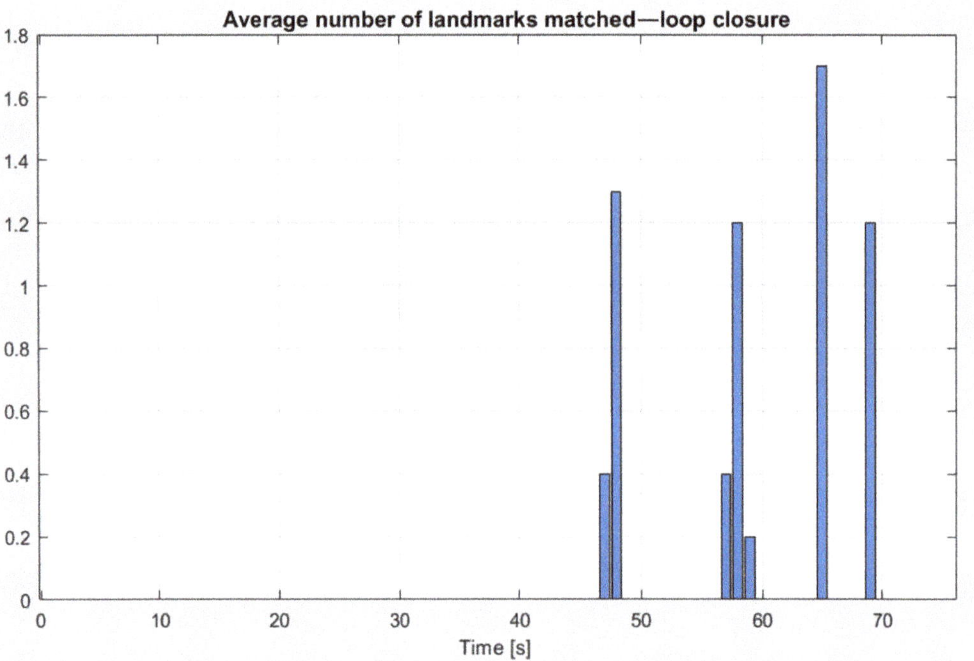

Figure 11. The average number of landmarks which were consequently matched after more than 5 s for the correct closure of all four loops.

Moreover, the total number of resampling procedures which took place during the flight was compared. Every set of the filter settings was evaluated 10 times. However, only those runs which ended in the successful closing of all four loops are included in the mean calculations.

The overall filter performance, interpreted as the ability to follow a given path accurately and evaluated using the amount of correctly closed loops, points to our novel stratification method as the most reliable one. Even for as little as 20 particles, the algorithm using our approach was able to reach the end of the trajectory accurately enough to close the last loop almost every time. On the contrary, the approach, which limited weight adjustment only to the addition of penalties in accordance to Equation (24,) resulted in the inability to correctly follow the true trajectory, even once, when using 20 particles. The increase in the number of particles improved its effectiveness, but even the employment of 80 particles led to successful closures of all loops in only 30% of runs. Although the third approach performed slightly better than the previous one, it was still erroneous most of the time. Similar conclusions can be drawn when analyzing the RMS errors, as the novel stratification method proves to be significantly more accurate.

To present the manner in which different SLAM algorithms diverged from the reference trajectory, three examples are presented in Figures 12–14.

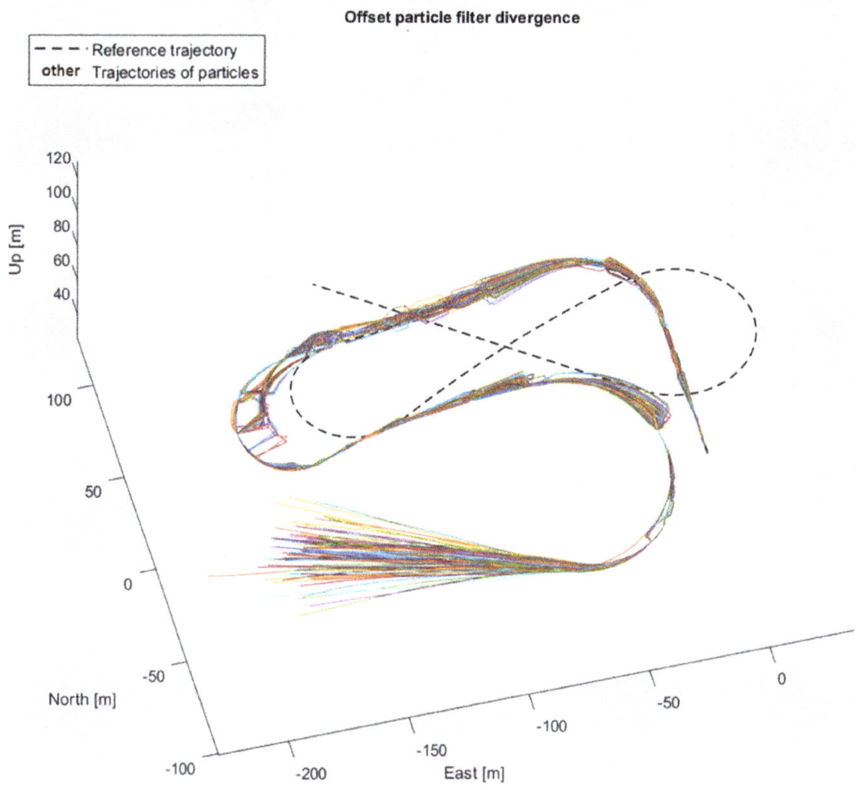

Figure 12. The example of filter divergence (method with adding penalty for unmatched landmarks).

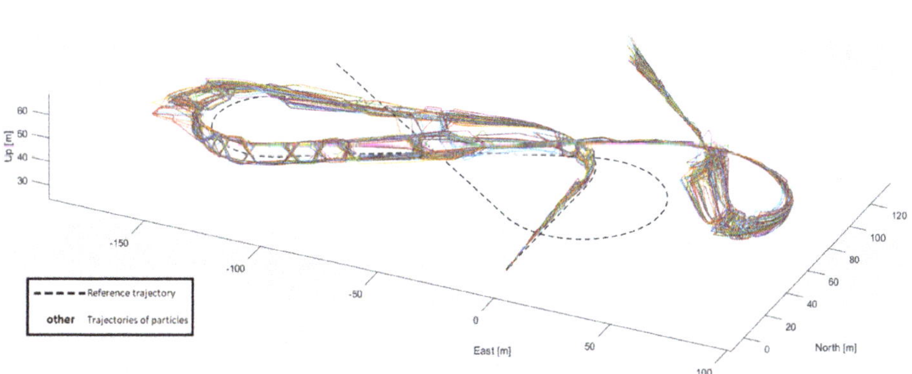

Figure 13. The example of filter divergence (method with weighting particles using only the lowest number of matched landmarks).

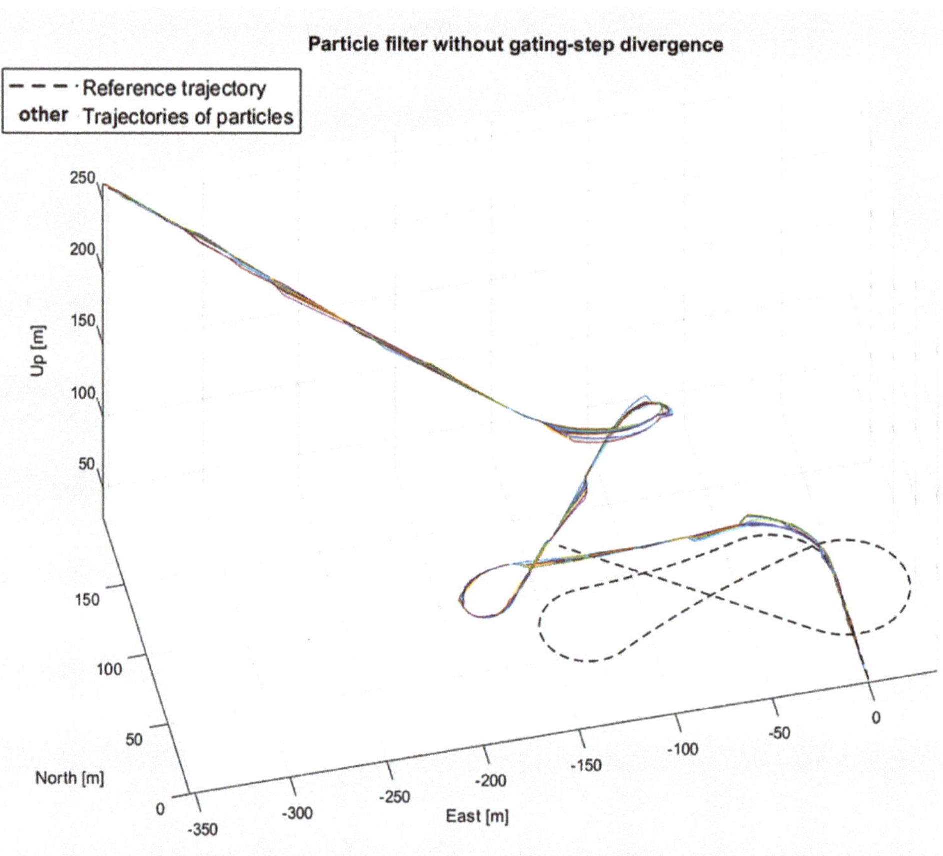

Figure 14. The example of filter divergence (method without gating for the outliers removal).

4.2. Real-World Data

To evaluate the utility of our approach more profoundly, we compared the SLAM procedures using real-world data from a UAV. The data were collected using the DJI Matrice M100, with a Raspberry Pi as an onboard computer—shown in Figure 15. Images were recorded using a Zenmuse X3 camera, while the onboard IMU was the source of kinematic data—one significantly less accurate than the sensor simulated in the previously described experiment.

The flight took place in the area where the aerial picture for the Gazebo simulation was taken. It lasted about 35 s, during which the UAV's altitude varied between 10 and 13 m above ground.

The mapping concept is similar to the one for the simulation experiment, however, we assumed that the starting point of the UAV is point (0,0,0) in the local ENU coordinate frame. As the data were registered outdoors, no ground-truth trajectory was available. Hence, we compared the results of our SLAM algorithm to the data from an INS/GNSS integrated navigation system installed onboard the UAV. Such a trajectory can be used to detect significant errors (e.g., filter divergence), but the RMS trajectory error, calculated with respect to it, can be treated as a crude estimate of positioning error only.

Figure 15. DJI Matrice 100 with an on-board computer.

First, to show the trajectory and the mapping procedure, we present the result of an exemplary SLAM procedure run in Figure 16. The areas of potential loop closure are marked with blue ellipses.

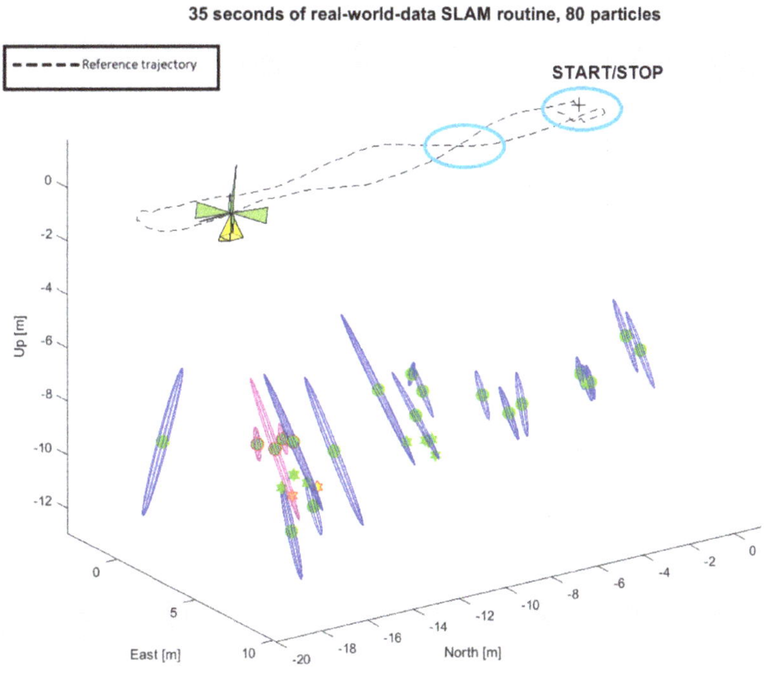

Figure 16. An exemplary map building procedure for real-world data.

Below, the results of comparisons between the SLAM procedures are presented. The algorithm was examined using nine sets of parameters, analogously to the previous examination. However, only two potential loop closures are possible for the assumed flight trajectory. Moreover, the beginning and the end of the trajectory are localized very close in space. The data collected during the runs are presented in Table 2.

Table 2. Comparison of real-world data processing results.

Number of particles		20			40			80		
Weighting and gating approach		1	2	3	1	2	3	1	2	3
Percent of correct loop closures	loop 2	0	0	0	20	-	30	60	0	40
	loop 1	30	0	0	50	-	40	90	0	70
Number of resamplings		-	-	-	85.5	-	72.5	76.25	-	67.17
Root-mean-square error [m]		-	-	-	10.44	-	12.28	5.94	-	7.16

The most apparent conclusion is that the solution of the SLAM problem for this set of data proved to be significantly more difficult than for the simulated data. However, the results are similar, i.e. our stratified-filter approach performed more robustly and more accurately than the others. It was the only filter variant to correctly identify the turning-back maneuver—the first loop closure—using as few as 20 particles. Secondly, the performance of the approach with penalties for unmatched landmarks was least effective. We conducted an additional simulation to see whether this method would close at least the first loop correctly with 160 particles. Still, every run of the algorithm was unsuccessful.

In Figure 17, the trajectories of single particles, together GNSS reference trajectory, are shown, using horizontal projection for image clarity.

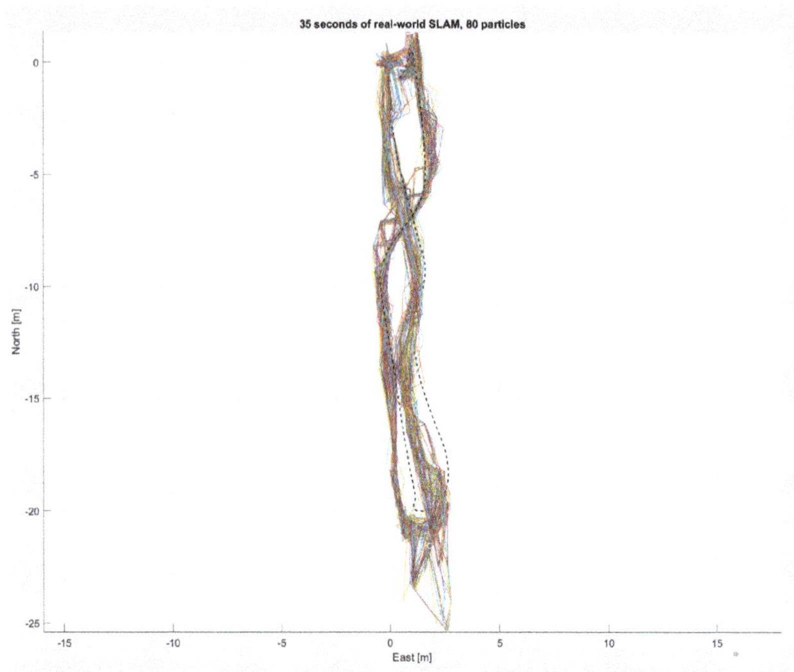

Figure 17. Real-world-data-SLAM routine for 80 particles.

In Figure 18, the average track for this run is presented. The altitude is shown separately in the right subplot.

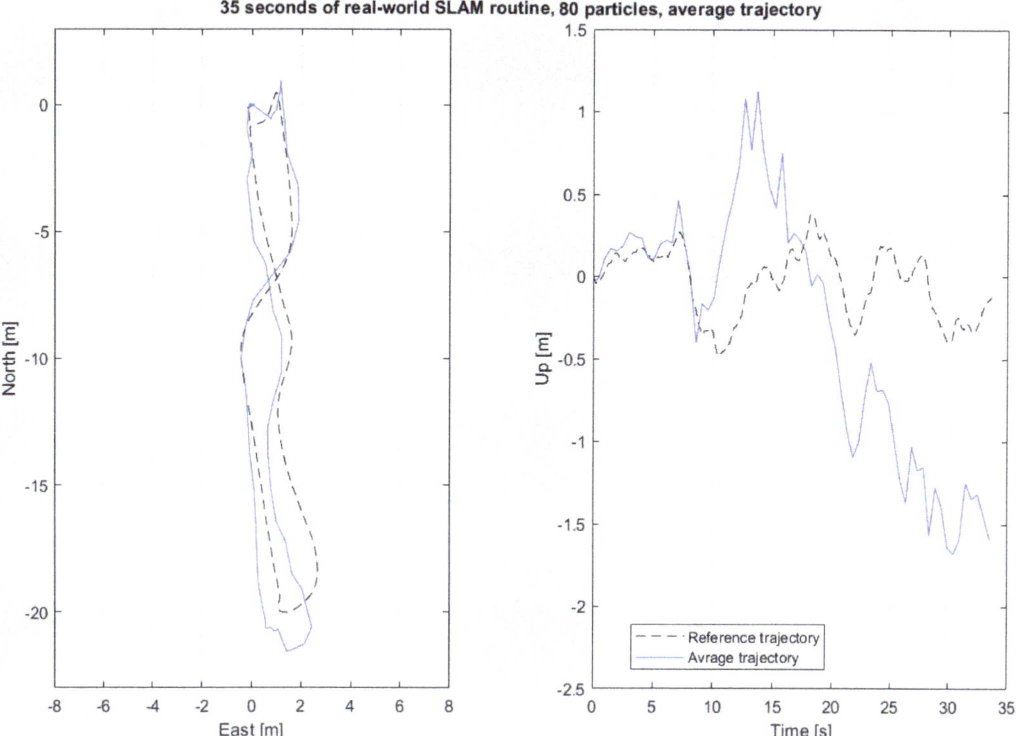

Figure 18. Average trajectory for the real-world-data-SLAM routine for 80 particles.

5. Discussion

The simulations and experiments compiled for compare different weighting and gating approaches for a PF SLAM provide consistent results that justify the implementation of the described stratification algorithm. First of all, its accuracy, measured in terms of RMSE, is superior. Secondly, the robustness of the algorithm is demonstrated by its having the highest percentage of successful runs. Last but not least, the amount of resampling procedures performed in different variants suggests that the method of data processing, in the stratified approach, provided the largest amount of information, as it led to more distinct differences in sample weights and more frequent resampling.

As mentioned earlier, the method of real-world data registration provided no ground-truth trajectory. This can be considered an additional source of uncertainty, in terms of accurate RMSE calculation, however, its impact should not be overestimated, because the INS/GNSS trajectory was available. Still, the removal of such an uncertainty can be pointed out as a future research direction, and we plan to implement our algorithm on a platform which will be capable of performing real-time kinematic (RTK) surveying. This will allow centimeter-level accuracy of positioning in providing reference trajectories.

6. Conclusions

This article discusses a particle-filter-SLAM algorithm that introduces a novel approach to the particle-weighting procedure.

Theoretical analysis and experimental evaluation were conducted for multiple simulated and real-world flights. As a result, the usage of Mahalanobis gating with weight stratification by the number of matched landmarks, was identified to be a beneficial and desirable element of monocular-particle-filter-SLAM algorithms.

The experiments proved that, overall, performance of a particle filter's simultaneous localization-and-mapping algorithm is better when the presented approach is implemented. The stratified particle filter is more robust and accurate than other filter variants. Furthermore, the loop closure is performed more effectively and the particles are resampled more often.

Consequently, the application of the presented algorithm allows to reduce the number of particles—lowering the computational complexity.

Author Contributions: Conceptualization, P.S.; methodology, P.S. and P.K.; software, P.S.; validation, P.S. and P.K.; formal analysis, P.S. and P.K.; investigation, P.S. and P.K.; resources, P.S.; data curation, P.S.; writing—original draft preparation, P.S.; writing—review and editing, P.K.; visualization, P.S.; supervision, P.K.; project administration, P.K.; funding acquisition, P.K. Both authors have read and agreed to the published version of the manuscript.

Funding: This work was financed/co-financed by Military University of Technology under research project UGB 22-856.

Data Availability Statement: The data presented in this study are available on request from the corresponding author.

Conflicts of Interest: The authors declare no conflict of interest.

References

1. Cadena, C.; Carlone, L.; Carrillo, H.; Latif, Y.; Scaramuzza, D.; Neira, J.; Reid, I.; Leonard, J.J. Past, present, and future of simultaneous localization and mapping: Toward the robust-perception age. *IEEE Trans. Robot.* **2016**, *32*, 1309–1332. [CrossRef]
2. Younes, G.; Asmar, D.; Shammas, E.; Zelek, J. Keyframe-based monocular SLAM: Design, survey, and future directions. *Rob. Auton. Syst.* **2017**, *98*, 67–88. [CrossRef]
3. Solà, J.; Vidal-Calleja, T.; Civera, J.; Montiel, J.M.M. Impact of landmark parametrization on monocular EKF-SLAM with points and lines. *Int. J. Comput. Vis.* **2012**, *97*, 339–368. [CrossRef]
4. Durrant-Whyte, H.; Bailey, T. Simultaneous localization and mapping: Part I. *IEEE Robot. Autom. Mag.* **2006**, *13*, 99–108. [CrossRef]
5. Wang, H.; Fu, G.; Li, J.; Yan, Z.; Bian, X. An Adaptive UKF Based SLAM Method for Unmanned Underwater Vehicle. *Math. Probl. Eng.* **2013**, *2013*, 605981. [CrossRef]
6. Montemerlo, M.; Thrun, S.; Koller, D.; Wegbreit, B. FastSLAM: A factored solution to the simultaneous localization and mapping problem. In Proceedings of the National Conference on Artificial Intelligence, Edmonton, AB, Canada, 28 July–1 August 2002; pp. 593–598.
7. Thrun, S.; Montemerlo, M.; Koller, D.; Wegbreit, B.; Nieto, J.; Nebot, E. Fastslam: An efficient solution to the simultaneous localization and mapping problem with unknown data association. *J. Mach. Learn. Res.* **2004**, *4*, 380–407.
8. Grisetti, G.; Kummerle, R.; Stachniss, C.; Burgard, W. A Tutorial on Graph-Based SLAM. *IEEE Intell. Transp. Syst. Mag.* **2010**, *2*, 31–43. [CrossRef]
9. Strasdat, H.; Montiel, J.M.M.; Davison, A.J. Real-time monocular SLAM: Why filter? In Proceedings of the IEEE International Conference on Robotics and Automation, Anchorage, AK, USA, 3–7 May 2010; pp. 2657–2664. [CrossRef]
10. Minkler, G.; Minkler, J. *Theory and Application of Kalman Filtering*; Magellan Book Co.: Palm Bay, FL, USA, 1993; ISBN 0962161829.
11. Mur-Artal, R.; Montiel, J.M.M.; Tardos, J.D. ORB-SLAM: A Versatile and Accurate Monocular SLAM System. *IEEE Trans. Robot.* **2015**, *31*, 1147–1163. [CrossRef]
12. Altermatt, M.; Martinelli, A.; Tomatis, N.; Siegwart, R. SLAM with comer features based on a relative map. In Proceedings of the 2004 IEEE/RSJ International Conference on Intelligent Robots and Systems (IROS) (IEEE Cat. No. 04CH37566), Sendai, Japan, 28 September–2 October 2004; Volume 2, pp. 1053–1058.
13. An, S.-Y.; Kang, J.-G.; Lee, L.-K.; Oh, S.-Y. SLAM with salient line feature extraction in indoor environments. In Proceedings of the 2010 11th International Conference on Control Automation Robotics & Vision, Singapore, 7–10 December 2010; pp. 410–416.
14. Pillai, S.; Leonard, J. Monocular SLAM Supported Object Recognition. In Proceedings of the Robotics: Science and Systems XI, Rome, Italy, 13–17 July 2015; Volume 11.
15. Eade, E.; Drummond, T. Scalable Monocular SLAM Simultaneous Localization and Mapping. In Proceedings of the IEEE Computer Society Conference on Computer Vision and Pattern Recognition, New York, NY, USA, 17–22 June 2006; pp. 469–476. [CrossRef]
16. Civera, J.; Lee, S.H. RGB-D Odometry and SLAM. In *Advances in Computer Vision and Pattern Recognition*; Springer: Berlin/Heidelberg, Germany, 2019; pp. 117–144. ISBN 9783030286033.

17. Bresson, G.; Alsayed, Z.; Yu, L.; Glaser, S. Simultaneous Localization and Mapping: A Survey of Current Trends in Autonomous Driving. *IEEE Trans. Intell. Veh.* **2017**, *2*, 194–220. [CrossRef]
18. Saeedi, S.; Trentini, M.; Li, H.; Seto, M. Multiple-robot Simultaneous Localization and Mapping-A Review 1 Introduction 2 Simultaneous Localization and Mapping: Problem statement. *J. F. Robot.* **2016**, *33*, 3–46. [CrossRef]
19. Kaniewski, P.; Słowak, P. Simulation and Analysis of Particle Filter Based Slam System. *Annu. Navig.* **2019**, *25*, 137–153. [CrossRef]
20. Murphy, K.P. Bayesian map learning in dynamic environments. In *Advances in Neural Information Processing Systems*; MIT Press: Cambridge, MA, USA, 2000; pp. 1015–1021.
21. Qian, K.; Ma, X.; Dai, X.; Fang, F. Improved Rao-Blackwellized particle filter for simultaneous robot localization and person-tracking with single mobile sensor. *J. Control Theory Appl.* **2011**, *9*, 472–478. [CrossRef]
22. Carlone, L.; Kaouk Ng, M.; Du, J.; Bona, B.; Indri, M. Simultaneous localization and mapping using rao-blackwellized particle filters in multi robot systems. *J. Intell. Robot. Syst. Theory Appl.* **2011**, *63*, 283–307. [CrossRef]
23. Xuexi, Z.; Guokun, L.; Genping, F.; Dongliang, X.; Shiliu, L. SLAM algorithm analysis of mobile robot based on lidar. In Proceedings of the 2019 Chinese Control Conference (CCC), Guangzhou, China, 27–30 July 2019; pp. 4739–4745. [CrossRef]
24. Kwok, N.M.; Dissanayake, G. Bearing-only SLAM in indoor environments using a modified particle filter. In Proceedings of the Australasian Conference on Robotics Automation 2003, Brisbane, Australia, 1–3 December 2003.
25. Pupilli, M.L.; Calway, A.D. Real-Time Camera Tracking Using a Particle Filter. In Proceedings of the British Machine Vision Conference, Oxford, UK, 5–8 September 2005; British Machine Vision Association: Durham, UK, 2005; pp. 519–528.
26. Sim, R.; Elinas, P.; Griffin, M.; Little, J.J. Vision-based SLAM using the rao-blackwellised particle filter. In Proceedings of the IJCAI-05 Workshop Reasoning with Uncertainty in Robotics (RUR-05), Edinburgh, UK, 30 July 2005; Volume 14, pp. 9–16.
27. Lowe, D.G. Object recognition from local scale-invariant features. In Proceedings of the IEEE International Conference on Computer Vision, Kerkyra, Greece, 20–27 September 1999; Volume 2, pp. 1150–1157.
28. Lemaire, T.; Lacroix, S.; Sola, J. A practical 3D bearing-only SLAM algorithm. In Proceedings of the 2005 IEEE/RSJ International Conference on Intelligent Robots and Systems, Edmonton, AB, Canada, 2–6 August 2005; pp. 2449–2454.
29. Strasdat, H.; Stachniss, C.; Bennewitz, M.; Burgard, W. Visual Bearing-Only Simultaneous Localization and Mapping with Improved Feature Matching. In Proceedings of the Autonome Mobile Systeme 2007, 20. Fachgespräch, Kaiserslautern, Germany, 18–19 October 2007; pp. 15–21.
30. Bay, H.; Ess, A.; Tuytelaars, T.; Van Gool, L. Speeded-Up Robust Features (SURF). *Comput. Vis. Image Underst.* **2008**, *110*, 346–359. [CrossRef]
31. Kwon, J.; Lee, K.M. Monocular SLAM with locally planar landmarks via geometric rao-blackwellized particle filtering on Lie groups. In Proceedings of the 2010 IEEE Computer Society Conference on Computer Vision and Pattern Recognition, San Francisco, CA, USA, 13–18 June 2010; pp. 1522–1529.
32. Lee, S.-H. Real-time camera tracking using a particle filter combined with unscented Kalman filters. *J. Electron. Imaging* **2014**, *23*, 013029. [CrossRef]
33. Ababsa, F.; Mallem, M. Robust camera pose tracking for augmented reality using particle filtering framework. *Mach. Vis. Appl.* **2011**, *22*, 181–195. [CrossRef]
34. Harris, C.; Stephens, M. A Combined Corner and Edge Detector. In Proceedings of the Alvey Vision Conference, Manchester, UK, 31 August–2 September 1988; pp. 147–151.
35. Çelik, K.; Somani, A.K. Monocular Vision SLAM for Indoor Aerial Vehicles. *J. Electr. Comput. Eng.* **2013**, *2013*, 374165. [CrossRef]
36. Vidal, F.S.; Barcelos, A.D.O.P.; Rosa, P.F.F. SLAM solution based on particle filter with outliers filtering in dynamic environments. In Proceedings of the IEEE 24th International Symposium on Industrial Electronics (ISIE), Buzios, Brazil, 3–5 June 2015; pp. 644–649. [CrossRef]
37. Hoseini, S.; Kabiri, P. A Novel Feature-Based Approach for Indoor Monocular SLAM. *Electronics* **2018**, *7*, 305. [CrossRef]
38. Zhou, Y.; Maskell, S. RB2-PF: A novel filter-based monocular visual odometry algorithm. In Proceedings of the 2017 20th International Conference on Information Fusion (Fusion), Xi'an, China, 10–13 July 2017. [CrossRef]
39. Chen, X.; Zhang, H.; Lu, H.; Xiao, J.; Qiu, Q.; Li, Y. Robust SLAM system based on monocular vision and LiDAR for robotic urban search and rescue. In Proceedings of the 2017 IEEE International Symposium on Safety, Security and Rescue Robotics (SSRR), Shanghai, China, 11–13 October 2017; pp. 41–47. [CrossRef]
40. Wang, S.; Kobayashi, Y.; Ravankar, A.A.; Ravankar, A.; Emaru, T. A Novel Approach for Lidar-Based Robot Localization in a Scale-Drifted Map Constructed Using Monocular SLAM. *Sensors* **2019**, *19*, 2230. [CrossRef]
41. Acevedo, J.J.; Messias, J.; Capitan, J.; Ventura, R.; Merino, L.; Lima, P.U. A Dynamic Weighted Area Assignment Based on a Particle Filter for Active Cooperative Perception. *IEEE Robot. Autom. Lett.* **2020**, *5*, 736–743. [CrossRef]
42. Deng, X.; Mousavian, A.; Xiang, Y.; Xia, F.; Bretl, T.; Fox, D. PoseRBPF: A Rao–Blackwellized Particle Filter for 6-D Object Pose Tracking. *IEEE Trans. Robot.* **2021**. [CrossRef]
43. Yuan, D.; Lu, X.; Li, D.; Liang, Y.; Zhang, X. Particle filter re-detection for visual tracking via correlation filters. *Multimed. Tools Appl.* **2019**, *78*, 14277–14301. [CrossRef]
44. Zhang, Q.; Wang, P.; Chen, Z. An improved particle filter for mobile robot localization based on particle swarm optimization. *Expert Syst. Appl.* **2019**, *135*, 181–193. [CrossRef]
45. Montiel, J.M.M.; Civera, J.; Davison, A.J. Unified inverse depth parametrization for monocular SLAM. *Robot. Sci. Syst.* **2007**, *2*, 81–88. [CrossRef]

46. Tareen, S.A.K.; Saleem, Z. A comparative analysis of SIFT, SURF, KAZE, AKAZE, ORB, and BRISK. In Proceedings of the 2018 International Conference on Computing, Mathematics and Engineering Technologies (iCoMET), Sukkur, Pakistan, 3–4 March 2018; pp. 1–10. [CrossRef]
47. Koenig, N.; Howard, A. Design and use paradigms for gazebo, an open-source multi-robot simulator. In Proceedings of the 2004 IEEE/RSJ International Conference on Intelligent Robots and Systems (IROS) (IEEE Cat. No.04CH37566), Sendai, Japan, 28 September–2 October 2004; Volume 3, pp. 2149–2154.
48. Ivaldi, S.; Peters, J.; Padois, V.; Nori, F. Tools for simulating humanoid robot dynamics: A survey based on user feedback. In Proceedings of the 2014 IEEE-RAS International Conference on Humanoid Robots, Madrid, Spain, 18–20 November 2014; pp. 842–849.
49. Gong, Z.; Liang, P.; Feng, L.; Cai, T.; Xu, W. Comparative Analysis Between Gazebo and V-REP Robotic Simulators. In Proceedings of the ICMREE 2011 International Conference on Materials for Renewable Energy and Environment, Shanghai, China, 20–22 May 2011; Volume 2, pp. 1678–1683. [CrossRef]
50. Körber, M.; Lange, J.; Rediske, S.; Steinmann, S.; Glück, R. Comparing Popular Simulation Environments in the Scope of Robotics and Reinforcement Learning. *arXiv* **2021**, arXiv:2103.04616.

MDPI

Article

Infrared and Visible Image Object Detection via Focused Feature Enhancement and Cascaded Semantic Extension

Xiaowu Xiao [1,*], Bo Wang [1], Lingjuan Miao [1], Linhao Li [1], Zhiqiang Zhou [1], Jinlei Ma [2] and Dandan Dong [3]

1 School of Automation, Beijing Institute of Technology, Beijing 100081, China; wangbo@bit.edu.cn (B.W.); miaolingjuan@bit.edu.cn (L.M.); lilinhao@bit.edu.cn (L.L.); zhzhzhou@bit.edu.cn (Z.Z.)
2 China Helicopter Research and Development Institute, Tianjin 300300, China; majl027@avic.com
3 College of Petroleum, China University of Petroleum, Karamay 834000, China; ddd@cupk.edu.cn
* Correspondence: 3120170438@bit.edu.cn

Abstract: Infrared and visible images (multi-sensor or multi-band images) have many complementary features which can effectively boost the performance of object detection. Recently, convolutional neural networks (CNNs) have seen frequent use to perform object detection in multi-band images. However, it is very difficult for CNNs to extract complementary features from infrared and visible images. In order to solve this problem, a difference maximum loss function is proposed in this paper. The loss function can guide the learning directions of two base CNNs and maximize the difference between features from the two base CNNs, so as to extract complementary and diverse features. In addition, we design a focused feature-enhancement module to make features in the shallow convolutional layer more significant. In this way, the detection performance of small objects can be effectively improved while not increasing the computational cost in the testing stage. Furthermore, since the actual receptive field is usually much smaller than the theoretical receptive field, the deep convolutional layer would not have sufficient semantic features for accurate detection of large objects. To overcome this drawback, a cascaded semantic extension module is added to the deep layer. Through simple multi-branch convolutional layers and dilated convolutions with different dilation rates, the cascaded semantic extension module can effectively enlarge the actual receptive field and increase the detection accuracy of large objects. We compare our detection network with five other state-of-the-art infrared and visible image object detection networks. Qualitative and quantitative experimental results prove the superiority of the proposed detection network.

Keywords: infrared and visible image object detection; convolutional neural network; difference maximum loss function; focused feature enhancement module; cascaded semantic extension module

Citation: Xiao, X.; Wang, B.; Miao, L.; Li, L.; Zhou, Z.; Ma, J.; Dong, D. Infrared and Visible Image Object Detection via Focused Feature Enhancement and Cascaded Semantic Extension. *Remote Sens.* **2021**, *13*, 2538. https://doi.org/10.3390/rs13132538

Academic Editors: Stefano Mattoccia, Piotr Kaniewski and Mateusz Pasternak

Received: 6 June 2021
Accepted: 23 June 2021
Published: 29 June 2021

1. Introduction

Infrared sensors can perform target detection in almost all weather conditions, and are not affected by night, occlusion, or fog. However, infrared images are usually lacking detailed information of the scene and objects within it. By contrast, visible images contain more detail features and are more convenient for visual perception. When it comes to bad weather, night, or occlusion, visible sensors usually tend to lose interesting objects. Combined infrared and visible image object detection (also known as multi-band/multi-sensor image object detection) aims to fuse the advantages of both infrared and visible images, producing more accurate object detection results for scene perception and intelligent decision-making.

In recent years, the convolutional neural network (CNN) has become the most effective way to implement object detection for natural images [1–5]. Inspired by this, more and more scholars have used CNNs to perform infrared and visible image object detection [6–10], in which the infrared image and the visible image have the same resolution and are already registered. Infrared and visible image object detection methods based on CNNs generally include two stages: a feature extraction stage and an object detection stage. For the object detection in the natural image, only one base CNN (e.g., VGG16 [11], ResNet [12], DenseNet [13]) is used to extract convolutional features. Unlike the detection in natural images, infrared and visible image object detection usually needs two base CNNs to respectively extract infrared features and visible features. The extracted features are then combined in one convolutional layer (usually the last layer), or multiple convolutional layers. In the object detection stage, combined features are utilized to classify and locate interesting objects. The combined features from only one convolutional layer usually face difficulties in detecting multi-scale objects in various scenes. In contrast, the combined features from multiple convolutional layers are more suitable for the detection of multi-scale objects.

In feature extraction stage, since the inputs of the multi-band detection network comprise two images (i.e., an infrared image and a visible image), two base CNNs are added to the whole detection network. The two CNNs are usually designed to be the same in many detection methods [8,9,14]. However, the features contained in the infrared image and the visible image are complementary and quite different. It would be very difficult for two identical CNNs to extract those diverse infrared and visible features. To solve this problem, we design a difference maximum loss function to extract diverse features. The designed loss function would punish similar features and reward different features in order to maximize the diversity and complementarity of the infrared features and the visible features in the extracted features.

In the object detection stage, the extracted features from multiple convolutional layers are usually used to detect multi-scale objects [15,16]. The convolutional layers are usually divided into two groups, that is, shallow layers and deep layers. The shallow layer is at the front of the CNN and has a relatively high resolution, while the deep layer is at the back of CNN and the resolution of the deep layer is relatively high. The features from relatively shallow convolutional layers are responsible for the detection of small objects. Large objects are recognized and located by features from deeper convolutional layers. However, directly using features from shallow or deep layers to implement object detection may result in some drawbacks.

For the detection of small objects, although the shallow detail features are used, these features are still relatively rough and not significant for some small objects. In this case, the shallow layers may not produce good detection results for these small objects. In this paper, we design a focused feature enhancement module to strengthen the shallow convolutional features of small objects, so as to make the detection of small objects easier. The designed module is achieved via the supervised training of semantic segmentation. The segmentation labels can be automatically generated on the basis of ground-truth detection labels (bounding box). In addition, the focused feature enhancement module is only added to the training stage, and not used in the testing stage. Hence, our designed module can effectively improve the detection accuracy of small objects without increasing the testing time.

On the other hand, large objects are usually detected and recognized by using the features from deeper convolutional layers. This is because deeper layers have a larger receptive field and thus contain more semantic and structural information, which is good for the detection of large objects. However, according to practical experience, the actual receptive field is usually smaller than the theoretical receptive field. In this case, convolutional features from deeper layers would not completely cover some large objects, resulting in the decrease of detection accuracy for large objects. In this paper, we propose a cascaded semantic extension module to enlarge the receptive field, in order to improve the detection performance for large objects. The proposed module utilizes multi-scale convolutional kernels and dilated convolutions, and can easily be integrated into original convolutional neural networks.

The rest of this paper is organized as follows. In Section 2, we describe the details of the proposed infrared and visible image object detection algorithm. In Section 3, experimental results and comparisons are given to verify the superiority of the proposed detection method. Some network analysis is given in Section 5. We conclude this paper in Section 6.

2. The Proposed Detection Network

Figure 1 shows the pipeline of the proposed infrared and visible image object detection network. The proposed detection network adopts the architecture of the classical detection method Faster R-CNN [1]. The two base networks (i.e., CNN1 and CNN2) both use the architecture of feature pyramid network (FPN) [15] (as shown in Figure 2) to effectively utilize multiple convolutional layers. The 50 convolutional layers in the ResNet-50 model [12] are used for CNN1 and CNN2. The resolutions of the second stage, the third stage, the fourth stage, the fifth stage are 1/4, 1/8, 1/16, and 1/32, respectively.

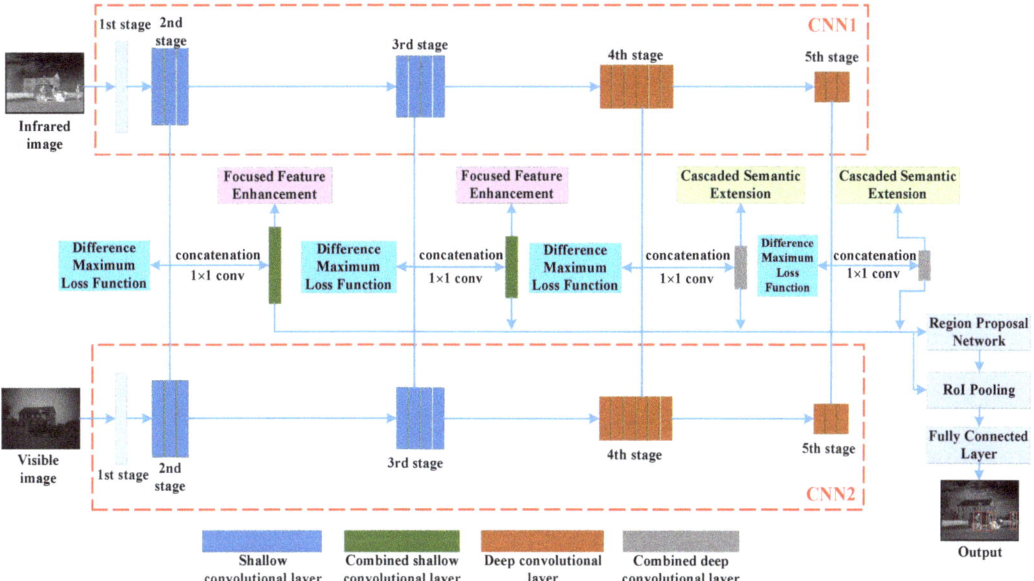

Figure 1. The pipeline of the proposed infrared and visible image object detection network.

Firstly, CNN1 and CNN2 are used to extract the multi-scale features of the infrared and the visible images from multiple convolutional layers. We design the difference maximum loss function to guide the learning directions of the two base networks in order to extract more complementary and diverse multi-band features. Then, infrared features and visible features in each stage (except for the first stage) are respectively combined via concatenation by the channel and a 1×1 convolution. Since the concatenation doubles the number of channels, a 1×1 convolution is added to reshape the channels to the original number.

We define the second stage and the third stage as shallow convolutional layers, and the fourth stage and the fifth stage as deep convolutional layers. The shallow/deep convolutional layers from the infrared image and the shallow/deep convolutional layers from the visible image are combined shallow/deep convolutional layers. We propose a focused feature enhancement module to enhance the shallow features of small objects. In this way, the detection performance of small objects can be effectively improved while keeping the computational cost unchanged. In addition, we design a cascaded semantic extension module to enlarge the receptive field of the deep convolutional layer. The large receptive field is able to increase the detection accuracy of large objects. With the proposed focused feature enhancement module and cascaded semantic extension module, the detection network can more accurately detect small and large objects.

The other components, such as the region proposal network, RoI pooling, and fully connected layer, are the same as those in Faster R-CNN [1]. More details can be found in [1].

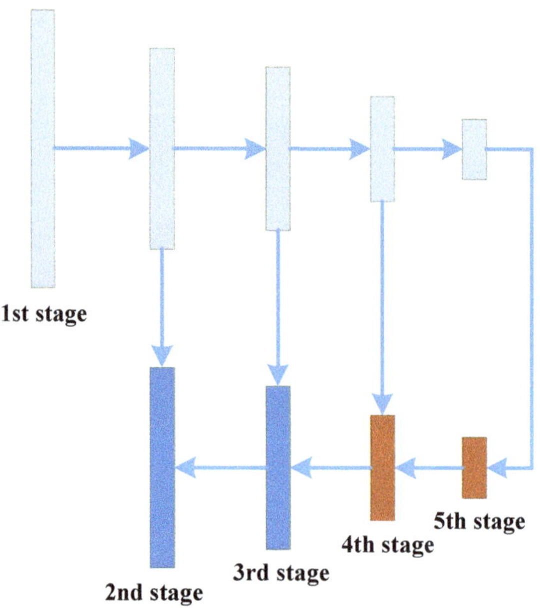

Figure 2. The architecture of the feature pyramid network (FPN).

2.1. Difference Maximum Loss

In recent years, convolutional neural networks (CNNs) have shown great advantages in object detection for natural images [17–19]. This encouraged researchers in related fields to detect and recognize multi-band (infrared and visible) images using CNNs. For infrared and visible image object detection methods based on CNNs, they usually design two base CNNs to respectively extract infrared features and visible features. The two base CNNs can be the same or different.

On the one hand, when two base CNNs are the same, although two CNNs are used to respectively extract features from two images, two networks may learn in the same direction in the training process. In this situation, the extracted features using two of the same networks would not be distinct and complementary. These features are not able to represent the respective advantages of the infrared image and the visible image, resulting in the reduction of the detection accuracy. On the other hand, when the two base CNNs are different, extracted features are usually distinct and complementary. However, the extracted features are complementary only on one or several levels (i.e., not both), because a network with a given structure can only extract one type or several types of features. There are many complementary features on other levels in infrared and visible images. Hence, it is also difficult for two different base CNNs to effectively extract the complementary features.

Although we do not know how many complementary features exist in the infrared image and the visible image, we can be sure that the complementary features must be distinct and varied, because only by combining distinct features can the object detection accuracy be improved. Based on this finding, in this paper we propose a difference maximum loss function. The loss function can guide two base CNNs in learning in different directions in the training process, in order to extract complementary features on more levels. As shown in Figure 1, the input of the difference maximum loss function is set as the last convolutional layer of each stage (except the first stage). By judging the similarity of two convolutional layers, the loss function guides the learning directions of the two base CNNs. Since the loss function is used to extract complementary features, the structures of the two base CNNs are set to be the same, that is, 50 convolutional layers in the ResNet-50 model [12]. We take advantage of Kullback–Leibler (KL) divergence [20] to define the difference maximum loss function:

$$L_d(p_1, p_2) = 1 - \frac{1}{N} \sum_{p_1 \in E_1, p_2 \in E_2} p_1 \log \frac{p_1}{p_2}, \quad (1)$$

where E_1 denotes features from the last convolutional layer of each stage in CNN1, E_2 denotes features from the last convolutional layer of each stage in CNN2, p_1 is the intensity value in each position of E_1, and p_2 is the intensity value in each position of E_2. p_1 and p_2 are computed via softmax function. N denotes the number of features from E_1 or E_2.

The second term in Equation (1) is KL divergence. When CNN1 and CNN2 are learning in different directions, the gap between p_1 and p_2 becomes large, resulting in the enlargement of the KL divergence. In this case, the loss function L_d becomes small, which implies that the learning directions of the two networks can meet the requirements of extracting complementary features. When CNN1 and CNN2 are learning in the same direction, the gap between p_1 and p_2 becomes small, resulting in the decrease of KL divergence. Then, the loss function becomes large, implying that the learning directions of the two networks are incorrect. A large loss function would guide the two networks to learn in different directions in subsequent iterations. Through continuous iterations in the training process, features from the two networks can be diverse and complementary on multiple levels.

2.2. Focused Feature Enhancement

Since R-CNN [21], SPP-Net [22], Fast R-CNN [23], and Faster R-CNN [1] have been used to perform object detection, researchers have found that these CNN-based detection algorithms can produce significantly higher classification and location accuracy than conventional detection algorithms like the Viola–Jones detector [24,25], HOG detector [26], deformable part-based models [27–31], and so on. However, these early CNN-based methods are not able to produce satisfactory detection results for small objects. This is mainly because detection methods like Faster R-CNN only utilize one convolutional layer to perform object detection, and the used convolutional features are rough and sparse, which is bad for the detection of small objects but good for the detection of large objects. To solve this problem, multiple convolutional layers are employed to improve the small-object detection performance in later CNN-based detection methods, including SSD [2], FPN [15], YOLOv4 [32], and so on.

Generally, multiple convolutional layers are usually divided into two groups: shallow layers and deep layers. The deep layer is at the back of CNN and has relatively low resolution. The deep layer contains more semantic features, which are good for the detection of large objects. The early detection methods, such as Fast R-CNN [23] and Faster R-CNN [1], take advantage of the deep convolutional layer to produce more accurate detection results than conventional detection methods. However, the deep layer lacks detail information, which is usually required for the detection of small objects. Hence, it is very difficult for the early detection methods to accurately detect small objects. On the other hand, the shallow layer is in the front of the CNN. The resolution of the shallow layer is relatively large. Thus, the shallow layer contains more detail features and fewer semantic features, which is beneficial for the detection of small objects. Based on the above characteristics, deep layers and shallow layers (i.e., multiple convolutional layers) are used at the same time to perform multi-scale object detection [33–37]. This strategy can produce higher detection accuracy for both small objects and large objects.

However, directly utilizing the shallow layer and the deep layer may present some shortcomings. When it comes to the detection of small objects using shallow convolutional layers, in order to make shallow layers contain more semantic features, the resolution of the initial shallow layer is usually set as 1/4 of the resolution of the input image [2,4]. In this situation, the shallow layer would contain very few features for some small objects, and these features would be rough and not significant. In addition, since some small objects are relatively vague and indistinguishable, the shallow layer may lose features for these small objects. Although shallow layers are more suited to the detection of small objects than deep layers, it is still difficult for shallow layers to accurately detect some small objects.

In order to solve this problem, in this paper, we propose a focused feature enhancement module to strengthen the convolutional features of small objects in shallow layers. A common way to strengthen convolutional features is to stacking many convolutional layers. Although this can be straightforward and effective, it is time- and resource-consuming. To overcome this difficulty, we introduce semantic segmentation to achieve focused feature enhancement. As shown in Figure 1, semantic segmentation is added to shallow convolutional layers (i.e., the second and third stages) of the detection network. Figure 3a,e shows the infrared image and the visible image, respectively. Figure 3b,f shows the ground-truth segmentation labels of subfigures (a) and (e). In the segmentation label, pixels in white regions denote positive samples (i.e., 1), and pixels in black regions denote negative samples (i.e., 0).

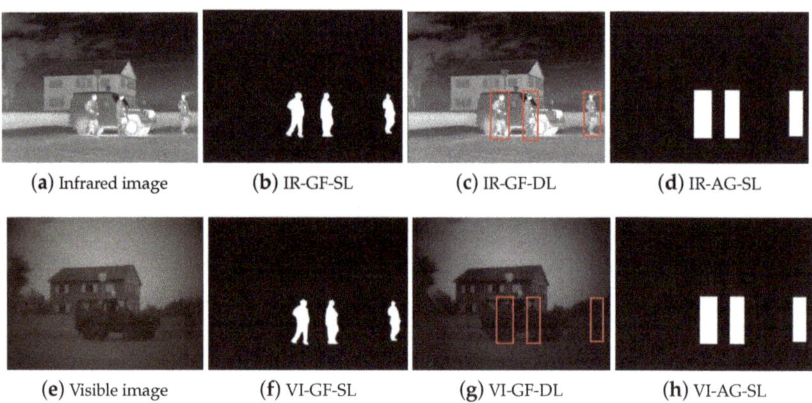

(**a**) Infrared image (**b**) IR-GF-SL (**c**) IR-GF-DL (**d**) IR-AG-SL

(**e**) Visible image (**f**) VI-GF-SL (**g**) VI-GF-DL (**h**) VI-AG-SL

Figure 3. (**a**) The infrared image; (**b**) the ground-truth segmentation label of (**a**), abbreviated IR-GF-SL; (**c**) the ground-truth detection label of (**a**), abbreviated IR-GF-DL; (**d**) the automatically generated segmentation label based on bounding boxes in (**c**), abbreviated IR-AG-SL. (**e**) The visible image; (**f**) the ground-truth segmentation label of (**e**), abbreviated VI-GF-SL; (**g**) the ground-truth detection label of (**e**), abbreviated VI-GF-DL; (**h**) the automatically generated segmentation label based on bounding boxes in (**g**), abbreviated VI-AG-SL.

In the training process, as shown in Figure 4, the last convolutional layer of the shallow layer outputs the segmentation result. The resolution of the last layer of the second stage is $w_2 \times h_2 \times c_2$ (width $\times$ height $\times$ channel). The last layer is then computed with a 3×3 convolution to produce the segmentation result, whose resolution is $w_2 \times h_2 \times 2$, where 2 denotes the number of the sample category (i.e., positive sample and negative sample). The softmax function is used to resize the feature values of the segmentation result to [0,1]. Finally, we use a cross-entropy loss function to compute the gap between the segmentation result and the ground-truth segmentation label. Based on the gap, the detection network guides shallow layers to be focused on enhancing the features of small objects. This process is also suitable for the third stage. Through the supervised training of semantic segmentation, the features of small objects in the shallow layers can be effectively enhanced in order to improve the small-object detection performance.

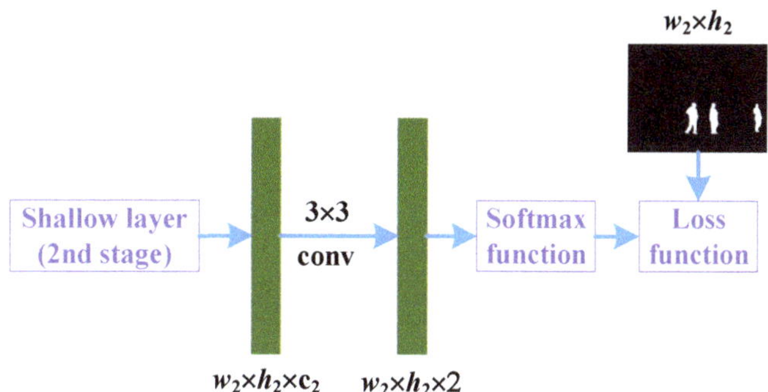

Figure 4. The focused feature enhancement for the second stage.

Since semantic segmentation is used, we need to manually annotate the ground-truth segmentation label for each training image using annotation tool LabelMe (https://github.com/CSAILVision/LabelMeAnnotationTool, accessed on 20 April 2019). However, this will consume too much time and effort. Figure 3c,g shows the ground-truth detection labels of subfigures (a) and (e), respectively. For the relief of the burden, we automatically generate segmentation labels (Figure 3d,h) based on the ground-truth detection labels (Figure 3c,g). This saves significant time and effort. From Figure 3b,d, we can see that the generated labels cover a relatively larger region than the ground-truth labels. Hence, the generated labels (Figure 3d,h) can also be focused on strengthening the features of small objects. The cross-entropy loss function used in semantic segmentation is defined as

$$L_f(h, p, q) = -\lambda \sum_{p \in I_+} \log h_p - \sum_{q \in I_-} \log h_q, \tag{2}$$

where I_+ and I_- denote the positive sample set and the negative sample set, respectively. h_p is the probability that pixel p is classified as a positive sample. h_q denotes the probability that pixel q is classified as a negative sample. h_p and h_q are computed with softmax function. For class balancing, we introduce the weight λ. λ is defined as $\frac{|I_-|}{|I_+|}$, where $|I_-|$ and $|I_+|$ are the number of negative and positive samples, respectively.

The proposed focused feature enhancement module based on semantic segmentation is only added to the training stage. In the testing stage, the shallow layers do not output segmentation results. In this way, the proposed module can effectively increase the detection rate of small objects without increasing the testing time.

2.3. Cascaded Semantic Extension

Compared with the shallow layer, the deep layer is at the back of CNN, and has a relatively lower resolution and larger receptive field. The deep layer contains more semantic features and structure information, which can contribute to a more accurate detection of large objects. The receptive field is defined as the region in the input image that the pixel in the convolutional layer can affect. The deeper the convolutional layer, the larger the receptive field. The larger receptive field can make the pixel in the convolutional layer affect a greater range, and contain more deep features. The reason that the deep convolutional layer is usually taken advantage of to detect large objects is that the deeper layer contributes a larger receptive field.

However, the actual receptive field only occupies a fraction of the theoretical receptive field [38–41]. The actual receptive field has a Gaussian distribution, and pixels at the center of a receptive field have a much larger impact on the output, and the impact of surrounding pixels generally decays quickly. Under this circumstance, convolutional features from the deep layer would not completely cover some large objects, and some important information would be left out when making the prediction. Therefore, this probably induces some decrease of detection accuracy and robustness.

A simple and natural way of enlarging the actual receptive field is to increase the number of convolutional layers. Unfortunately, this would lead to a high computational cost and limit the efficiency of the detection network. In this paper, we propose a cascaded semantic extension module to effectively enlarge the receptive field while still keeping the computational cost under control. As shown in Figure 1, the proposed cascaded semantic extension module is cascaded with deep convolutional layers (i.e., the fourth stage and the fifth stage) of the detection network. The structure of the proposed module is shown in Figure 5. We can see that the proposed module mainly makes use of a multi-branch convolutional layer with different kernel sizes, and each convolution is followed by a dilated convolution with a corresponding dilation rate. The outputs of the three branches are concatenated and then reshaped with a 1×1 convolution to produce the final enhanced features.

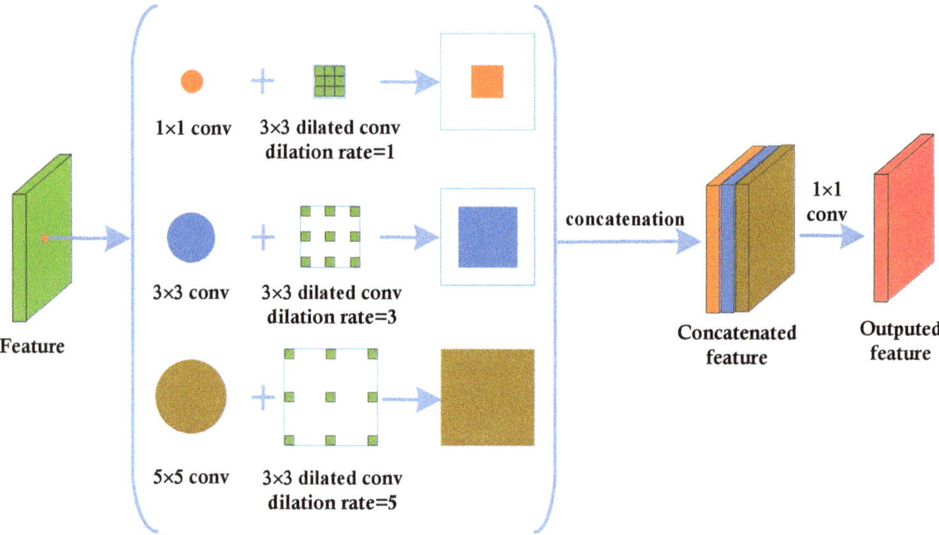

Figure 5. The structure of the proposed cascaded semantic extension module, in which 'conv' denotes convolution.

As shown in Figure 5, in the proposed module, we first design a three-branch convolutional layer (i.e., 1×1 convolution, 3×3 convolution, 5×5 convolution). The 1×1 and 3×3 convolutions are responsible for extracting relatively small-scale features, and the aim of the 5×5 convolution is to extract large-scale features. Through the three-branch convolutional layer, the deep features can be enhanced and the receptive field can be enlarged. To further enlarge the receptive field, we introduce dilated convolution following the three-branch convolutional layer. Dilated convolution [42–44], also known as atrous convolution [41,45,46], aims to generate a feature map with higher resolution, capturing information in a larger area with more context while minimally increasing the computation cost.

We set a 3 × 3 dilated convolution with dilation rate 1 followed by 1 × 1 convolution, a 3 × 3 dilated convolution with dilation rate 3 followed by 3 × 3 convolution, and a 3 × 3 dilated convolution with dilation rate 5 followed by 5 × 5 convolution. The larger the dilation rate, the larger the receptive field. The dilation rate is set as the same as the front convolutional kernel size in order to make different branches focus on enhancing the features with particular sizes. Then, we concatenate the outputs of the three branches by the channel. Finally, a 1 × 1 convolution is used to reshape the concatenated features to the original size.

In order to effectively enlarge the receptive field, we stack three cascaded semantic extension modules following each deep layer. In this way, the actual receptive field can be significantly enlarged and the deep features can also be enhanced. Therefore, the detection performance of large objects can be effectively boosted. Besides, since the proposed module only contains a three-branch convolutional layer, the increase of the computational cost can be very minor.

2.4. End-to-End Training

The proposed detection method adopts the detection architecture of Faster R-CNN, and we define the loss function of Faster R-CNN as L_{Faster}. In this paper, the loss functions of the newly proposed difference maximum loss and focused feature enhancement module are L_d (Equation (1)) and L_f (Equation (2)), respectively. Thus, the loss function of the whole detection network is defined as

$$L = L_{Faster} + L_d + L_f, \tag{3}$$

The base networks CNN1 and CNN2 are initialized with ResNet-50 pre-trained weights for ImageNet classification [12]. The stochastic gradient descent (SGD) optimizer [47] is used to optimize the network parameters. The detection network is trained end-to-end, which means that the proposed detection network can directly output the detection results based on the input images, without any other operations. We use an NVIDIA GTX 1080 Ti GPU to train and test the detection network. The weights of the network are updated with a learning rate of 10^{-4} for the first 50k iterations, and 10^{-5} for the next 50k iterations. The momentum, weight decay, and batch size are set as 0.9, 0.0005, and 2, respectively. The code of the proposed detection network is implemented based on PyTorch [48].

3. Experiments

3.1. Infrared and Visible Image Dataset

In this paper, the used infrared and visible image dataset is collected from [49,50]. Each pair of infrared and visible images are collected from aligned infrared and visible cameras, and each pair of images were already registered. Both the infrared and visible images are single-channel gray images. The used infrared images are far-infrared images. Table 1 lists the composition of the infrared and visible image dataset. We can see that the dataset contains a total of 3318 pairs of infrared and visible images, in which 1641 pairs of images are in the daytime and 1677 are in the night. We randomly select 668 pairs of infrared and visible images as the testing images, and the remaining 2650 pairs of infrared and visible images as the training images. In the 668 pairs of testing images, 352 pairs of images are in the daytime and 316 are in the night. In the 2650 pairs of training images, 1289 pairs of images are in the daytime, and 1361 are in the night. The image sizes include 640 × 471 (width × height) and 640 × 480, and we resize all infrared and visible images into 640 × 480. Data augmentation is introduced to avoid the over-fitting of the detection network. We use two augmentation strategies, that is, horizontal flip and Gaussian blur with standard deviation of 2, to increase the number of training images. Through data augmentation, we get 7950 pairs of infrared and visible images for the training of the detection network.

Table 1. The composition of the infrared and visible image dataset.

	Training Images	Testing Images	Sum
Daytime	1289	352	1641
Night	1361	316	1677
Sum	2650	668	3318

In the infrared and visible image dataset, there are several different object categories, including person, car, tree, building, and so on. The images of people in the person category include people that are still, walking, running, and carrying various things. In this paper, we only detect one category, that is, person. Figure 6 shows some examples in the infrared and visible image dataset. The images in the first row are visible images, and the images in the second row are infrared images. The first two columns show the images in the daytime, and the last three columns show the images in the night. We can see that although the objects in the visible images contain more detail information, the visible images are easily affected by low brightness (see Figure 6h,j), smoke (see Figure 6i), and noise (see Figure 6h). On the other hand, the contrast of the infrared objects is relatively high (see Figure 6c–e), while detail features in infrared objects are missing (see Figure 6c–e). Besides, from the first two columns, we can see that in the daytime, visible images may have better visual effects than infrared images.

(**a**) Infrared Image 1 (**b**) Infrared Image 2 (**c**) Infrared Image 3 (**d**) Infrared Image 4 (**e**) Infrared Image 5

(**f**) Visible Image 1 (**g**) Visible Image 2 (**h**) Visible Image 3 (**i**) Visible Image 4 (**j**) Visible Image 5

Figure 6. (**a**–**j**) Some examples of infrared and visible images from the used image dataset. The images in the first row are infrared images, and the images in the second row are visible images.

3.2. Method Comparison

The proposed detection method is compared with five other infrared and visible image detection methods, including MCDetection [51], FusionDetection [1,52], TwoFusion [53], TripleFusion [8], and IAF R-CNN [9]. The first two detection methods use pixel-level fusion, and the last three methods and our detection method use feature-level fusion. All six methods are respectively trained and tested with the same training images and testing images, and the common network parameters of the six methods are identical.

MCDetection: As shown in Figure 7a, MCDetection [51] first combines a single-channel infrared image and a single-channel visible image into a two-channel pseudo-color image. Then, the two-channel image is used as the input of the detection network Fast R-CNN. Fast R-CNN finally outputs the detection results for the infrared image and the visible image. Since the other five detection networks are designed based on Faster R-CNN, we change the Fast R-CNN in MCDetection to Faster R-CNN. The infrared features and the visible features are fused at the pixel level, and thus MCDetection uses one base CNN to extract infrared and visible features.

FusionDetection: As shown in Figure 7b, FusionDetection first utilizes the multi-band image fusion method HMSD [52] to fuse a single-channel infrared image and a single-channel visible image into a single-channel fused image. Note that other state-of-the-art multi-band image fusion methods [54–58] can also be used for FusionDetection. Then, the fused image is detected with the detection network Faster R-CNN [1]. For the detection method FusionDetection, infrared features and visible features are also fused at the pixel level, and thus one base CNN is used to extract multi-band features.

TwoFusion: As shown in Figure 7c, TwoFusion [53] uses two base CNNs to respectively extract infrared features and visible features, and then the extracted features are fused for the subsequent recognition and location of the objects of interest. The detection architecture of Faster R-CNN is adopted in TwoFusion, and the two base CNNs have the same structure.

TripleFusion: As shown in Figure 8a, TripleFusion [8] proposes a three-branch detection architecture, and takes advantage of infrared features, visible features, and their fused features to respectively perform classification and regression of the region proposal. Then, the output results of the three branches are combined with an accumulated probability fusion layer to produce more accurate detection results. Note that TripleFusion also uses two of the same base CNNs.

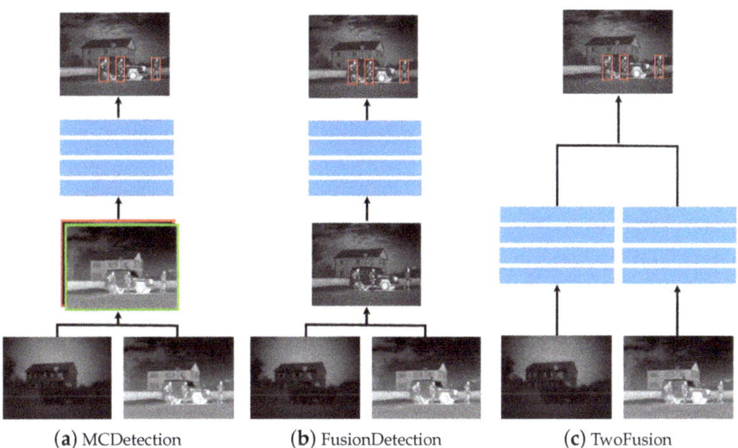

(**a**) MCDetection (**b**) FusionDetection (**c**) TwoFusion

Figure 7. (**a–c**) The pipelines for MCDetection, FusionDetection, and TwoFusion.

IAF R-CNN: As shown in Figure 8b, IAF R-CNN [9] discovers that the detection performance is correlated with illumination conditions. Therefore, IAF R-CNN first uses two detection networks (Faster R-CNN) to respectively produce detection results for the infrared image and the visible image. In this process, the fused features from the two networks are used to generate region proposals. Then, an illumination-aware network is introduced to measure the illumination of the visible image. According to the measured illumination value, the detection results from the two detection networks are adaptively merged to obtain the final detection outputs.

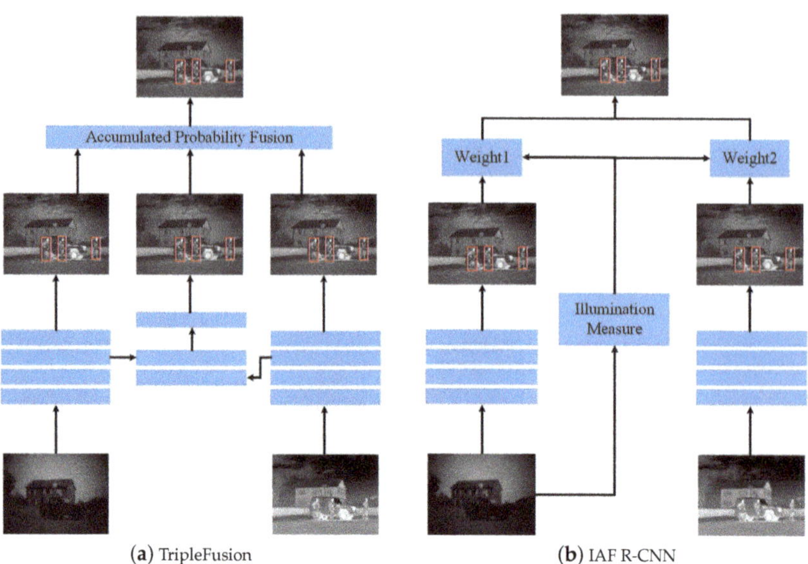

Figure 8. (**a**,**b**) The pipelines for TripleFusion and IAF R-CNN.

3.3. Comparison

Figure 9 shows the infrared and visible image detection results from the six different detection methods. In order to display the results more clearly, the detection results are shown on the fused images of the infrared and visible images, and the fusion method HMSD [52] is used to produce the fused images. Since MCDetection and FusionDetection simply combine an infrared image and a visible image into a fused image, the detection network cannot effectively extract complementary features from only a fused image. Therefore, they produce inaccurate location and classification results (Figure 9c,d). TwoFusion uses two of the same base CNNs to extract complementary and diverse infrared and visible features, and it is very difficult for this strategy to achieve the desired goal. Without complementary features, the detection network TwoFusion gives incorrect detection results (Figure 9e). Although TripleFusion and IAF R-CNN design more complex network structures, the detection results are still unsatisfactory (Figure 9f,g). Thanks to the carefully designed difference maximum loss function, focused feature enhancement module, and cascaded semantic extension module, our detection network gives more accurate detection results.

Figure 10 shows another comparison example. We can see that objects in the images are crowded and not easily distinguished, and some irrelevant objects are very similar to interesting objects to be detected. In this situation, the other five detection methods give inaccurate location or classification results (Figure 10c–g), while the proposed method outputs satisfactory detection results (Figure 10h). Figure 11 shows a similar example to Figure 10. Some objects in Figure 11 are very small, and some lighting can confuse detection networks. From Figure 11c–e, we can see that MCDetection, FusionDetection, and TwoFusion give quite inaccurate location and classification results. The detection results of TripleFusion and IAF R-CNN can also be improved (see Figure 11f,g). In contrast, the output of our detection network is more accurate.

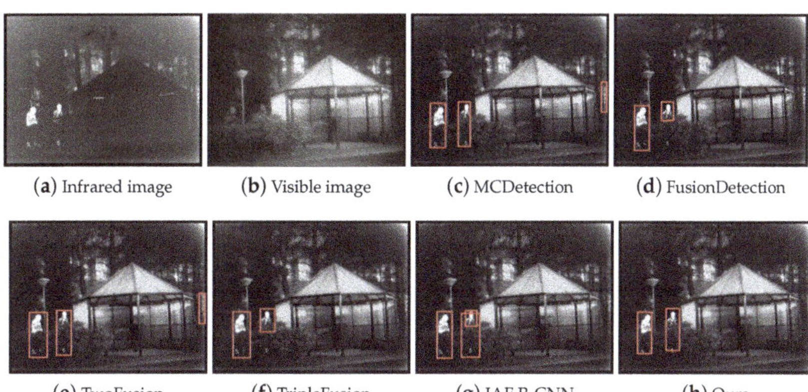

Figure 9. (**a–h**) The infrared and visible image detection results from six different detection methods. In order to better display the results, the detection results are shown on the fused images.

We use mean average precision (mAP) to quantitatively evaluate the detection performance of the six detection networks. Table 2 lists the mAPs of the different detection methods on the testing set. The first column shows the six detection methods, the second column shows the mAPs of the testing images in the daytime, the third column shows the mAPs of the testing images in the night, and the fourth column shows the mAPs of all testing images. MCDetection and FusionDetection only use one base CNN to extract diverse multi-band features, resulting in lower mAP compared with the other detection networks. Since TwoFusion uses the same two CNNs, its detection accuracy is also relatively low. Although TripleFusion and IAF R-CNN introduce well-designed network structures, their mAPs are still lower than that of our detection network.

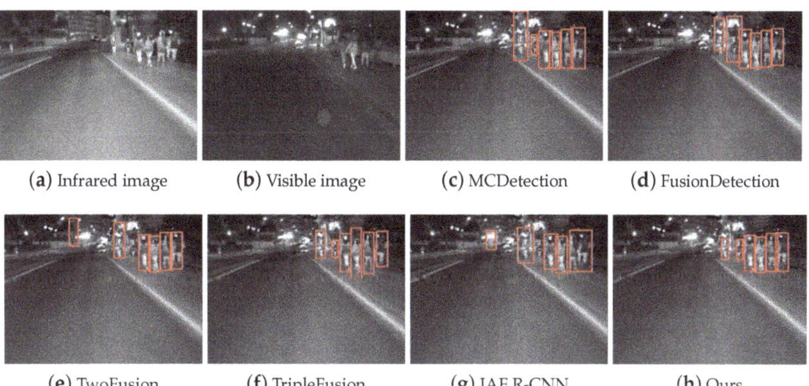

Figure 10. (**a–h**) The infrared and visible image detection results from six different detection methods. In order to better display the results, the detection results are shown on the fused images.

Table 2. The mAP for the different detection methods on the testing set.

Method	Daytime	Night	ALL
MCDetection	74.8%	73.9%	74.3%
FusionDetection	74.4%	75.5%	75.1%
TwoFusion	78.8%	77.5%	78.2%
TripleFusion	81.1%	80.1%	80.5%
IAF R-CNN	80.9%	81.8%	81.3%
Ours	84.3%	83.2%	83.7%

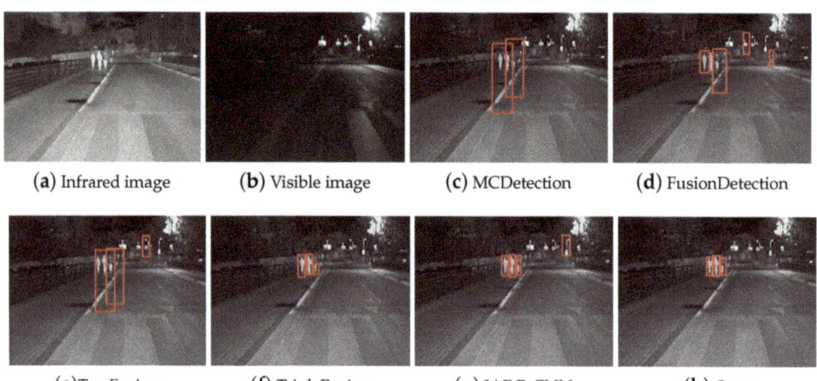

(**a**) Infrared image (**b**) Visible image (**c**) MCDetection (**d**) FusionDetection

(**e**)TwoFusion (**f**) TripleFusion (**g**) IAF R-CNN (**h**) Ours

Figure 11. (**a–h**) The infrared and visible image detection results from six different detection methods. In order to better display the results, the detection results are shown on the fused images.

4. Individual Results

Figure 12 gives some detection results from our detection network. The first column shows infrared images, the second column shows visible images, and the third column shows detection results. From the first three rows, we can see that our method is able to accurately recognize low-contrast objects. Owning to the focused feature enhancement module, the proposed network outputs good detection results for small objects (see the fourth row in Figure 12). Thanks to our proposed cascaded semantic extension module, the large object in the fifth row is accurately located. In the daytime (see the last row), our method still produces good detection performance.

(**a**) Infrared image (**b**) Visible image (**c**) Ours

Figure 12. (**a–c**) Some detection results from our detection network. In order to better display the results, the detection results are shown on the fused images. The images in the first three rows have very low contrast; the objects in the fourth row are relatively small; the object in the fifth row is relatively large; and the images in the last row are in the daytime.

5. Discussion

A. The effectiveness of the proposed difference maximum loss, focused feature enhancement, and cascaded semantic extension. In this paper, we mainly propose three novel modules (i.e., difference maximum loss function, focused feature enhancement module, and cascaded semantic extension module). In order to demonstrate the effectiveness of the three proposed modules, massive experiments are implemented and some results are listed in Table 3. × denotes that the module is not used in the detection network, and ✓ denotes that the module is used in the detection network. All testing images are used to produce the detection accuracy (mAP). We can see that without the three proposed modules, the accuracy of the detection network decreases to 78.9%. Using only the difference maximum loss function, the detection accuracy is 80.7%. Using only the focused feature enhancement module, the detection accuracy is 80.1%. Using only the cascaded semantic extension module, the detection accuracy is 81.6%. Hence, all three modules can improve the detection performance, and the cascaded semantic extension module has the largest benefit. When using two modules for the detection network, the detection accuracy can be further improved. When using all three modules, the detection accuracy reaches a maximum of 83.7%.

Table 3. The effectiveness of difference maximum loss, focused feature enhancement, and cascaded semantic extension.

Difference Maximum Loss	Focused Feature Enhancement	Cascaded Semantic Extension	mAP
×	×	×	78.9%
✓	×	×	80.7%
×	✓	×	80.1%
×	×	✓	81.6%
×	✓	✓	82.5%
✓	×	✓	82.8%
✓	✓	×	82.1%
✓	✓	✓	83.7%

B. The effectiveness for small objects and large objects. In order to demonstrate the detection effectiveness of small objects and large objects, we define objects with size smaller than 48 × 48 in testing images as small objects, and objects with sizes larger than 96 × 96 as large objects. Our method produces a mAP of 76.5% for small objects, while without the focused feature enhancement module the mAP for small objects drops to 74.2%. For large objects, the mAP from our method is 88.6%. Without the cascaded semantic extension module, the mAP for large objects drops to 87.1%.

C. The first stage in CNN1 and CNN2. In the proposed detection network, the first stage in CNN1 and CNN2 is not used for the difference maximum loss function and the focused feature enhancement module (see Figure 1). There are two reasons for this. Firstly, in many other state-of-the-art object detection networks [19,59], the first stage of the base CNN is not used for various designed modules. Secondly, the resolution of the first stage is one-half the resolution of the input image. Therefore, semantic features in the first stage are very few, and features in the first stage are usually edges and gradients. In this case, those features would be useless for the detection of small objects. Edges and gradients have very small differences between two base CNNs, and thus the difference maximum loss function also does not use the first stage. Table 4 lists the detection accuracy with and without the first stage. We can see that it would be good for the designed detection network not to use the first stage.

Table 4. The mAP with and without the first stage for difference maximum loss and focused feature enhancement.

With/Without	mAP
With the first stage	83.5%
Without the first stage	83.7%

D. Automatically generated segmentation labels for the focused feature enhancement. In the focused feature enhancement module, we use automatically generated segmentation labels instead of ground-truth segmentation labels to save time and effort (see Figure 3). Although the automatically generated segmentation labels are relatively inaccurate compared with the ground-truth segmentation labels, the automatically generated segmentation labels can meet the requirements of the focused feature enhancement module. The generated labels can make the enhancement module effectively concentrate on strengthening the features of small objects. Table 5 lists the detection accuracy based on the ground-truth segmentation labels and the automatically generated segmentation labels. We can see that our used strategy (i.e., automatically generated segmentation labels) can produce satisfactory detection results while effectively saving time and effort for annotation.

Table 5. The mAP for the ground-truth segmentation labels and the automatically generated segmentation labels.

Segmentation Labels	mAP
Ground-truth	83.8%
Automatically generated	83.7%

6. Conclusions

In this paper, a novel infrared and visible image object detection network is proposed. First, we design a difference maximum loss function to guide the learning directions of the two base CNNs. In this way, the extracted multi-band features from the two base CNNs can be complementary on more levels and diverse, which is beneficial for the multi-band object detection. Secondly, the proposed focused feature enhancement module is added to the shallow convolutional layer to improve the small-object detection performance. The proposed module is only employed in the training process without increasing the testing time. Finally, in order to enlarge the receptive field of the deep convolutional layer and increase the large-object detection accuracy, a cascaded semantic extension module is introduced. This module can be easily integrated into the detection network while minimally affecting the computational cost. Experimental results demonstrate that the proposed detection network can achieve superior performance compared with many other state-of-the-art detection methods. Further research will include infrared and visible image fusion and semantic segmentation for the infrared and visible images.

Author Contributions: Conceptualization, X.X.; methodology, X.X., B.W. and J.M.; software, X.X.; validation, X.X., B.W.; investigation, B.W. and L.M.; resources, L.L. and L.M.; data curation, X.X.; writing—original draft preparation, X.X., Z.Z. and J.M.; writing—review and editing, X.X., L.M. and J.M.; visualization, B.W., L.L. and D.D.; supervision, B.W. All authors have read and agreed to the published version of the manuscript.

Funding: This research received no external funding.

Institutional Review Board Statement: Not applicable.

Informed Consent Statement: Not applicable.

Data Availability Statement: Not applicable.

Conflicts of Interest: The authors declare no conflict of interest.

References

1. Ren, S.; He, K.; Girshick, R.; Sun, J. Faster R-CNN: Towards Real-Time Object Detection with Region Proposal Networks. *IEEE Trans. Pattern Anal. Mach. Intell.* **2017**, *39*, 1137. [CrossRef]
2. Liu, W.; Anguelov, D.; Erhan, D.; Szegedy, C.; Reed, S.; Fu, C.Y.; Berg, A.C. SSD: Single Shot MultiBox Detector. In *European Conference on Computer Vision*; Springer: Cham, Switzerland, 2016; pp. 21–37.
3. Lin, T.Y.; Goyal, P.; Girshick, R.; He, K.; Dollar, P. Focal Loss for Dense Object Detection. In Proceedings of the IEEE International Conference on Computer Vision, Venice, Italy, 22–29 October 2017; pp. 2999–3007.
4. Redmon, J.; Farhadi, A. Yolov3: An incremental improvement. *arXiv* **2018**, arXiv:1804.02767.
5. Wang, J.; Chen, K.; Yang, S.; Loy, C.C.; Lin, D. Region Proposal by Guided Anchoring. In Proceedings of the 2019 IEEE/CVF Conference on Computer Vision and Pattern Recognition (CVPR), Long Beach, CA, USA, 16–20 June 2019.
6. Alejandro, G.; Fang, Z.; Yainuvis, S.; Joan, S.; David, V.; Xu, J.; Antonio, L. Pedestrian Detection at Day/Night Time with Visible and FIR Cameras: A Comparison. *Sensors* **2016**, *16*, 820.
7. Konig, D.; Adam, M.; Jarvers, C.; Layher, G.; Teutsch, M. Fully Convolutional Region Proposal Networks for Multispectral Person Detection. In Proceedings of the IEEE Conference on Computer Vision and Pattern Recognition (CVPR) Workshops, Honolulu, HI, USA, 22–25 July 2017.
8. Park, K.; Kim, S.; Sohn, K. Unified multi-spectral pedestrian detection based on probabilistic fusion networks. *Pattern Recognit. J. Pattern Recognit. Soc.* **2018**, *80*, 143–155. [CrossRef]
9. Li, C.; Song, D.; Tong, R.; Tang, M. Illumination-aware Faster R-CNN for Robust Multispectral Pedestrian Detection. *Pattern Recognit.* **2019**, *85*, 161–171. [CrossRef]
10. Shopovska, I.; Jovanov, L.; Philips, W. Deep Visible and Thermal Image Fusion for Enhanced Pedestrian Visibility. *Sensors* **2019**, *19*, 3727. [CrossRef]
11. Simonyan, K.; Zisserman, A. Very deep convolutional networks for large-scale image recognition. *arXiv* **2014**, arXiv:1409.1556.
12. He, K.; Zhang, X.; Ren, S.; Jian, S. Deep Residual Learning for Image Recognition. In Proceedings of the IEEE Conference on Computer Vision and Pattern Recognition, Las Vegas, NV, USA, 27–30 June 2016.
13. Huang, G.; Liu, Z.; Laurens, V.D.M.; Weinberger, K.Q. Densely Connected Convolutional Networks. In Proceedings of the IEEE Conference on Computer Vision and Pattern Recognition, Las Vegas, NV, USA, 27–30 June 2016.
14. Hou, Y.L.; Song, Y.; Hao, X.; Shen, Y.; Qian, M. Multispectral pedestrian detection based on deep convolutional neural networks. In Proceedings of the 2017 IEEE International Conference on Signal Processing, Communications and Computing (ICSPCC), Macau, China, 21–24 August 2018.
15. Lin, T.Y.; Dollar, P.; Girshick, R.; He, K.; Hariharan, B.; Belongie, S. Feature Pyramid Networks for Object Detection. In Proceedings of the 2017 IEEE Conference on Computer Vision and Pattern Recognition (CVPR), Honolulu, HI, USA, 21–26 July 2017.
16. Liu, S.; Huang, D.; Wang, Y. Receptive Field Block Net for Accurate and Fast Object Detection. In Proceedings of the European Conference on Computer Vision, Munich, Germany, 8–14 September 2018.
17. Kaiming, H.; Georgia, G.; Piotr, D.; Ross, G. Mask R-CNN. In Proceedings of the IEEE International Conference on Computer Vision, Venice, Italy, 22–29 October 2017.
18. Cao, J.; Cholakkal, H.; Anwer, R.M.; Khan, F.S.; Shao, L. D2Det: Towards High Quality Object Detection and Instance Segmentation. In Proceedings of the 2020 IEEE/CVF Conference on Computer Vision and Pattern Recognition (CVPR), Seattle, WA, USA, 14–19 June 2020.
19. Dong, Z.; Li, G.; Liao, Y.; Wang, F.; Ren, P.; Qian, C. CentripetalNet: Pursuing High-quality Keypoint Pairs for Object Detection. In Proceedings of the IEEE/CVF Conference on Computer Vision and Pattern Recognition, Seattle, WA, USA, 13–19 June 2020
20. Rubner, Y.; Puzicha, J.; Tomasi, C.; Buhmann, J.M. Empirical Evaluation of Dissimilarity Measures for Color and Texture. *Comput. Vis. Image Underst.* **2001**, *84*, 25–43. [CrossRef]
21. Girshick, R.; Donahue, J.; Darrell, T.; Malik, J. Rich Feature Hierarchies for Accurate Object Detection and Semantic Segmentation. In Proceedings of the Computer Vision and Pattern Recognition, Columbus, OH, USA, 23–28 June 2014.
22. He, K.; Zhang, X.; Ren, S.; Sun, J. Spatial Pyramid Pooling in Deep Convolutional Networks for Visual Recognition. *IEEE Trans. Pattern Anal. Mach. Intell.* **2014**, *37*, 1904–1916. [CrossRef]
23. Girshick, R. Fast r-cnn. In Proceedings of the IEEE International Conference on Computer Vision, Santiago, Chile, 7–13 December 2015; pp. 1440–1448.
24. Viola, P. Rapid Object Detection using a Boosted Cascade of Simple Features. In Proceedings of the 2001 IEEE Computer Society Conference on Computer Vision and Pattern Recognition, Kauai, HI, USA, 8–14 December 2001.
25. Viola, P.; Jones, M.J. Robust Real-Time Face Detection. *Int. J. Comput. Vis.* **2004**, *57*, 137–154. [CrossRef]
26. Dalal, N. Histograms of oriented gradients for human detection. In Proceedings of the 2005 IEEE Conference on Computer Vision and Pattern Recognition, San Diego, CA, USA, 20–26 June 2005.
27. Felzenszwalb, P.F.; Mcallester, D.A.; Ramanan, D. A discriminatively trained, multiscale, deformable part model. In Proceedings of the 2008 IEEE Conference on Computer Vision and Pattern Recognition, Anchorage, AK, USA, 23–28 June 2008.
28. Felzenszwalb, P.F.; Girshick, R.B.; Mcallester, D.A. Cascade object detection with deformable part models. In Proceedings of the 2010 IEEE Computer Society Conference on Computer Vision and Pattern Recognition, San Francisco, CA, USA, 13–18 June 2010.
29. Felzenszwalb, P.F.; Girshick, R.B.; McAllester, D.; Ramanan, D. Object Detection with Discriminatively Trained Part-Based Models. *IEEE Trans. Pattern Anal. Mach. Intell.* **2010**, *32*, 1627–1645. [CrossRef] [PubMed]

30. Girshick, R.B.; Felzenszwalb, P.F.; Mcallester, D. Object detection with grammar models. *Adv. Neural Inf. Process. Syst.* **2011**, *24*, 442–450.
31. Girshick, R.B. *From Rigid Templates to Grammars: Object Detection with Structured Models*; University of Chicago: Chicago, IL, USA, 2012.
32. Bochkovskiy, A.; Wang, C.Y.; Liao, H. YOLOv4: Optimal Speed and Accuracy of Object Detection. *arXiv* **2020**, arXiv:2004.10934.
33. Zhou, P.; Geng, C.; Transmission. Scale-Transferrable Object Detection. In Proceedings of the 2018 IEEE/CVF Conference on Computer Vision and Pattern Recognition (CVPR), Salt Lake City, UT, USA, 18–23 June 2018.
34. Singh, B.; Davis, L.S. An Analysis of Scale Invariance in Object Detection—SNIP. In Proceedings of the 2018 IEEE/CVF Conference on Computer Vision and Pattern Recognition (CVPR), Salt Lake City, UT, USA, 18–23 June 2018.
35. Singh, B.; Najibi, M.; Davis, L.S. SNIPER: Efficient Multi-Scale Training. *arXiv* **2018**, arXiv:1805.09300.
36. Zhu, C.; He, Y.; Savvides, M. Feature Selective Anchor-Free Module for Single-Shot Object Detection. In Proceedings of the 2019 IEEE/CVF Conference on Computer Vision and Pattern Recognition (CVPR), Long Beach, CA, USA, 16–20 June 2019.
37. Hong, S.; Roh, B.; Kim, K.H.; Cheon, Y.; Park, M. Pvanet: Lightweight deep neural networks for real-time object detection. *arXiv* **2016**, arXiv:1611.08588.
38. Glorot, X.; Bengio, Y. Understanding the difficulty of training deep feedforward neural networks. *J. Mach. Learn. Res.* **2010**, *9*, 249–256.
39. He, K.; Zhang, X.; Ren, S.; Sun, J. Delving Deep into Rectifiers: Surpassing Human-Level Performance on ImageNet Classification. In Proceedings of the 2015 IEEE International Conference on Computer Vision (ICCV), Santiago, Chile, 7–13 December 2015.
40. Luo, W.; Li, Y.; Urtasun, R.; Zemel, R. Understanding the Effective Receptive Field in Deep Convolutional Neural Networks. In Proceedings of the IEEE Conference on Computer Vision and Pattern Recognition, Honolulu, HI, USA, 21–26 July 2017.
41. Chen, L.C.; Papandreou, G.; Kokkinos, I.; Murphy, K.; Yuille, A.L. DeepLab: Semantic Image Segmentation with Deep Convolutional Nets, Atrous Convolution, and Fully Connected CRFs. *IEEE Trans. Pattern Anal. Mach. Intell.* **2018**, *40*, 834–848. [CrossRef]
42. Yu, F.; Koltun, V. Multi-Scale Context Aggregation by Dilated Convolutions. *arXiv* **2015**, arXiv:1511.07122.
43. Zhang, X.; Zou, Y.; Wei, S. Dilated convolution neural network with LeakyReLU for environmental sound classification. In Proceedings of the 2017 22nd International Conference on Digital Signal Processing (DSP), London, UK, 23–25 August 2017.
44. Qiao, Z.; Cui, Z.; Niu, X.; Geng, S.; Yu, Q. Image Segmentation with Pyramid Dilated Convolution Based on ResNet and U-Net. In *International Conference on Neural Information Processing*; Springer: Cham, Switzerland, 2017.
45. Chen, L.C.; Papandreou, G.; Schroff, F.; Adam, H. Rethinking Atrous Convolution for Semantic Image Segmentation. In Proceedings of the Computer Vision and Pattern Recognition, Columbus, OH, USA, 23–28 June 2014.
46. Chen, J.; Wang, C.; Tong, Y. AtICNet: Semantic segmentation with atrous spatial pyramid pooling in image cascade network. *EURASIP J. Wirel. Commun. Netw.* **2019**, *2019*, 1–7. [CrossRef]
47. Krizhevsky, A.; Sutskever, I.; Hinton, G. ImageNet Classification with Deep Convolutional Neural Networks. *Adv. Neural Inf. Process. Syst.* **2012**, *25*, 1097–1105. [CrossRef]
48. Paszke, A.; Gross, S.; Chintala, S.; Chanan, G.; Yang, E.; DeVito, Z.; Lin, Z.; Desmaison, A.; Antiga, L.; Lerer, A. Automatic Differentiation in Pytorch. 2017. Available online: https://openreview.net/forum?id=BJJsrmfCZ (accessed on 14 March 2019).
49. Toet, A.; Hogervorst, M.A.; Pinkus, A.R. The TRICLOBS Dynamic Multi-Band Image Data Set for the Development and Evaluation of Image Fusion Methods. *PLoS ONE* **2016**, *11*, e0165016. [CrossRef]
50. CVC14 Dataset. Available online: http://adas.cvc.uab.es/elektra/enigma-portfolio/cvc-14-visible-fir-day-night-pedestrian-sequence-dataset (accessed on 25 July 2019).
51. Liu, S.; Liu, Z. Multi-Channel CNN-based Object Detection for Enhanced Situation Awareness. In Proceedings of the Computer Vision and Pattern Recognition, Honolulu, HI, USA, 21–26 July 2017.
52. Zhou, Z.; Wang, B.; Li, S.; Dong, M. Perceptual fusion of infrared and visible images through a hybrid multi-scale decomposition with Gaussian and bilateral filters. *Inf. Fusion* **2016**, *30*, 15–26. [CrossRef]
53. Liu, J.; Zhang, S.; Wang, S.; Metaxas, D.N. Multispectral Deep Neural Networks for Pedestrian Detection. *arXiv* **2016**, arXiv:1611.02644.
54. Li, S.; Kang, X.; Hu, J. Image Fusion With Guided Filtering. *IEEE Trans. Image Process.* **2013**, *22*, 2864–2875.
55. Adu, J.; Gan, J.; Wang, Y.; Huang, J. Image fusion based on nonsubsampled contourlet transform for infrared and visible light image. *Infrared Phys. Technol.* **2013**, *61*, 94–100. [CrossRef]
56. Liu, Y.; Liu, S.; Wang, Z. A general framework for image fusion based on multi-scale transform and sparse representation. *Inf. Fusion* **2015**, *24*, 147–164. [CrossRef]
57. Zhang, Q.; Maldague, X. An adaptive fusion approach for infrared and visible images based on NSCT and compressed sensing. *Infrared Phys. Technol.* **2016**, *74*, 11–20. [CrossRef]
58. Ma, J.; Zhou, Z.; Wang, B.; Zong, H. Infrared and visible image fusion based on visual saliency map and weighted least square optimization. *Infrared Phys. Technol.* **2017**, *82*, 8–17. [CrossRef]
59. Aslam, A. Object Detection for Unseen Domains while Reducing Response Time using Knowledge Transfer in Multimedia Event Processing. In Proceedings of the ICMR 20 International Conference on Multimedia Retrieval, Dublin, Ireland, 8–11 June 2020.

MDPI

Article

Sentinel-1 and 2 Time-Series for Vegetation Mapping Using Random Forest Classification: A Case Study of Northern Croatia

Dino Dobrinić [1], Mateo Gašparović [2,*] and Damir Medak [1]

[1] Chair of Geoinformatics, Faculty of Geodesy, University of Zagreb, 10000 Zagreb, Croatia; ddobrinic@geof.unizg.hr (D.D.); dmedak@geof.unizg.hr (D.M.)
[2] Chair of Photogrammetry and Remote Sensing, Faculty of Geodesy, University of Zagreb, 10000 Zagreb, Croatia
* Correspondence: mgasparovic@geof.unizg.hr; Tel.: +385-1-4639-223

Abstract: Land-cover (LC) mapping in a morphologically heterogeneous landscape area is a challenging task since various LC classes (e.g., crop types in agricultural areas) are spectrally similar. Most research is still mostly relying on optical satellite imagery for these tasks, whereas synthetic aperture radar (SAR) imagery is often neglected. Therefore, this research assessed the classification accuracy using the recent Sentinel-1 (S1) SAR and Sentinel-2 (S2) time-series data for LC mapping, especially vegetation classes. Additionally, ancillary data, such as texture features, spectral indices from S1 and S2, respectively, as well as digital elevation model (DEM), were used in different classification scenarios. Random Forest (RF) was used for classification tasks using a proposed hybrid reference dataset derived from European Land Use and Coverage Area Frame Survey (LUCAS), CORINE, and Land Parcel Identification Systems (LPIS) LC database. Based on the RF variable selection using Mean Decrease Accuracy (MDA), the combination of S1 and S2 data yielded the highest overall accuracy (OA) of 91.78%, with a total disagreement of 8.22%. The most pertinent features for vegetation mapping were GLCM Mean and Variance for S1, NDVI, along with Red and SWIR band for S2, whereas the digital elevation model produced major classification enhancement as an input feature. The results of this study demonstrated that the aforementioned approach (i.e., RF using a hybrid reference dataset) is well-suited for vegetation mapping using Sentinel imagery, which can be applied for large-scale LC classifications.

Keywords: classification; CORINE; feature selection; LUCAS; MDA; random forest; SAR; sentinel

Citation: Dobrinić, D.; Gašparović, M.; Medak, D. Sentinel-1 and 2 Time-Series for Vegetation Mapping Using Random Forest Classification: A Case Study of Northern Croatia. *Remote Sens.* **2021**, *13*, 2321. https://doi.org/10.3390/rs13122321

Academic Editor: Piotr Kaniewski

Received: 12 May 2021
Accepted: 10 June 2021
Published: 13 June 2021

1. Introduction

Vegetation mapping is essential for sustainable forest management, deforestation, agricultural, and silvicultural planning [1,2]. Remotely sensed optical imagery is a common tool for straightforward land-cover classification and vegetation monitoring [3,4]. However, in complex land-cover areas, it is difficult to map multiple classes that are spectrally similar. Therefore, time-series of low to medium-resolution optical satellite imagery (e.g., MODIS, Landsat) have been extensively used for vegetation monitoring since the 1970s [5–8].

In the last decade, time-series imagery provided by the Sentinel-2 (S2) satellites offered a unique opportunity for vegetation mapping [3,9–12]. The S2 satellite, developed from the Copernicus Programme of the European Space Agency (ESA), has a three-day revisit time and a 10 m spatial resolution. However, the acquisition of optical images in key monitoring periods may be limited because of their vulnerability to rainy or cloudy weather. In this context, as a form of active remote sensing that is mostly independent of solar illumination and cloud cover, synthetic aperture radar (SAR) can be used as an important alternative or complementary data source [13]. SAR systems register the amplitude and phase of the backscattered signal, which depends on the physical and electrical properties of the imaged object (e.g., terrain roughness, permittivity) [14]. Recently, multitemporal C-band SAR imagery has been investigated for vegetation monitoring. Gašparović and Dobrinić [15] used single-date and multitemporal (MT) Sentinel-1 (S1) imagery for urban vegetation mapping. Various machine learning methods were used for classification, and the research confirmed the possibility of MT C-band SAR imagery for vegetation mapping.

Recently, integration of SAR and optical (i.e., S1 and S2) data has been mostly used for flood and wetland monitoring or forest disturbance mapping caused by the important abiotic (e.g., fire, drought, wind, snow) and biotic (insects and pathogens) disturbance effects [16–18]. In the research from Gašparović and Dobrinić [15], Figure 1 shows that SAR imagery is neglected for vegetation mapping in land-cover classification tasks compared to optical data and usage of MT series compared to the single-date imagery. Thus, time-series of S1 and S2 imagery provide great potential for vegetation monitoring, and this research investigated the potential of S1, S2, and combined S1 and S2 data for vegetation mapping.

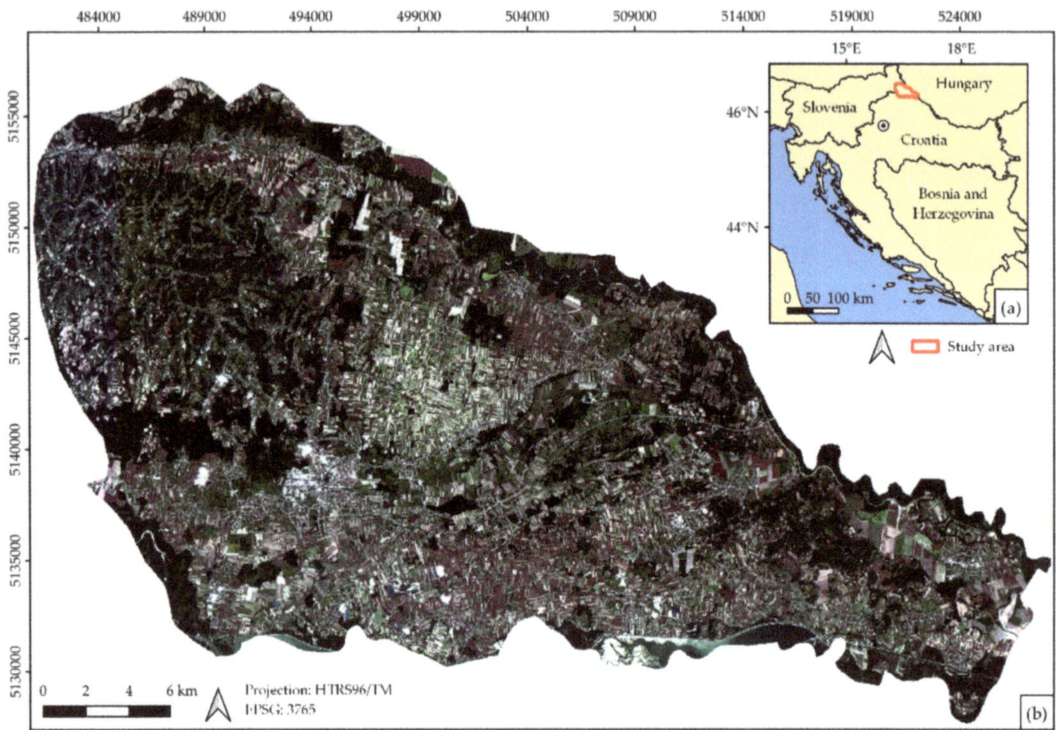

Figure 1. (**a**) Location of the study area and (**b**) overview of the study area (background: true-color composite of Sentinel-2 imagery; bands: B4-B3-B2, acquisition date: 28th September 2018).

Besides using MT optical or SAR imagery for vegetation mapping, recent studies have used vegetation indices and textural features to obtain phenological vegetation information. Jin et al. [19] used normalized difference vegetation index (NDVI) time-series data and textural features computed from the Grey Level Co-occurrence Matrix (GLCM) for land-cover mapping in central Shandong. The highest overall accuracy (OA) of 89% was achieved using multitemporal Landsat 5 TM imagery, topographic (digital elevation model—DEM), NDVI time-series, and textural variables. Furthermore, the influence of the NDVI time-series variables had a greater impact on OA than the influence of textural variables. Gašparović and Dobrinić [20] investigated the impact of different pre-processing steps for SAR imagery when applied to pixel-based classification. Classification using GLCM texture bands (Mean and Variance) increased OA by 19.38% compared to the classification on vertical–vertical (VV) and vertical–horizontal (VH) polarization bands.

Additionally, Lee's spatial filter with a 5 × 5 window size proved the most effective filter for speckle reduction [19].

The use of multi-source and MT remote sensing data creates high-dimensional datasets for classification tasks. Many features in the aforementioned datasets are highly correlated, which causes noise that hinders the classification itself [21]. Although deep learning techniques, especially convolutional neural networks (CNNs), have the ability to extract high-level abstract features for complex image classification tasks, a large training set representative of the considered study area is required [22,23]. Therefore, various feature selection (FS) methods are proposed to address these challenging classification tasks [24]. Following Saeys et al. [25], FS techniques can be organized into three categories: filter methods, wrapper methods, and embedded methods. Filter methods rank the relevance of individual features by their correlation with the dependent variable. Wrapper methods use feature subsets and evaluate them based on the classifier performance [26]. This method is computationally very expensive due to the repeated model classifications and cross-validations. Embedded methods perform FS during the model training, and they combine the qualities of filter and wrapper methods. These methods are mostly embedded within the algorithm, such as Random Forest (RF). A RF classifier, introduced by Breiman in 2001 [27], is a very popular algorithm in a remote sensing community due to the ability to deal with noise, high dimensional, and unbalanced datasets. RF belongs to an ensemble learning algorithm built on decision trees and is increasingly being applied in vegetation mapping using multispectral and radar satellite sensor imagery [4,28–30]. As mentioned before, FS can be done during the modelling algorithm's execution, based on the following indices for variable importance: Mean Decrease Accuracy (MDA) and Mean Decrease Gini (MDG) [24].

Besides using state-of-the-art machine learning methods for vegetation mapping, the overall accuracy of the classified image depends on the quality, quantity, and semantic distribution of the reference data [31]. Balzter et al. [32] investigated SAR imagery for land-cover classification using the CORINE land-cover mapping scheme. The CORINE was initiated in 1985 and consists of a land-cover inventory in 44 classes. In [32], 17 land-cover classes from hybrid CORINE level 2/3 were used as training data for the RF classifier. Besides CORINE, European Land Use and Coverage Area Frame Survey (LUCAS) was carried out by EUROSTAT for identifying land-use and land-cover (LULC) changes across the European Union. Weigand et al. [31] investigated spatial and semantic effects of LUCAS samples using S2 imagery for land-cover (LC) mapping and proposed pre-processing schemes for LUCAS data. RF classifier was used for discriminating the proposed LC class hierarchy of LUCAS samples, and the results indicated that LUCAS data can be used for LULC classifications using S2 data. Belgiu and Csillik [30] used LUCAS data for study areas in Europe for cropland mapping. Depending on the study area, six or seven LC classes were discriminated using a RF classifier. Therefore, suitable reference data for classification tasks must be used to ensure the research's reproducibility and combined with the sampling design and "good practice" recommendations presented in [33,34].

This research aims to assess the classification accuracy using SAR and optical imagery for different scenarios and evaluate the addition of textural features for S1 and spectral indices for S2 imagery. Hence, the use of S1 and S2 time-series, which contain most of the phenological changes, was investigated for vegetation mapping. Moreover, the performance of the RF classifier in a morphologically heterogeneous landscape of northern Croatia was evaluated by using a hybrid reference dataset derived from CORINE, LUCAS, and national Land Parcel Identification Systems (LPIS) LC datasets.

2. Study Area and Datasets

2.1. Study Area

The Međimurje County, illustrated in Figure 1, is one of the main crop product regions and is the northernmost part of Croatia. The study area covers over 700 km^2, from which around 360 km^2 are used in agriculture, which mostly includes fields of cereals, maize, potato, orchards, and vineyards. According to the Koppen–Geiger climate classification system [35], this region has a temperate oceanic climate (Cfb), characterized by warm summer. The mean annual temperature for 2018 in the study area is 10.2 °C with precipitation of 846 mm/year [36].

2.2. Satellite Datasets

For vegetation mapping in heterogeneous land-cover (LC) areas, multitemporal (MT) satellite imagery is used to characterize phenological changes in vegetation LC classes, instead of using single-date imagery [19]. Therefore, SAR and optical satellite imagery from each temperate season have been used in this research. The SAR data are described in Section 2.2.1; the optical satellite imagery is elaborated in Section 2.2.2; ancillary features derived from SAR and optical imagery are introduced in Section 2.2.3, and topographic data used in different classification scenarios is described in Section 2.2.4.

2.2.1. Sentinel-1 Data

Sentinel-1 (S1) Ground Range Detected (GRD) products in dual-polarization mode (VV + VH) were used in this research (Table 1). Data were downloaded from the European Space Agency (ESA) Data Hub and, as a GRD product, imagery has already been detected, multi-looked, and projected to ground range using an Earth ellipsoid model [37]. Additionally, in ESA SNAP software, data were calibrated to sigma naught (σ^0) backscatter intensities, speckle filtered using Lee filter [38], and terrain-correction was made using the shuttle radar topography mission (SRTM) one-arcsecond tiles.

Table 1. Sentinel-1 (S1) and Sentinel-2 (S2) imagery used in this research.

	S1	S1 Orbit	S2	S2 Cloud Cover [%]	ΔS1-S2 [Days]
Date	04-12-2017	ASC	19-12-2017	0.0	15
	27-04-2018	ASC	28-04-2018	0.1	1
	19-08-2018	ASC	16-08-2018	3.6	3
	12-10-2018	ASC	10-10-2018	0.2	2
	29-11-2018	ASC	04-12-2018	1.1	5

2.2.2. Sentinel-2 Data

The Sentinel-2 (S2) constellation includes two identical satellites (S2A and S2B), which carry a multispectral instrument (MSI) for the acquisition of optical imagery at high spatial resolution (i.e., four spectral bands at 10 m, six bands at 20 m, and three bands at 60 m resolution). The S2 sensor acquired optical imagery during the same periods as S1 (Table 1), and the cloud-free tiles were downloaded in Level-2A (L2A), which provides orthorectified Bottom-Of-Atmosphere (BOA) reflectance, with sub-pixel multispectral registration. For this research, bands with 60 m spatial resolution were not considered due to their sensitivity to aerosol and clouds, whereas 20 m spectral bands were resampled to 10 m using the nearest neighbor method to preserve the pixels' original values [39].

2.2.3. SAR Texture Features and Multispectral Indices

Since radar backscatter is strongly influenced by the roughness, geometric shape, and dielectric properties of the observed target, radar-derived texture information represents valuable information for classification tasks. Introduced by Haralick et al. [40], grey-level co-occurrence matrix (GLCM), depending on a given direction and a certain distance in the image, estimates the local patterns in image pixel intensities and spatial arrangement. Among many developed texture measures for vegetation mapping, GLCM, combined with the original radar image, is one of the most trustworthy methods for improving mapping accuracy. In this research, a set of nine texture features, derived from the GLCM, were calculated in the SNAP 8.0 software and used for vegetation mapping: Angular Second Moment (ASM), Contrast, Correlation, Dissimilarity, Energy, Entropy, Homogeneity, Mean, and Variance.

Satellite-based indices are commonly calculated from the spectral reflectance of two or more bands [41]. Using these indices indicates the relative abundance of features of interest, such as canopy chlorophyll content estimations, vegetation cover, and leaf area (Normalized Difference Vegetation Index—NDVI; Enhanced Vegetation Index—EVI; Soil Adjusted Vegetation Index—SAVI; Pigment Specific Simple Ratio—$PSSR_a$) or water surfaces (Normalized Difference Water Index—NDWI). Moreover, modified and refined versions of the aforementioned indices were used (Modified Chlorophyll Absorption in Reflectance—MCARI; Green Normalized Vegetation Index—GNDVI; Modified Soil Adjusted Vegetation Index—MSAVI), as well as indices that use narrower red edge bands from S2 (Normalized Difference Index 45—NDI45; Inverted Red-Edge Chlorophyll Index—IRECI). Table 2 shows the multispectral indices employed in this research.

Table 2. Sentinel-1 (S1) and Sentinel-2 (S2) imagery used in this research.

Spectral Index	Equation *	S2 Bands Used	Reference
NDVI	$\frac{NIR - R}{NIR + R}$	$\frac{B8 - B4}{B8 + B4}$	[42]
NDWI	$\frac{NIR - G}{NIR + G}$	$\frac{B8 - B3}{B8 + B3}$	[43]
EVI	$2.5 * \frac{NIR - R}{NIR + 6.0 * R - 7.5 * B + L}$	$2.5 * \frac{B8 - B4}{B8 + 6.0 * B4 - 7.5 * B2 + 0.5}$	[44]
SAVI	$\frac{NIR - R}{NIR + R + L} * (1 + L)$	$\frac{B8 - B4}{B8 + B4 + 0.5} * 1.5$	[45]
NDI45	$\frac{RE1 - R}{RE1 + R}$	$\frac{B5 - B4}{B5 + B4}$	[46]
MCARI	$[(RE1 - R) - 0.2 * (RE1 - G)] * (RE1 - R)$	$[(B5 - B4) - 0.2 * (B5 - B3)] * (B5 - B4)$	[47]
GNDVI	$\frac{G - R}{G + R}$	$\frac{B3 - B4}{B3 + B4}$	[48]
MSAVI	$\frac{NIR - R}{NIR + R + L} * (1 + L)$	$\frac{B8 - B4}{B8 + B4 + L} * (1 + 0.5)$	[49]
$PSSR_a$	$\frac{RE3}{R}$	$\frac{B7}{B4}$	[50]
IRECI	$\frac{RE3 - R}{RE1 / RE2}$	$\frac{B7 - B4}{B5 / B6}$	[51]

* NIR: Near-infrared band; R: Red band; G: Green band; B: Blue band; RE1, RE2, and RE3 represent Red-edge 1, 2, and 3 band, respectively. L is the adjusted factor that depends on terrain conditions and vegetation cover, where 0 indicates dense vegetation cover, and 1 represents areas without vegetation. In this research, L factor was set to 0.5 [52].

2.2.4. Topographic Data

Some research indicated that using topographic variable data, such as the digital elevation model (DEM) produces major classification enhancements [32,53]. Therefore, shuttle radar topography mission (SRTM) one-arcsecond DEM tiles were resampled to a 10-m spatial resolution by the bilinear interpolation. Additionally, during the test phase of the research, slope and aspect were derived from the SRTM DEM, but their presence as input features brought noise in the dataset, which led to lower classification accuracy. Therefore, the aforementioned features (i.e., slope and aspect) were not further considered as input features for vegetation mapping.

2.3. Reference Data

To reflect the major land-cover classes that are present in the area, reference data were derived, based on our expert knowledge and information, from CORINE, LUCAS, and LPIS land-cover database. Since reference data in the aforementioned databases vary in spatial and semantic consistency, a hybrid classification scheme was devised for this research. Therefore, higher thematic levels from CORINE, LUCAS, and LPIS database were visually interpreted from a time-series of Landsat and Google Earth high-resolution imagery and reduced to the following eight major land-cover classes which were sampled in the study area: cropland, forest, water, built-up, bare soil, grassland, orchard, and vineyard (Table 3). Training and validation pixels were selected at random from the polygon-eroded CORINE and LPIS land-cover maps, and a maximum pixel threshold of 300 pixels per class was set. Afterwards, signatures of the proposed hybrid land-cover classes were checked with LUCAS sample points and visually confirmed from a time-series of high-resolution imagery. This threshold was set following the recommendation from Jensen and Lulla [54] that a number of training pixels should be 10 times the number of the variable used in the classification model. This hybrid approach was proposed in this research, in order to ensure the reproducibility and optimal number of LC classed were chosen since the difference in the number of distinct classes can affect the classification accuracy [55].

Table 3. Description of the major LC classes used in this research, with included codes of CORINE Level 2/3 and LUCAS classification scheme.

ID	Class	CORINE	LUCAS
1	Cropland	2.1 Arable land 2.4 Heterogeneous agricultural areas	B00 Cropland (except B70)
2	Forest	3.1 Forests	C00 Woodland
3	Water	4.1 Inland wetlands 5.1 Inland waters	G00 Water areas
4	Built-up	1.1 Urban fabric	A00 Artificial land
5	Bare soil	3.3 Open spaces with little or no vegetation	F00 Bare land and lichens/moss
6	Grassland	2.3 Pastures 3.2 Scrub and/or herbaceous vegetation associations	D00 Shrubland E00 Grassland
7	Orchard	2.2.2 Fruit trees and berry plantations	B70 Permanent crops: Fruit trees
8	Vineyard	2.2.1 Vineyards	B82 Vineyards

3. Methods

The analysis of potentially separable LC classes was conducted using the time-series of optical NDVI values and radar polarization bands. A total of seven cloud-free scenes of the S2 L2A were used to calculate NDVI profiles to identify areas containing different vegetation and agriculture characteristics [56].

3.1. *Jeffries–Matusita (JM) Distance*

To evaluate the spectral similarity between the LC classes in the reference dataset used for this research, Jeffries–Matusita (*JM*) distance was calculated [57,58]. This spectral separability measure compares distances between the distribution of classes (e.g., A_1 and A_2), which are then ranked according to this distance, following the equation:

$$JM_{A_1A_2} = 2 * \left(1 - e^{-B}\right) \tag{1}$$

where *B* represents the Bhattacharyya distance [56]:

$$B_{A_1A_2} = \frac{1}{8}\left(\overline{m}_{A_1} - \overline{m}_{A_2}\right)^2 \frac{2}{S_{A_1}^2 + S_{A_2}^2} + \frac{1}{2}\ln\left[\frac{S_{A_1}^2 + S_{A_2}^2}{2S_{A_1} + S_{A_2}}\right] \tag{2}$$

where $\overline{m}_{A_i}$ are average values of LC classes A_1 and A_2, and S_{A_i} are their covariance matrices.

3.2. *Random Forest (RF) Classification*

For this research, RF classifier was chosen due to the simple parametrization, feature importance estimation in the classification, and short calculation time [3]. Therefore, optimization of the RF hyperparameters and feature importance estimation as input for vegetation mapping will be explained in Sections 3.2.1 and 3.2.2, respectively.

3.2.1. Hyperparameter Tuning

RF consists of several hyperparameters, which allow users to control the structure and size of the forest (*ntree*) and its randomness (e.g., number of random variables used in each tree—*mtry*). Default values for the *ntree* and *mtry* parameters are 500 and the square root of the number of input variables, respectively. Therefore, a grid search approach with cross-validation was used in this research for hyperparameter tuning, and optimal parameter values were determined as those that produced the highest classification accuracy.

3.2.2. Feature Importance and Selection

During the training phase, RF classifier constructs a bootstrap sample from 2/3 samples of the training dataset, whereas the remaining samples, which are not included in the training subset, are used for internal error estimation called out-of-bag (OOB) error [59]. The random sampling procedure was repeated ten times, allowing to compute average performances with confidence intervals. Afterwards, by evaluating the OOB error of each decision tree when the values of the feature are randomly permuted, the relative importance of each feature can be evaluated [60]. In such a way, mean decrease in accuracy (*MDA*) can be expressed as [24]:

$$MDA_j = \frac{1}{n}\sum_{t=1}^{n}(MP_{tj} - M_{tj}) \tag{3}$$

where n is equal to *ntree*, and M_{tj} and MP_{tj} denote the OOB error of tree t before and after permuting the values of predictor variable X_j, respectively [61]. MDA value of zero indicates that there is no connection between the predictor and the response feature, whereas the larger positive of MDA value, the more important the feature is for the classification.

Another measure for calculating the feature importance is based on the mean decrease in Gini (MDG), which measures the impurity at each tree node split of a predictor feature, normalized by the number of trees. Similar to MDA, the higher the MDG, the more important the feature is. Similar research in the remote sensing community is not united on which measure to use for feature selection using RG in classification tasks. Belgiu and Dragut [62] reported that most studies in their review used MDA, whereas Cánovas-García and Alonso-Sarría [60] obtained the highest accuracy for all of the classification algorithms using MDG. Since former research used pixel-based and latter object-based classification, MDA was used as a measure for feature selection in this research.

3.3. *Accuracy Assessment*

The ability to discriminate LC classes was first assessed using optical NDVI profiles and radar backscatter (i.e., VV and VH) coefficients and JM distance, which measures statistical separability between two distributions.

Afterwards, the validation protocol of different classification scenarios used a stratified random sample of 70% of the reference pixels for training and 30% for the validation [33]. Mean overall accuracy (OA) with confidence intervals was reported in this research since twenty random splits of training and validation data were performed [11]. Besides OA, two simple measures (i.e., quantity disagreement (QD) and allocation disagreement (AD) [63]) were used in this research. As reported in the paper by Stehman and Foody [64], Kappa coefficient is highly correlated with OA, and therefore, we opted for QD and AD. The former measure refers to a difference in a number of pixels of the same class and the latter measure refers to a spatial location mismatch for every LC class between the training and test dataset [65]. Additionally, the user's accuracy (UA), as a measure of the reliability of the map, and producer's accuracy (PA), as a measure of how well the reference pixels were classified, were computed for individual LC classes [66].

In this research, various classification scenarios (package "randomForest" [67]) and accuracy assessments (package 'caret' [68]) were conducted using the R programming language, version 4.0.3., through RStudio version 1.3.1093.

4. Results and Discussion

4.1. *Optical NDVI and Radar Backscatter (VV, VH) Time-Series*

As shown in Figure 2, water and forest can easily be detected throughout the whole season, whereas built-up and bare soil class show a similar pattern, except in August, which can easily be resolved by using additional spectral indices (e.g., SAVI, normalized difference built-up index (NDBI) [69]). Since the cropland class in the investigated study area is consisted mostly of single cropping plant systems (e.g., cereals, maize, and potato), characteristic crop phenology pattern can be recognized, which consists of the sowing (March), growth (from April to August), and harvest (September). The biggest inter-class overlap for separating the vegetation occurs between grassland, orchard, and vineyard. Therefore, in this research, Jeffries–Matusita (JM) distance was used as a spectral separability measure [10,70].

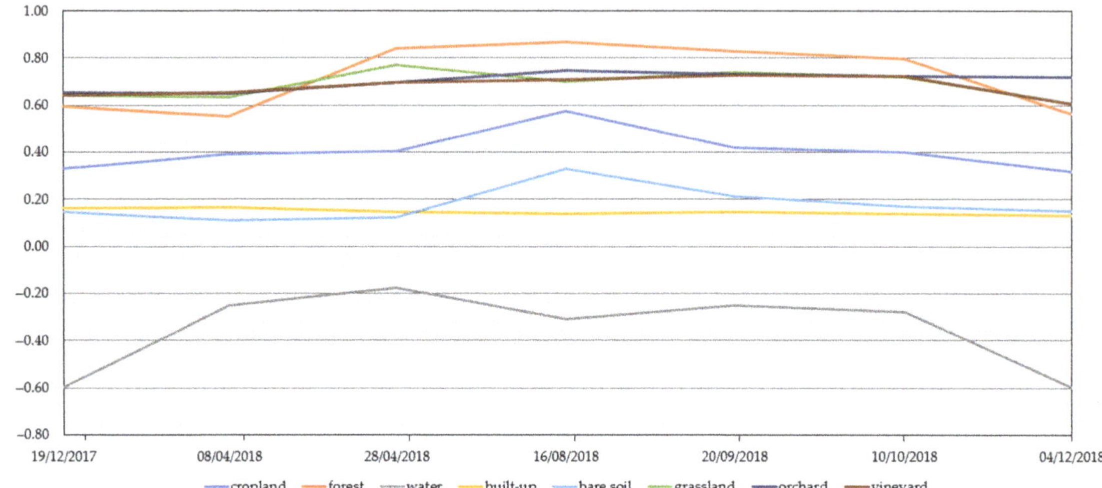

Figure 2. Temporal behavior of optical NDVI time-series profiles for LC class analysis.

Since the backscatter signal is affected by soil moisture, surface roughness, and terrain topography, the VV and VH polarization bands analysis is presented in Figure 3. Overall, the lowest VV and VH values have water class, since only a very small proportion of backscatter is returned to the sensor due to the side-looking geometry [71]. On the other hand, the highest mean VV and VH values consist of the built-up class, due to the double-bounce effect in the urban areas [72]. Vegetation classes tend to overlap within the VV and VH bands due to the volume scattering, whereas the backscatter values are higher in the VV than VH due to a combination of single bounce (e.g., leaves, stems) and bare soil double-bounce backscatter [73].

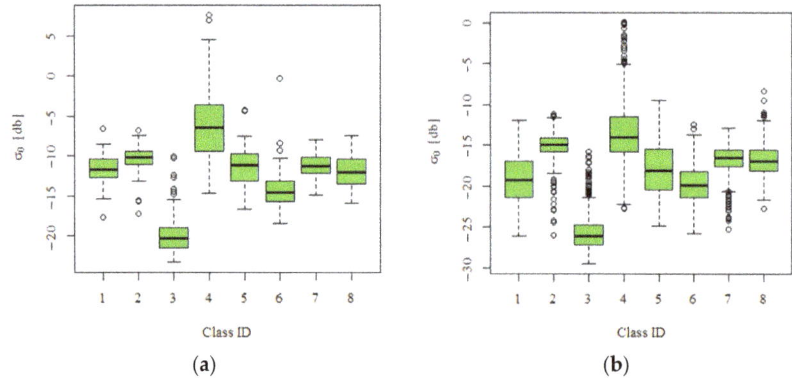

Figure 3. Mean (**a**) VV and (**b**) VH backscatter values, for each LC class investigated. The classes are represented as follows: 1 = Cropland; 2 = Forest; 3 = Water; 4 = Built-up; 5 = Bare soil; 6 = Grassland; 7 = Orchard; 8 = Vineyard.

4.2. Jeffries–Matusita (JM) Distance Variability Results of Each Class

The JM distance results for the similarity of each class and each sensor calculated are shown in Table 4. For both sensors used in this research (i.e., S1 and S2), the water class was the only LC class identified with JM values above 1.7, which indicates good separability with other classes. This class separability was also confirmed with calculated NDVI profiles (Figure 2). Furthermore, fairly good separation can be found for the forest class using the S1 polarization bands, whereas bare soil class separability is noticeable for the S2 bands. Similar to the NDVI profiles and radar backscatter (VV, VH) values, vegetation classes yielded low JM distance values, indicating that additional features (e.g., spectral indices, GLCM textures) should be used for better class differentiation.

Table 4. JM * distance values of each LC class used in this research calculated for S1 (blue color) and S2 (green color) sensors.

ID #	1	2	3	4	5	6	7	8
1	-	0.56	1.35	0.63	0.07	0.14	0.34	0.25
2	0.31	-	1.82	0.37	0.28	0.52	0.11	0.20
3	1.91	1.93	-	1.86	1.36	1.27	1.68	1.59
4	0.12	0.52	1.96	-	0.30	0.69	0.41	0.49
5	0.27	0.52	1.99	0.17	-	0.16	0.18	0.14
6	0.13	0.59	1.97	0.08	0.37	-	0.26	0.16
7	0.08	0.31	1.96	0.08	0.36	0.06	-	0.05
8	0.08	0.65	1.96	0.07	0.35	0.04	0.08	-

* JM values are in the range from 0 to 2, where distance values greater than 1.7 indicate a good separability between the LC classes. # The ID represents following classes: 1 = Cropland; 2 = Forest; 3 = Water; 4 = Built-up; 5 = Bare soil; 6 = Grassland; 7 = Orchard; 8 = Vineyard.

Since this research used reference data from higher thematic levels of CORINE, LUCAS, and LPIS database, the aforementioned eight LC classes were used for different classification scenarios and comparison with similar research. According to Dabboor et al. [74], the JM distance measure is wide in the case of high-dimensional feature space, mostly when hyperspectral imagery is used. In this research, different texture measures were used for increasing the class separability, as noted in the research by Klein et al. [75].

4.3. Random Forest Hyperparameter Tuning Results

For optimization of the RF hyperparameters, a grid search approach with *k*-fold cross-validation was performed (Table 5), and *k* was set to 5. Although Cánovas-García and Alonso-Sarría [57] mentioned that RF is not very sensitive to its hyperparameters, *ntree* and *mtry* values were set to 1000 and a one half of the input variables for each classification scenario, respectively. A larger number of trees of the forest led to a more stable classification, albeit it can increase computational time for vegetation mapping at regional to global scales.

Table 5. The cross-validated grid search relationship between the overall accuracy (%) and hyperparameters (*mtry* and *ntree*) of the RF classifier.

		ntree		
		100	**500**	**1000**
mtry	**10**	93.85	93.78	93.40
	20	95.51	96.07	96.20
	30	96.30	96.41	96.58

4.4. Importance and Selection of S1 and S2 Input Features for Vegetation Mapping

Before any classification scenario was conducted, the feature selection was performed for SAR (i.e., S1) and optical (i.e., S2) time-series data, as well as their ancillary features. As shown in Figure 4, major improvements in the overall accuracy are perceptible up to 50 features. An increase from the aforementioned number of features in the classification model provides a negligible improvement in the OA in relation to the computational cost and processing requirements. Therefore, one-fourth most important features from the overall number of input features available for each classification scenario were used in this research.

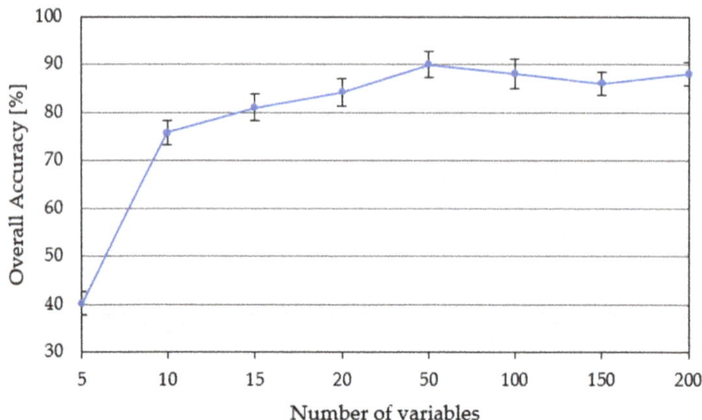

Figure 4. Mean overall accuracy (OA) for combined S1 and S2 time-series as a function of the various number of input features.

According to the feature importance approach described in Section 3.2.2, Figure 5 shows the 50 first features sorted by the decreasing MDA. Color coding was used, depending on the source of the input feature (e.g., S1 or S2 band, derived ancillary features from S1 and S2). The digital elevation model (DEM) was the most important input feature for the classification, followed by the summer B4 (i.e., Red) S2 band, and winter MSAVI and NDWI S2 indices. Overall, for S1, the VH polarization band was the most important feature among the first 50 features, whereas GLCM Mean, Variance, and Correlation were the most important features among the nine textural features used in this research. The former variable (i.e., VH) is expected to be included in the final classification model, since it contains volume scattering information [76], whereas the latter GLCM features have already been proven for vegetation mapping [20,41].

In terms of S2, B12, B11, and B5 (i.e., SWIR2, SWIR1, and RE1) are the most present S2 spectral bands. These results coincide with similar research [77], e.g., Abdi [78] where nearly half of the input S2 variables belonged to the RE and SWIR bands, included in scenes from spring and summer dates. The high importance of the RE1 band could be associated with the mapping of different crop types [79], whereas SWIR bands were found to be important for mapping the forest class [80]. In the research by Immitzer et al. [81], the aforementioned S2 bands were most important for tree species and crop type mapping using single-date S2 imagery. In this research, the spectral indices were represented the most, with 28 of them among the 50 input features. NDVI, NDWI, SAVI, and MSAVI were represented the most in the classification model within this feature group, whereas NDI45, MCARI, and GNDVI were not included at all in the model. This is expected, since vegetation phenology in the time-series can be greatly represented with NDVI [82], and other indices provided good separation between other LC classes. In terms of the relevance of time periods, the spring dates are the most present for features that are connected with the vegetation classes, whereas December appeared to be the most important month for discriminating other non-vegetation LC classes.

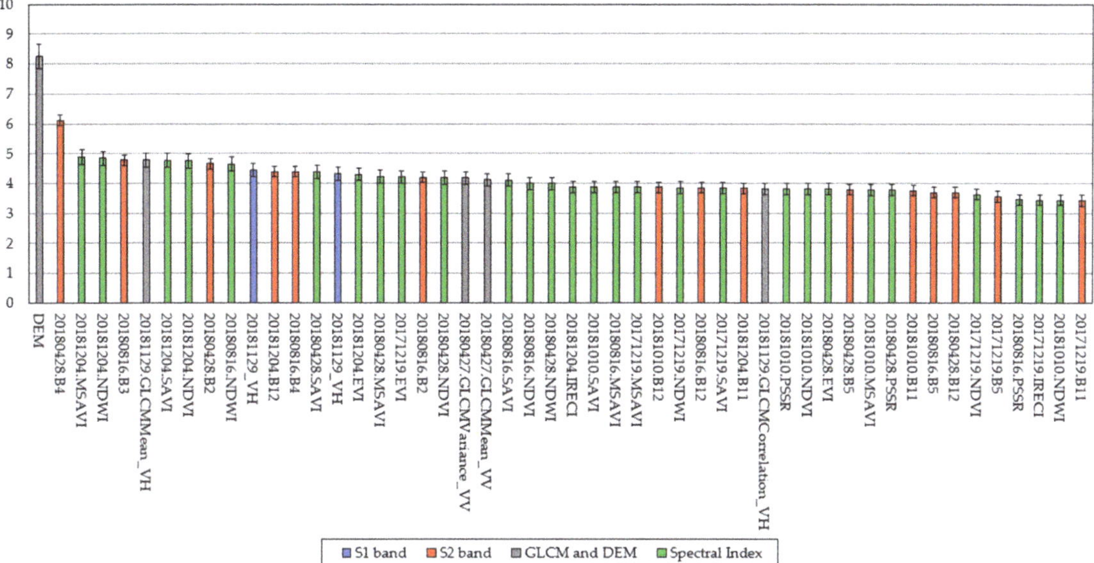

Figure 5. S1 and S2 time-series feature importance sorted by the decreasing MDA. Error bars indicate 95% confidence intervals.

The aforementioned feature selection results coincide with similar research. Jin et al. [19] evaluated the variable importance of Landsat imagery through MDA and Gini index using RF for LC classification. Summer NDVI and NIR band, DEM, GLCM mean, and contrast were the key input features for classification, which achieved OA of 88.9%. Abdi [78] classified boreal landscape using S2 imagery and RF, support vector machine (SVM), extreme gradient boosting (XGB), and deep learning (DL) classifiers. RE and SWIR S2 bands were the most important features, and interestingly, none of the spectral indices were ranked highly in his research, probably because of the high correlation with the red edge bands. RF achieved an overall accuracy of 73.9%. Tavares et al. [77] used S1 and S2 data for urban area classification. Red and SWIR S2 bands were identified as the most significant contributors to the classification, whereas VV and GLCM mean were the most important features from S1 and texture features, respectively. The authors agree that DEM should be included for major classification enhancements, which was done in our research. The integration of S1 and S2 data yielded the highest OA of 91.07% [74].

4.5. Accuracy Assessment

Confusion or error matrix [83] was computed for six different classification scenarios in a time-series (i.e., S1 and S2 alone, S1 with GLCM, S2 with Indices, S1 and S2, and all features together). Overall accuracy (OA), quantity disagreement (QD), and allocation disagreement (AD) were derived from the confusion matrix (Section 3.3).

Table 6 shows that the highest OA of 91.78% was achieved using combined S1 with S2 features, whereas classification using only S1 bands yielded the lowest OA of 79.45%. Interestingly, classification using all available features achieved lower OA than S1 with S2, leading to a conclusion that ancillary S1 and S2 features (i.e., GLCM and spectral indices) brought additional noise in the data, leading to a decrease in the classification. On the other hand, QD was lower for classification using all features compared to S1 with S2 classification, which means that the level of difference in the number of pixels between the reference map versus the classified map is different for the classification using all features, but the number of misallocated pixels is higher (i.e., more pixels are omitted from a particular LC class). The addition of texture features (i.e., GLCM) for S1 and spectral indices for S2 yielded an increase of 2.26% and 1.33% of OA, respectively. Similar results were obtained in the research from Sun et al. [84], where the RF classifier provided the best crop-type mapping results. Classification using S1 and S2 sensors yielded OA of 92%, whereas the index features were most dominant in the classification results.

Table 6. Mean overall accuracy (OA), quantity, and allocation disagreement (QD and AD) calculated from 10 random trials ranked in ascending order of OA.

Class. Scenario	OA (%)	QD (%)	AD (%)
S1	79.45	7.31	13.24
S1 with GLCM	81.71	8.57	9.72
S2	89.04	2.74	8.22
S2 with Indices	90.37	1.83	7.80
All	90.78	1.38	7.84
S1 with S2	91.78	1.83	6.39

To assess the ability of differentiation between the LC classes, UA and PA values for each classification scenario is presented in Figure 6. As indicated in Table 6, classification using S1 features only poorly predicted LC classes, except for the water class (Figure 6a), which achieved very high UA values in each scenario, irrespective of the sensor used. As seen in Figure 6b, GLCM features improved the classification accuracy in vegetation classes, and similar to [85], texture features improved the supervised classification for urban areas. Although S2 and S2 with Indices (Figure 6c,d, respectively) produced similar results in this research, cropland and bare soil classes were better differentiated when the spectral indices were used. Therefore, it is confirmed that for vegetation mapping, when enough optical imagery is available for the time-series analysis, it outperforms LC classifications solely using SAR data in many agricultural applications. In order to mitigate this obstacle, Holtgrave et al. [86] compared S1 and S2 data and indices for agricultural monitoring. In their study, radar vegetation index (RVI) and VH backscatter had the strongest correlation with the spectral indices, whereas the soil more influences VV backscatter in general. Therefore, SAR indices need to be investigated for vegetation mapping in future research.

For the best classification scenario (i.e., S1 with S2), forest and water class achieved the highest UA values to measure map reliability. From other vegetation LC classes, UA was highest for orchard, whereas PA was highest for the vineyard. Cropland was mostly committed to the bare soil or orchard class, while bare soil was the most underestimated LC class (i.e., high omission error) due to the confusion with the built-up class. In order to reduce the misclassification between built-up and bare soil class, built-up indices, such as normalized difference built-up index (NDBI), built-up index (BUI), built-up area extraction index (BAEI), etc., should be included in the classification [38]. Similar to our research, Jin et al. [19] yielded UA higher than 90% for vegetation classes, and the confusion of grassland, forest, and cropland could be associated with their accuracy errors. Sonobe et al. [10] investigated the potential of SAR (i.e., S1) and optical (i.e., S2) data for crop classification. Accuracy metrics for RF classifier were 95.70%, 2.83%, and 1.47%, in terms of OA, AD, and QD, respectively. Most of the misclassified fields were below 200 a, mostly for the grassland and maize class. Overall, the large potential of S1 and S2 data was proven for crop mapping, mostly of their high temporal resolution and free of charge availability.

Figure 7 represents the best supervised pixel-based classification scenario (i.e., S1 with S2) using a RF classifier. Water and forest were in good agreement with testing data, whereas some bare soil pixels were classified as cropland or orchard. Having added S2 imagery extracted the urban area more accurately, including main traffic roads. The confusion of orchards, cropland, and bare soil was the main cause of their misclassification errors. Vineyards are located in the northwestern part of the study areas and are mostly situated on the slopes of hills. Due to the large terrain slopes or SAR shadowing [87], it can be seen in Figure 7 that some confusion between built-up and vineyards occurred. This effect can be removed with the help of high-quality DEM or GLCM textural features [88].

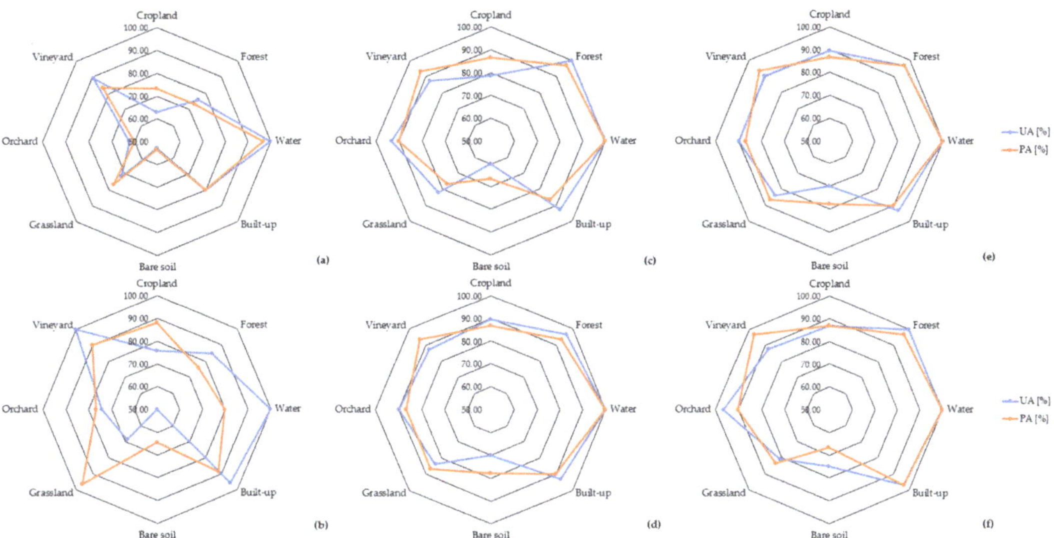

Figure 6. Spider chart representing the User's (UA) and Producer's accuracy (PA) for each LC class in the: (**a**) S1; (**b**) S1 with GLCM; (**c**) S2; (**d**) S2 with Indices; (**e**) All; (**f**) S1 with S2 classification scenario.

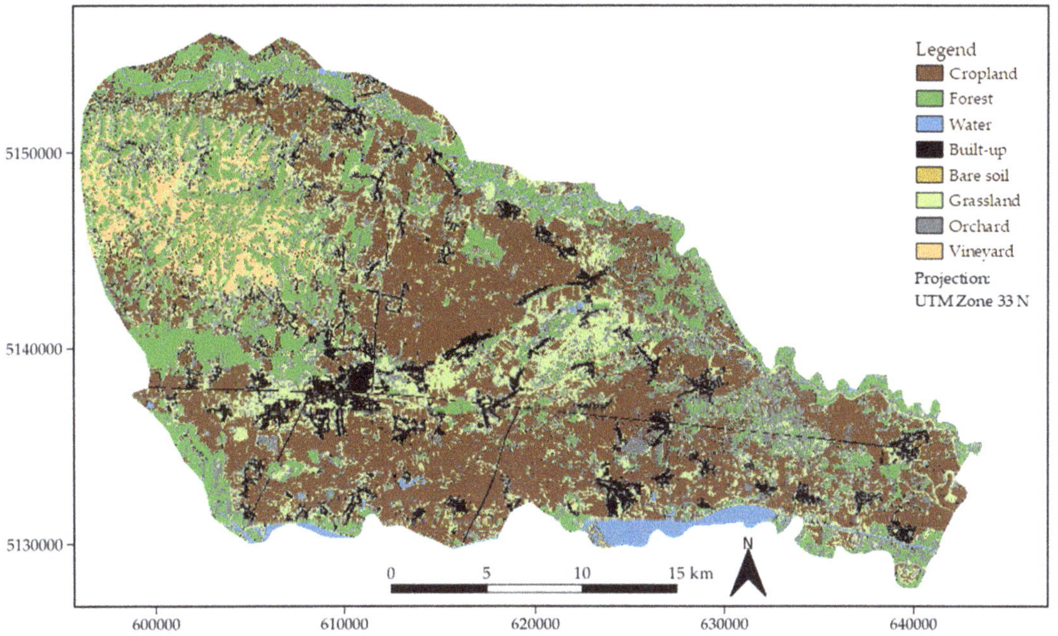

Figure 7. Classification map of the Međimurje County produced by RF using S1 with S2 imagery.

4.6. Impact of the Reference Dataset on Classification Accuracies

This research used a hybrid reference dataset derived from CORINE, LUCAS, and LPIS land-cover datasets, which collect in situ data every six, three, and one year, respectively. The goal of the hybrid dataset was to take the best of each representation, where only an agreement is targeted [89]. As noted in the research by Baudoux et al. [28], within this approach, two main limitations can arise—spatial [31] and semantic consistency [90]. A former limitation was solved using a GRID location of the LUCAS sample points since a difference between GRID and GPS locations exists [31]. Since the nomenclatures across different LC databases are not standardized, the latter limitation was resolved using $n\text{-} > 1$ associations between each class of each nomenclature (as described in Section 2.3), which resulted in identifying eight major LC classes. This proved to be a good trade-off between the overall classification accuracy and the spectral difference between the LC classes since through an analysis of 64 similar research, Van Thinh et al. [91] noted that a significant decrease in OA occurs when increasing the number of classes. Moreover, variations in the performance of the RF classifier, in terms of OA, could occur due to the imbalanced and mislabeled training datasets. The former obstacle could be mitigated using a weighted confusion matrix, which provides confidence estimates associated with correctly and misclassified instances in the RF classification model [92], whereas the latter obstacle is little influenced for low random noise levels up to 25%–30% [93]. This research used a balanced training dataset, which, as presented in [92], resulted in the lowest overall error rates for classification scenarios.

In this research, S1 and S2 imagery along with RF classifier were used for vegetation mapping on a proposed hybrid reference dataset. Compared to similar research, Dabija et al. [94] compared SVM and RF for 14 CORINE classes using multitemporal S2 and Landsat 8 imagery. SVM with radial kernel yielded the highest OA, whereas RF achieved OA of 80%. Close et al. [95] used S2 imagery and the LUCAS reference dataset for LC mapping in Belgium. Single-date and multitemporal classifications of five LC classes were tested for different seasons, and RF yielded an OA of 88%. In their research, the size of the training samples was also investigated, and the highest OA was achieved with approximately 400 sample points of a balanced training dataset. Balzter et al. [32] used S1 imagery and RF classifier for mapping CORINE land cover. Additional texture features were derived from S1 imagery, and in addition, SRTM was used as an input feature for landscape topography. Hybrid CORINE Level 2/3 classification scheme was proposed in the research, which reduced 44 LC classes to 27. The highest classification result, in terms of OA, of 68.4% was achieved using S1, texture bands, and DEM data. As noted in the review paper by Phiri et al. [96], RF and SVM classifiers provide the highest accuracies in the range from 89% to 92% for land cover/use mapping using S2 imagery, which was confirmed in our research.

5. Conclusions

This research aimed to evaluate the classification accuracy of multi-source time-series data (i.e., radar and optical imagery) with a high temporal and spatial resolution for vegetation mapping.

Sentinel-1 SAR time-series were combined with Sentinel-2 imagery showing that an improvement in classification accuracy can be obtained in regard to the results with each sensor independently. In this research, Random Forest was used as a classifier for vegetation mapping, due to the ability to deal with high-dimensional data through feature importance strategy. The aforementioned measure allowed us to use one-fourth of input variables as a trade-off between model complexity and overall accuracy. Therefore, in this research, the highest OA of 91.78% was achieved using S1 with S2, with a total disagreement of 8.22%.

For vegetation mapping, the most pertinent features derived from S1 imagery were GLCM Mean and Variance, along with the VH polarization band. Considering the spectral indices derived from S2 imagery, NDVI, NDWI, SAVI, and MSAVI contained most of the information needed for vegetation mapping, along with Red and SWIR S2 spectral bands. Overall, SRTM DEM produced major classification enhancement as an input feature for vegetation mapping.

Within this research, a hybrid classification scheme was derived from European (i.e., LUCAS and CORINE) and national (LPIS) land-cover (LC) databases. The results of this study demonstrated that the aforementioned approach is well-suited for vegetation mapping using Sentinel imagery, which can be applied for large-scale LC classifications.

Future research should focus on more advanced deep learning techniques (e.g., convolutional neural networks), which can exploit relations between pixels and objects on the image. Furthermore, these deep learning methods need many training samples, which can be derived from the proposed hybrid classification scheme and combined with S1 and S2 time-series imagery.

Author Contributions: Conceptualization, D.D. and M.G.; methodology, M.G.; software, D.D.; validation, M.G., D.D., and D.M.; formal analysis, D.D. and M.G.; investigation, D.D.; resources, M.G. and D.M.; data curation, D.D. and M.G.; writing—original draft preparation, D.D. and M.G.; writing—review and editing, D.D., M.G, and D.M.; visualization, D.D. and M.G.; supervision, M.G. and D.M.; project administration, D.M.; funding acquisition, M.G. and D.M. All authors have read and agreed to the published version of the manuscript.

Funding: This research was funded by the Croatian Science Foundation for the GEMINI project: "Geospatial Monitoring of Green Infrastructure by Means of Terrestrial, Airborne and Satellite Imagery" (Grant No. HRZZIP-2016-06-5621); and by the University of Zagreb for the project: "Advanced photogrammetry and remote sensing methods for environmental change monitoring" (Grant No. RS4ENVIRO).

Institutional Review Board Statement: Not applicable.

Informed Consent Statement: Not applicable.

Data Availability Statement: All data are publicly available online: S1 and S2 imagery were acquired from the Copernicus Open Access Hub (https://scihub.copernicus.eu, accessed on 8 January 2021), Corine Land Cover data from Copernicus Land Monitoring Service (https://land.copernicus.eu/pan-european/corine-land-cover/clc2018, accessed on 8 January 2021), Land Use and Coverage Area Frame Survey (LUCAS) data from Copernicus Land Monitoring Service (https://land.copernicus.eu/imagery-in-situ/lucas/lucas-2018, accessed on 17 March 2021), Land Parcel Identification System (LPIS) data from National Spatial Data Infrastructure (https://registri.nipp.hr/izvori/view.php?id=401, accessed on 17 March 2021).

Conflicts of Interest: The authors declare no conflict of interest.

References

1. Gong, P.; Wang, J.; Yu, L.; Zhao, Y.; Zhao, Y.; Liang, L.; Niu, Z.; Huang, X.; Fu, H.; Liu, S.; et al. Finer Resolution Observation and Monitoring of Global Land Cover: First Mapping Results with Landsat TM and ETM+ Data. *Int. J. Remote Sens.* **2013**, *34*, 2607–2654. [CrossRef]
2. Liu, X.; Liang, X.; Li, X.; Xu, X.; Ou, J.; Chen, Y.; Li, S.; Wang, S.; Pei, F. A Future Land Use Simulation Model (FLUS) for Simulating Multiple Land Use Scenarios by Coupling Human and Natural Effects. *Landsc. Urban Plan.* **2017**, *168*, 94–116. [CrossRef]
3. Mercier, A.; Betbeder, J.; Rumiano, F.; Baudry, J.; Gond, V.; Blanc, L.; Bourgoin, C.; Cornu, G.; Ciudad, C.; Marchamalo, M.; et al. Evaluation of Sentinel-1 and 2 Time Series for Land Cover Classification of Forest–Agriculture Mosaics in Temperate and Tropical Landscapes. *Remote Sens.* **2019**, *11*, 979. [CrossRef]
4. Dobrinić, D.; Medak, D.; Gašparović, M. Integration Of Multitemporal Sentinel-1 And Sentinel-2 Imagery For Land-Cover Classification Using Machine Learning Methods. Int. Arch. Photogramm. *Remote Sens. Spat. Inf. Sci.* **2020**, *43*, 91–98. [CrossRef]
5. Zhang, X.; Friedl, M.A.; Schaaf, C.B.; Strahler, A.H.; Hodges, J.C.F.; Gao, F.; Reed, B.C.; Huete, A. Monitoring Vegetation Phenology Using MODIS. *Remote Sens. Environ.* **2003**, *84*, 471–475. [CrossRef]
6. Gao, F.; Hilker, T.; Zhu, X.; Anderson, M.; Masek, J.; Wang, P.; Yang, Y. Fusing Landsat and MODIS Data for Vegetation Monitoring. IEEE Geosci. *Remote Sens. Mag.* **2015**, *3*, 47–60. [CrossRef]
7. Schultz, M.; Clevers, J.G.P.W.; Carter, S.; Verbesselt, J.; Avitabile, V.; Quang, H.V.; Herold, M. Performance of Vegetation Indices from Landsat Time Series in Deforestation Monitoring. *Int. J. Appl. Earth Obs. Geoinf.* **2016**, *52*, 318–327. [CrossRef]
8. Bhandari, S.; Phinn, S.; Gill, T. Preparing Landsat Image Time Series (LITS) for Monitoring Changes in Vegetation Phenology in Queensland, Australia. *Remote Sens.* **2012**, *4*, 1856–1886. [CrossRef]
9. Vuolo, F.; Neuwirth, M.; Immitzer, M.; Atzberger, C.; Ng, W.T. How Much Does Multi-Temporal Sentinel-2 Data Improve Crop Type Classification? *Int. J. Appl. Earth Obs. Geoinf.* **2018**, *72*, 122–130. [CrossRef]
10. Sonobe, R.; Yamaya, Y.; Tani, H.; Wang, X.; Kobayashi, N.; Mochizuki, K.-I. Assessing the Suitability of Data from Sentinel-1A and 2A for Crop Classification. *GISci. Remote Sens.* **2017**, *54*, 918–938. [CrossRef]
11. Inglada, J.; Vincent, A.; Arias, M.; Marais-Sicre, C. Improved Early Crop Type Identification by Joint Use of High Temporal Resolution Sar and Optical Image Time Series. *Remote Sens.* **2016**, *8*, 362. [CrossRef]

12. Gašparović, M.; Klobučar, D. Mapping Floods in Lowland Forest Using Sentinel-1 and Sentinel-2 Data and an Object-Based Approach. *Forests* **2021**, *12*, 553. [CrossRef]
13. Tomiyasu, K. Tutorial Review of Synthetic-Aperture Radar (SAR) with Applications to Imaging of the Ocean Surface. *Proc. IEEE* **1978**, *66*, 563–583. [CrossRef]
14. Moreira, A.; Prats-iraola, P.; Younis, M.; Krieger, G.; Hajnsek, I.; Papathanassiou, K.P. A tutorial on synthetic aperture radar. *IEEE Geosci. Remote Sens. Mag.* **2013**, *1*, 6–43. [CrossRef]
15. Gašparović, M.; Dobrinić, D. Comparative Assessment of Machine Learning Methods for Urban Vegetation Mapping Using Multitemporal Sentinel-1 Imagery. *Remote Sens.* **2020**, *12*, 1952. [CrossRef]
16. Chauhan, S.; Darvishzadeh, R.; Lu, Y.; Boschetti, M.; Nelson, A. Understanding Wheat Lodging Using Multi-Temporal Sentinel-1 and Sentinel-2 Data. *Remote Sens. Environ.* **2020**, *243*, 111804. [CrossRef]
17. Frantz, D.; Schug, F.; Okujeni, A.; Navacchi, C.; Wagner, W.; van der Linden, S.; Hostert, P. National-Scale Mapping of Building Height Using Sentinel-1 and Sentinel-2 Time Series. *Remote Sens. Environ.* **2021**, *252*, 112128. [CrossRef]
18. Zhang, W.; Brandt, M.; Wang, Q.; Prishchepov, A.V.; Tucker, C.J.; Li, Y.; Lyu, H.; Fensholt, R. From Woody Cover to Woody Canopies: How Sentinel-1 and Sentinel-2 Data Advance the Mapping of Woody Plants in Savannas. *Remote Sens. Environ.* **2019**, *234*, 111465. [CrossRef]
19. Jin, Y.; Liu, X.; Chen, Y.; Liang, X. Land-Cover Mapping Using Random Forest Classification and Incorporating NDVI Time-Series and Texture: A Case Study of Central Shandong. *Int. J. Remote Sens.* **2018**, *39*, 8703–8723. [CrossRef]
20. Gašparović, M.; Dobrinić, D. Green Infrastructure Mapping in Urban Areas Using Sentinel-1 Imagery. *Croat. J. For. Eng.* **2021**, *42*, 1–20. [CrossRef]
21. Isaac, E.; Easwarakumar, K.S.; Isaac, J. Urban Landcover Classification from Multispectral Image Data Using Optimized AdaBoosted Random Forests. *Remote Sens. Lett.* **2017**, *8*, 350–359. [CrossRef]
22. Feng, Q.; Yang, J.; Zhu, D.; Liu, J.; Guo, H.; Bayartungalag, B.; Li, B. Integrating Multitemporal Sentinel-1/2 Data for Coastal Land Cover Classification Using a Multibranch Convolutional Neural Network: A Case of the Yellow River Delta. *Remote Sens.* **2019**, *11*, 1006. [CrossRef]
23. Paris, C.; Weikmann, G.; Bruzzone, L. Monitoring of Agricultural Areas by Using Sentinel 2 Image Time Series and Deep Learning Techniques. *Proc. SPIE.* **2020**, *11533*, 115330K. [CrossRef]
24. Han, H.; Guo, X.; Yu, H. Variable Selection Using Mean Decrease Accuracy and Mean Decrease Gini Based on Random Forest. In Proceedings of the 2016 7th IEEE International Conference on Software Engineering and Service Science (ICSESS), Beijing, China, 26–28 August 2016; pp. 219–224. [CrossRef]
25. Saeys, Y.; Inza, I.; Larrañaga, P. A Review of Feature Selection Techniques in Bioinformatics. *Bioinformatics* **2007**, *23*, 2507–2517. [CrossRef] [PubMed]
26. Jović, A.; Brkić, K.; Bogunović, N. A Review of Feature Selection Methods with Applications. In Proceedings of the 2015 38th International Convention on Information and Communication Technology, Electronics and Microelectronics, MIPRO 2015-Proceedings, Opatija, Croatia, 25–29 May 2015; pp. 1200–1205.
27. Breiman, L. Random Forests. *Mach. Learn.* **2001**, *45*, 5–32. [CrossRef]
28. Baudoux, L.; Inglada, J.; Mallet, C. Toward a Yearly Country-Scale CORINE Land-Cover Map without Using Images: A Map Translation Approach. *Remote Sens.* **2021**, *13*, 1060. [CrossRef]
29. Van Tricht, K.; Gobin, A.; Gilliams, S.; Piccard, I. Synergistic Use of Radar Sentinel-1 and Optical Sentinel-2 Imagery for Crop Mapping: A Case Study for Belgium. *Remote Sens.* **2018**, *10*, 1642. [CrossRef]
30. Belgiu, M.; Csillik, O. Sentinel-2 Cropland Mapping Using Pixel-Based and Object-Based Time-Weighted Dynamic Time Warping Analysis. *Remote Sens. Environ.* **2018**, *204*, 509–523. [CrossRef]
31. Weigand, M.; Staab, J.; Wurm, M.; Taubenböck, H. Spatial and Semantic Effects of LUCAS Samples on Fully Automated Land Use/Land Cover Classification in High-Resolution Sentinel-2 Data. *Int. J. Appl. Earth Obs. Geoinf.* **2020**, *88*, 102065. [CrossRef]
32. Balzter, H.; Cole, B.; Thiel, C.; Schmullius, C. Mapping CORINE Land Cover from Sentinel-1A SAR and SRTM Digital Elevation Model Data Using Random Forests. *Remote Sens.* **2015**, *7*, 14876–14898. [CrossRef]
33. Olofsson, P.; Foody, G.M.; Herold, M.; Stehman, S.V.; Woodcock, C.E.; Wulder, M.A. Good Practices for Estimating Area and Assessing Accuracy of Land Change. *Remote Sens. Environ.* **2014**, *148*, 42–57. [CrossRef]
34. Stehman, S.V. Sampling Designs for Accuracy Assessment of Land Cover. *Int. J. Remote Sens.* **2009**, *30*, 5243–5272. [CrossRef]
35. Beck, H.E.; Zimmermann, N.E.; McVicar, T.R.; Vergopolan, N.; Berg, A.; Wood, E.F. Present and Future Köppen-Geiger Climate Classification Maps at 1-Km Resolution. *Sci. Data* **2018**, *5*, 1–12. [CrossRef] [PubMed]
36. World Weather Online. Available online: https://www.worldweatheronline.com/cakovec-weather-history/medimurska/hr.aspx (accessed on 15 January 2021).
37. Torres, R.; Snoeij, P.; Geudtner, D.; Bibby, D.; Davidson, M.; Attema, E.; Potin, P.; Rommen, B.Ö.; Floury, N.; Brown, M.; et al. GMES Sentinel-1 Mission. *Remote Sens. Environ.* **2012**, *120*, 9–24. [CrossRef]
38. Lee, J.S. Digital Image Enhancement and Noise Filtering by Use of Local Statistics. *IEEE Trans. Pattern Anal. Mach. Intell.* **1980**, *2*, 165–168. [CrossRef]
39. Osgouei, P.E.; Kaya, S.; Sertel, E.; Alganci, U. Separating Built-up Areas from Bare Land in Mediterranean Cities Using Sentinel-2A Imagery. *Remote Sens.* **2019**, *11*, 345. [CrossRef]

40. Haralick, R.M.; Shanmugam, K.; Dinstein, I. Textural Features for Image Classification. *IEEE Trans. Syst. Man Cybern.* **1973**, *3*, 610–621. [CrossRef]
41. Clerici, N.; Valbuena Calderón, C.A.; Posada, J.M. Fusion of Sentinel-1a and Sentinel-2A Data for Land Cover Mapping: A Case Study in the Lower Magdalena Region, Colombia. *J. Maps* **2017**, *13*, 718–726. [CrossRef]
42. Tucker, C.J. Red and Photographic Infrared Linear Combinations for Monitoring Vegetation. *Remote Sens. Environ.* **1979**, *8*, 127–150. [CrossRef]
43. McFeeters, S.K. The Use of the Normalized Difference Water Index (NDWI) in the Delineation of Open Water Features. *Int. J. Remote Sens.* **1996**, *17*, 1425–1432. [CrossRef]
44. Huete, A.; Didan, K.; Miura, T.; Rodriguez, E.P.; Gao, X.; Ferreira, L.G. Overview of the Radiometric and Biophysical Performance of the MODIS Vegetation Indices. *Remote Sens. Environ.* **2002**, *83*, 195–213. [CrossRef]
45. Huete, A.R. A Soil-Adjusted Vegetation Index (SAVI). *Remote Sens. Environ.* **1988**, *25*, 295–309. [CrossRef]
46. Delegido, J.; Verrelst, J.; Alonso, L.; Moreno, J. Evaluation of Sentinel-2 Red-Edge Bands for Empirical Estimation of Green LAI and Chlorophyll Content. *Sensors* **2011**, *11*, 7063–7081. [CrossRef]
47. Daughtry, C.S.T.; Walthall, C.L.; Kim, M.S.; De Colstoun, E.B.; McMurtrey, J.E. Estimating Corn Leaf Chlorophyll Concentration from Leaf and Canopy Reflectance. *Remote Sens. Environ.* **2000**, *74*, 229–239. [CrossRef]
48. Gitelson, A.A.; Merzlyak, M.N. Remote Sensing of Chlorophyll Concentration in Higher Plant Leaves. *Adv. Sp. Res.* **1998**, *22*, 689–692. [CrossRef]
49. Qi, J.; Chehbouni, A.; Huete, A.R.; Kerr, Y.H.; Sorooshian, S. A Modified Soil Adjusted Vegetation Index. *Remote Sens. Environ.* **1994**, *48*, 119–126. [CrossRef]
50. Blackburn, G.A. Spectral Indices for Estimating Photosynthetic Pigment Concentrations: A Test Using Senescent Tree Leaves. *Int. J. Remote Sens.* **1998**, *19*, 657–675. [CrossRef]
51. Frampton, W.J.; Dash, J.; Watmough, G.; Milton, E.J. Evaluating the Capabilities of Sentinel-2 for Quantitative Estimation of Biophysical Variables in Vegetation. *ISPRS J. Photogramm. Remote Sens.* **2013**, *82*, 83–92. [CrossRef]
52. Gonzalez-Piqueras, J.; Calera, A.; Gilabert, M.A.; Cuesta, A.; De la Cruz Tercero, F. Estimation of crop coefficients by means of optimized vegetation indices for corn. *Remote Sens. Agric. Ecosyst. Hydrol. V* **2004**, *5232*, 110. [CrossRef]
53. Chatziantoniou, A.; Psomiadis, E.; Petropoulos, G. Co-Orbital Sentinel 1 and 2 for LULC Mapping with Emphasis on Wetlands in a Mediterranean Setting Based on Machine Learning. *Remote Sens.* **2017**, *9*, 1259. [CrossRef]
54. Jensen, J.R.; Lulla, K. Introductory Digital Image Processing: A Remote Sensing Perspective. *Geocarto Int.* **1987**, *2*, 65. [CrossRef]
55. Ma, L.; Li, M.; Ma, X.; Cheng, L.; Du, P.; Liu, Y. A Review of Supervised Object-Based Land-Cover Image Classification. *ISPRS J. Photogramm. Remote Sens.* **2017**, *130*, 277–293.
56. Choudhary, K.; Shi, W.; Boori, M.S.; Corgne, S. Agriculture Phenology Monitoring Using NDVI Time Series Based on Remote Sensing Satellites: A Case Study of Guangdong, China. *Opt. Mem. Neural Netw.* **2019**, *28*, 204–214. [CrossRef]
57. Cánovas-García, F.; Alonso-Sarría, F. Optimal Combination of Classification Algorithms and Feature Ranking Methods for Object-Based Classification of Submeter Resolution Z/I-Imaging DMC Imagery. *Remote Sens.* **2015**, *7*, 4651–4677. [CrossRef]
58. Melgani, F.; Bruzzone, L. Classification of Hyperspectral Remote Sensing Images With Support Vector Machines. *IEEE Trans. Geosci. Remote Sens.* **2004**, *42*, 1778–1790. [CrossRef]
59. Htitiou, A.; Boudhar, A.; Lebrini, Y.; Hadria, R.; Lionboui, H.; Elmansouri, L.; Tychon, B.; Benabdelouahab, T. The Performance of Random Forest Classification Based on Phenological Metrics Derived from Sentinel-2 and Landsat 8 to Map Crop Cover in an Irrigated Semi-Arid Region. *Remote Sens. Earth Syst. Sci.* **2019**, *2*, 208–224. [CrossRef]
60. Behnamian, A.; Millard, K.; Banks, S.N.; White, L.; Richardson, M.; Pasher, J. A Systematic Approach for Variable Selection with Random Forests: Achieving Stable Variable Importance Values. *IEEE Geosci. Remote Sens. Lett.* **2017**, *14*, 1988–1992. [CrossRef]
61. Janitza, S.; Tutz, G.; Boulesteix, A.L. Random Forest for Ordinal Responses: Prediction and Variable Selection. *Comput. Stat. Data Anal.* **2016**, *96*, 57–73. [CrossRef]
62. Belgiu, M.; Drăgut, L. Random Forest in Remote Sensing: A Review of Applications and Future Directions. *ISPRS J. Photogramm. Remote Sens.* **2016**, *114*, 24–31. [CrossRef]
63. Pontius, R.G.; Millones, M. Death to Kappa: Birth of Quantity Disagreement and Allocation Disagreement for Accuracy Assessment. *Int. J. Remote Sens.* **2011**, *32*, 4407–4429. [CrossRef]
64. Stehman, S.V.; Foody, G.M. Key Issues in Rigorous Accuracy Assessment of Land Cover Products. *Remote Sens. Environ.* **2019**, *231*, 111199. [CrossRef]
65. Massetti, A.; Sequeira, M.M.; Pupo, A.; Figueiredo, A.; Guiomar, N.; Gil, A. Assessing the Effectiveness of RapidEye Multispectral Imagery for Vegetation Mapping in Madeira Island (Portugal). *Eur. J. Remote Sens.* **2016**, *49*, 643–672. [CrossRef]
66. Story, M.; Congalton, R.G. Remote Sensing Brief Accuracy Assessment: A User's Perspective. *Photogramm. Eng. Remote Sens.* **1986**, *52*, 397–399.
67. Liaw, A.; Wiener, M. Classification and Regression by randomForest. *R News* **2002**, *2*, 18–22.
68. Kuhn, M. Building Predictive Models in R Using the Caret Package. *J. Stat. Softw.* **2008**, *28*, 1–26. [CrossRef]
69. Abdullah, A.Y.M.; Masrur, A.; Gani Adnan, M.S.; Al Baky, M.A.; Hassan, Q.K.; Dewan, A. Spatio-Temporal Patterns of Land Use/Land Cover Change in the Heterogeneous Coastal Region of Bangladesh between 1990 and 2017. *Remote Sens.* **2019**, *11*, 790. [CrossRef]

70. Seo, B.; Bogner, C.; Poppenborg, P.; Martin, E.; Hoffmeister, M.; Jun, M.; Koellner, T.; Reineking, B.; Shope, C.L.; Tenhunen, J. Deriving a Per-Field Land Use and Land Cover Map in an Agricultural Mosaic Catchment. *Earth Syst. Sci. Data* **2014**, *6*, 339–352. [CrossRef]
71. Huang, W.; DeVries, B.; Huang, C.; Lang, M.; Jones, J.; Creed, I.; Carroll, M. Automated Extraction of Surface Water Extent from Sentinel-1 Data. *Remote Sens.* **2018**, *10*, 797. [CrossRef]
72. Koppel, K.; Zalite, K.; Voormansik, K.; Jagdhuber, T. Sensitivity of Sentinel-1 Backscatter to Characteristics of Buildings. *Int. J. Remote Sens.* **2017**, *38*, 6298–6318. [CrossRef]
73. Cable, J.W.; Kovacs, J.M.; Jiao, X.; Shang, J. Agricultural Monitoring in Northeastern Ontario, Canada, Using Multi-Temporal Polarimetric RADARSAT-2 Data. *Remote Sens.* **2014**, *6*, 2343–2371. [CrossRef]
74. Dabboor, M.; Howell, S.; Shokr, M.; Yackel, J. The Jeffries–Matusita Distance for the Case of Complex Wishart Distribution as a Separability Criterion for Fully Polarimetric SAR Data. *Int. J. Remote Sens.* **2014**, *35*, 6859–6873. [CrossRef]
75. Klein, D.; Moll, A.; Menz, G. Land Cover/Use Classification in a Semiarid Environment in East Africa Using Multi-Temporal Alternating Polarization ENVISAT ASAR Data. In Proceedings of the 2004 Envisat & ERS Symposium (ESA SP-572), Salzburg, Austria, 6–10 September 2004.
76. Harfenmeister, K.; Spengler, D. Analyzing Temporal and Spatial Characteristics of Crop Parameters Using Sentinel-1 Backscatter Data. *Remote Sens.* **2019**, *11*, 1569. [CrossRef]
77. Tavares, P.A.; Beltrão, N.E.S.; Guimarães, U.S.; Teodoro, A.C. Integration of Sentinel-1 and Sentinel-2 for Classification and LULC Mapping in the Urban Area of Belém, Eastern Brazilian Amazon. *Sensors* **2019**, *19*, 1140. [CrossRef]
78. Abdi, A.M. Land Cover and Land Use Classification Performance of Machine Learning Algorithms in a Boreal Landscape Using Sentinel-2 Data. GISci. *Remote Sens.* **2019**, *57*, 1–20. [CrossRef]
79. Forkuor, G.; Dimobe, K.; Serme, I.; Tondoh, J.E. Landsat-8 vs. Sentinel-2: Examining the Added Value of Sentinel-2's Red-Edge Bands to Land-Use and Land-Cover Mapping in Burkina Faso. GISci. *Remote Sens.* **2018**, *55*, 331–354. [CrossRef]
80. Eklundh, L.; Harrie, L.; Kuusk, A. Investigating Relationships between Landsat ETM+ Sensor Data and Leaf Area Index in a Boreal Conifer Forest. *Remote Sens. Environ.* **2001**, *78*, 239–251. [CrossRef]
81. Immitzer, M.; Atzberger, C.; Koukal, T. Tree Species Classification with Random Forest Using Very High Spatial Resolution 8-Band WorldView-2 Satellite Data. *Remote Sens.* **2012**, *4*, 2661–2693. [CrossRef]
82. Veloso, A.; Mermoz, S.; Bouvet, A.; Le Toan, T.; Planells, M.; Dejoux, J.F.; Ceschia, E. Understanding the Temporal Behavior of Crops Using Sentinel-1 and Sentinel-2-like Data for Agricultural Applications. *Remote Sens. Environ.* **2017**, *199*, 415–426. [CrossRef]
83. Congalton, R.G.; Oderwald, R.G.; Mead, R.A. Assessing Landsat Classification Accuracy Using Discrete Multivariate Analysis Statistical Techniques. *Photogramm. Eng. Remote Sens.* **1983**, *27*, 83–92.
84. Sun, C.; Bian, Y.; Zhou, T.; Pan, J. Using of Multi-Source and Multi-Temporal Remote Sensing Data Improves Crop-Type Mapping in the Subtropical Agriculture Region. *Sensors* **2019**, *19*, 2401. [CrossRef] [PubMed]
85. Zakeri, H.; Yamazaki, F.; Liu, W. Texture Analysis and Land Cover Classification of Tehran Using Polarimetric Synthetic Aperture Radar Imagery. *Appl. Sci.* **2017**, *7*, 452. [CrossRef]
86. Holtgrave, A.; Röder, N.; Ackermann, A.; Erasmi, S.; Kleinschmit, B. Comparing Sentinel-1 and -2 Data and Indices for Agricultural Land Use Monitoring. *Remote Sens.* **2020**, *12*, 2919. [CrossRef]
87. Bouvet, A.; Mermoz, S.; Ballère, M.; Koleck, T.; Le Toan, T. Use of the SAR Shadowing Effect for Deforestation Detection with Sentinel-1 Time Series. *Remote Sens.* **2018**, *10*, 1250. [CrossRef]
88. Xiang, D.; Tang, T.; Hu, C.; Fan, Q.; Su, Y. Built-up Area Extraction from Polsar Imagery with Model-Based Decomposition and Polarimetric Coherence. *Remote Sens.* **2016**, *8*, 685. [CrossRef]
89. Fritz, S.; See, L. Comparison of Land Cover Maps Using Fuzzy Agreement. *Int. J. Geogr. Inf. Sci.* **2005**, *19*, 787–807. [CrossRef]
90. Pérez-Hoyos, A.; Udías, A.; Rembold, F. Integrating Multiple Land Cover Maps through a Multi-Criteria Analysis to Improve Agricultural Monitoring in Africa. *Int. J. Appl. Earth Obs. Geoinf.* **2020**, *88*, 102064. [CrossRef]
91. Thinh, T.V.; Duong, P.C.; Nasahara, K.N.; Tadono, T. How Does Land Use/Land Cover Map's Accuracy Depend on Number of Classification Classes? *SOLA* **2019**, *15*, 28–31. [CrossRef]
92. Mellor, A.; Boukir, S.; Haywood, A.; Jones, S. Exploring Issues of Training Data Imbalance and Mislabelling on Random Forest Performance for Large Area Land Cover Classification Using the Ensemble Margin. *ISPRS J. Photogramm. Remote Sens.* **2015**, *105*, 155–168. [CrossRef]
93. Pelletier, C.; Valero, S.; Inglada, J.; Champion, N.; Sicre, C.M.; Dedieu, G. Effect of Training Class Label Noise on Classification Performances for Land Cover Mapping with Satellite Image Time Series. *Remote Sens.* **2017**, *9*, 173. [CrossRef]
94. Dabija, A.; Kluczek, M.; Zagajewski, B.; Raczko, E.; Kycko, M.; Al-Sulttani, A.H.; Tardà, A.; Pineda, L.; Corbera, J. Comparison of Support Vector Machines and Random Forests for Corine Land Cover Mapping. *Remote Sens.* **2021**, *13*, 777. [CrossRef]
95. Close, O.; Benjamin, B.; Petit, S.; Fripiat, X.; Hallot, E. Use of Sentinel-2 and LUCAS Database for the Inventory of Land Use, Land Use Change, and Forestry in Wallonia, Belgium. *Land* **2018**, *7*, 154. [CrossRef]
96. Phiri, D.; Simwanda, M.; Salekin, S.; Nyirenda, V.; Murayama, Y.; Ranagalage, M. Sentinel-2 Data for Land Cover/Use Mapping: A Review. *Remote Sens.* **2020**, *12*, 2291. [CrossRef]

MDPI

Article

An Interval Temporal Logic for Time Series Specification and Data Integration

Piotr Kosiuczenko

Institute of Information Systems, Military University of Technology, 00-908 Warsaw, Poland; piotr.kosiuczenko@wat.edu.pl

Abstract: The analysis of temporal series—in particular, analysis of multisensor data—is a complex problem. It depends on the application domain, the way the data have to be used, and sensors available, among other factors. Various models, algorithms, and technologies have been designed for this goal. Temporal logics are used to describe temporal properties of systems. The properties may specify the occurrence and the order of events in time, recurring patterns, complex behaviors, and processes. In this paper, a new interval logic, called duration calculus for functions (DC4F), is proposed for the specification of temporal series corresponding to multisensor data. DC4F is a natural extension of the well-known duration calculus, an interval temporal logic for the specification of process duration. The adequacy of the proposed logic is analyzed in the case of multisensor data concerning volcanic eruption monitoring. It turns out that the relevant behavior concerns time intervals, not only accumulated history as it is described in other kinds of temporal logics. The examples analyzed demonstrate that a description language is required to specify time series of various kind relative to time intervals. The duration calculus cannot be successfully applied for this task. The proposed calculus allows one to specify temporal series and complex interval-dependent behaviors, and to evaluate the corresponding data within a unifying logical framework. It allows to formulate hypotheses concerning volcano eruption phenomena. However, the expressivity of DC4F comes at the cost of its decidability.

Keywords: duration calculus; data models; temporal logic; temporal series; data fusion; data evaluation; multisensor data; signal and data processing; interval logic

Citation: Kosiuczenko, P. An Interval Temporal Logic for Time Series Specification and Data Integration. *Remote Sens.* **2021**, *13*, 2236. https://doi.org/10.3390/rs13122236

Academic Editors: Stefano Mattoccia, Piotr Kaniewski and Mateusz Pasternak

Received: 15 April 2021
Accepted: 2 June 2021
Published: 8 June 2021

1. Introduction

Remote sensing allows one to acquire information from a distance from cameras, sensors, microphones, and other external devices. The data may originate from satellites, aeroplanes, and sonar systems, among other sources. Satellite-based instruments are commonly used to monitor various parameters of the Earth's surface, such as temperatures in various infrared frequencies, and to take images. Global coverage is offered with frequency being as low as once per day. Stationary satellites are even capable of providing continuous monitoring of specific locations. The main problems associated with satellite data include their heterogeneity and large volumes. The data need to be analyzed on a daily basis, sometimes in real time, if timely response is needed.

In general, multisensor data are voluminous, heterogeneous, and sometimes incomplete. Prior to being analyzed, they need to be processed and represented in a specific form. They may be represented as discrete data by temporal series, i.e., by sequences of data usually taken in equal time intervals, or as continuous time-dependent functions and stochastic processes. Numerous algorithms have been developed, for example, concerning volcano monitoring (cf., e.g., [1]).

In the literature, heterogeneous means are used, such as means diagrams, functions, tables, and textual descriptions, to specify those heterogeneous types of phenomena. Dependencies between factors of different kind, such as similarities, regularities, and periodicity, are quite often described with the use of text only, meaning that various problems stemming from such an approach, such as imprecision and the lack of a unifying framework for specification, validation, and reasoning need to be dealt with. Textual specifications are inherently imprecise, as this is the feature of natural languages. They do not rely on formal semantics in the form of comprehensive mathematical models. Consequently, they do not allow for precise evaluation or a formal reasoning, and thus the reasoning lacks due formality.

Logics provide uniform formal languages, models, and reasoning methods, and thus provide a solution enabling to address above mentioned problems. Logics are associated with well-defined classes of models constituting their semantics. They provide precise formal languages for the description of the models and capabilities for correct reasoning abut properties of the models. In general, a logic is a system consisting of a formal language for specification, a class of models corresponding to the language, and a set of sound rules for a correct reasoning about the about the models.

Temporal logics are used to express change over time, properties of behaviors, and sequences of actions. Their languages provide temporal modalities for specifying future or past events. They are used also to define and synthesize system controllers. On the other hand, they provide rules for correct reasoning about temporal properties. Various kinds of temporal logics exist (see [2] for an overview).

Interval temporal logics (cf., e.g., [2,3]) are used for specifying time-dependent processes relative to time intervals. Duration calculus (DC) is an interval logic that is widely used for the specifying, modeling, and reasoning about discrete and continuous processes. It allows to specify propositional functions with Boolean values changing over time. There is an operator corresponding to the integral (cf., e.g., [4]) that measures how long such a propositional function remains true. It may be used to study periodicity of system states.

In this paper, we present duration calculus for functions (DC4F), a natural extension of the duration calculus for dealing with general integrable functions, not only Boolean-valued ones, as is the case of DC. The idea is simple; we take the integral operator on Riemann integrable functions and use it within a frame of an interval logic such as DC. Thus, the integral operator is used within a well-suited logical framework. DC4F, in addition to the expressive capabilities of DC, allows one to characterize the behavior of functions over time intervals in terms of their integrals. Consequently, we can characterize not only the duration of a certain property, as in the case of DC, but characterize the behavior of functions over time intervals in terms of integrals. The proposed extension is conservative in the logical sense: the DC part is unchanged meaning that its valid formulas remain valid and all its invalid formulas remain invalid. Even though the extension of DC is natural, we are, to our best knowledge, the first ones who propose it.

To evaluate the proposed logic, we investigated various phenomena, multisensor data, and facts concerning volcano monitoring, as this is a popular research topic and the degree of complexity of data is significant. Periodic degassing and temperature increases are common characteristic of active volcanoes. Distinct periodicity patterns concerning measurable parameters of volcanoes' activity have been widely identified. The timescales are ranging from seconds to weeks and months. The development of temperature and gas measurement techniques is aimed at enabling a robust quantification of high-frequency processes. Paper [5] presents an overview of the current state of knowledge regarding periodic volcanic degassing and evaluates the methods aiming at detecting periodicity. It summarizes and statistically analyzes published studies. Periodicity analysis of volcano activity (cf., e.g., [5]) is one of the challenges.

It turns out that such phenomena and their dependencies may be conveniently specified using DC4F, as it provides a expressive and uniform language for expressing various phenomena in a precise way, models for various data, and evaluation and reasoning capabilities. It provides a convenient specification language to express hypothesis concerning expected temporal properties. If data (in particular, multisensor data) are provided, then DC4F formulas may be validated for them and, thus, their truthfulness may be checked.

On the other hand, the reasoning rules provide for convenient reasoning possibilities. Thus, DC4F may be perceived as a unifying logical framework for data integration and for formulating hypotheses, their evaluation, and reasoning.

However, high expressivity of languages always comes at the cost of complexity of reasoning rules. The higher the expressivity, the more difficult the reasoning. More precisely, the question if a formula follows from a set of other formulas has high computational complexity. DC4F, like DC, is an expressive language and is, therefore, undecidable, i.e., there is no general algorithm for deciding the question mentioned above. Nonetheless, this does not hinder its use for specification, nor does it hinder the validation of formulas for concrete data.

The paper is organized as follows. In Section 2, we discuss related works. Section 3 contains a brief presentation of duration calculus and the way it is used. In Section 4, we present the main idea of DC4F and some examples of its application. In Section 5, we define its formal syntax and outline its semantics in an informal and exemplary way; we also list some of its properties and show how it applies to the multidimensional case. Section 6 is devoted to applications of DC4F in the area of volcano monitoring, in particular to the validation of the proposed concepts. In particular, we present an exemplary reasoning in DC4F. We conclude the paper with Section 7. The paper also contains an Appendix in which we define the formal semantics of DC4F.

2. Related Works

Data fusion means the integration of data and knowledge from different sources. It is a widely studied topic (cf. the overview papers [6,7] and the references there). Data fusion approaches may be classified in various ways—for example, data association, state estimation, and decision fusion. Data fusion has been studied in the context of temporal series as well. In paper [8], the authors study the fusion of long-term data in the form of dense time series from the Moderate Resolution Imaging Spectroradiometer (MODIS) and Landsat imagery. They investigate a spatiotemporal adaptive fusion algorithm in a regionalization study in which MODIS was used. They show the correlation of the time series achieved by different observation methods. In paper [8], the authors propose a data fusion method for producing high spatiotemporal resolution values for the normalized difference vegetation index in the case of time series.

We approach this topic from the perspective of temporal logic, rather than from the practical angle, as in the papers mentioned above and various others. Temporal logic (TL) describes behavior over time (cf., e.g., [2]). Their languages offer temporal modalities for specifying past or future events, as well as their temporal relations, such as: event A will happen or event A may happen in the future, property A has to be always present, property A holds until property B will be satisfied, and so on. They are also used to define and synthesize system controllers; the controllers are then guaranteed to monitor and control the underlying systems according to the specific requirements. On the other hand, they are used to ensure that conclusions drawn based on the assumptions made are correct, that the implementation satisfies the specifications, and so on.

There are different types of temporal logics (see [2] for the overview). The linear temporal logic describes behaviors by referring to linear sequences of events. The computation tree logic (CTL) describes the possibilities and, more precisely, the branching time structures. The interval temporal logic (ITL) (cf., e.g., [2,3]) describes behaviors relative to time intervals, stating, for example, that a property is valid during the entire time period or that it occurs within a certain subperiod. It is also possible to express the fact that a property holds within a given time period, which is then followed by another time period over which another property holds.

Duration calculus (DC) is a popular and widely studied kind of interval logic (cf., e.g., [4]). Duration calculus may be considered to be a form of ITL. In addition to the above-mentioned potential of ITL, DC contains an integral operator that allows one to express durations of properties in a quantitative manner.

In [9], automata-based semantics of DC have been defined covering data, real-time, and communication-related aspects. A model-checking algorithm has been presented for a subset of DC that may be model-checked. The algorithm has been implemented and its use demonstrated.

DC and ITL are expressive but not decidable, i.e., no procedure exists for figuring out whether a given formula is always valid or not. Thus, as usual, expensivity is at the cost of complexity—in this case, decidability. A restricted form of DC, known as RDC, has decidable inference relation [10]. In fact, formulas of this type may be reduced to the regular expressions. The problem of DC model checking is a topic of ongoing research (cf. e.g., [11] and the references there). In [12], the authors propose a method for solving minimal and maximal reachability problems for the multipriced timed automata. The automata are an extension of timed automata with multiple cost variables that may evolve according to specified rates.

The range of applications varies from the above-mentioned real-time systems (cf. e.g., [13] and the references there) to air traffic control (cf. e.g., [14]) and hybrid systems [15].

Duration calculus is often used for the synthesis of real-time system controllers (see, for example, [16]) and the references there). It is capable of specifying time constraints of dynamic systems. Various methods are used—for example, integer linear problem-solving and optimization problem-solving methods (cf. [16]). DC is used also for traffic system specification (cf., e.g., [13]). There is also a wide range of tools supporting DC (see, for example, [10] and the references there).

In [17], a variant of duration calculus was presented for system discounting, i.e., the idea that something happening earlier is more important than similar events happening later. The idea was introduced earlier into temporal logics such as LTL and CTL. The authors demonstrated decidability of the model-checking for timed automata and a dedicated fragment of discounted duration calculus.

We illustrate our ideas with data concerning volcano monitoring. In fact, volcano monitoring is one of the prime examples for the application of multisensor systems placed in satellites, aeroplanes, balloons, and on the ground. They provide huge amounts of data that have to be preprocessed, analyzed, evaluated, and reasoned about. Thus, it is a proper area to test ideas such as ours. Of particular interest are the Strombolian effects, i.e., periodic volcanic activity phenomena such as temperature picks, gas bursts, and lava eruptions (cf., e.g., [18–20]).

In [1], Koeppen et al. propose a fully automated interactive algorithm called MODVOLC for the analysis of thermal satellite time-series data. The algorithm is aimed at detecting and quantifying the excess energy radiated from the thermal anomalies such as active volcanoes. The algorithm enhances the previously developed approaches (see the references in [1]). It is characterized by the law rate of false positives (see [1,21]). It flags thermal anomalies—in particular, volcanic eruptions. The algorithm was tested for different localizations such as the Anatahan volcano, the Kīlauea volcano, and the Cantarell oil field in the Gulf of Mexico.

In [22], Laiolo et al. compared thermal satellite images of the Etna eruption that took place in 2018 with ground-based geophysical data of summit craters. The Moderate Resolution Imaging Spectroradiometer (MODIS) provided infrared images and helped to identify pixels including the possible hot spots.

Temporal series corresponding to Strombolian effects may be compared using similarity measures. In fact, various similarity measures are used to compare temporal series and functions in general (cf., e.g., [23] and the references there). One can use the dynamic time warping algorithm (cf., e.g., [24] and the references there). This algorithm is used in temporal series analysis for measuring similarity between temporal series that may vary in terms of speed. Algorithms of this type were applied to compare graphical data representing temporal series, but also in comparisons of audio and video materials (cf., e.g., [25,26]).

Dyea et al. [19] proposed a method of detecting Strombolian eruptions based on training a convolutional neural network. The method automatically categorizes eruptions based on infrared images taken at the rim of a crater atop Mount Erebus. The authors show that machine learning may be effectively used to classify the characteristics of Strombolian eruptions, to facilitate the process of studying their origins, and to assess the hazards posed by volcanic eruptions.

3. Duration Calculus

In this section, we present the basic features of duration calculus (DC). We present the basic idea behind temporal specification relative to time intervals. We also list and explain the temporal modalities it provides and explain how they can be used.

3.1. Using Duration Calculus

In this subsection, we show the basic properties of duration calculus (DC) in its original form. DC allows to specify Boolean-valued functions, which play the role of time-dependent formulas, and to use the DC integral operator to measure the time, or equivalently the duration, when they hold. We also consider an operator for splitting intervals.

DC allows to treat durations of processes. More precisely, it expresses integrals of Boolean-valued functions with values 1 and 0 representing true and false, respectively. Such functions are called state functions since they indicate whether the system is in a given state or not. It should be noted that the properties are defined not in respect to specific points in time but in respect to intervals and in terms of durations.

We start with a simple example of system with states modeled by the $sin(t)$ function and lasting for the time interval $[0, 2\pi]$ (see Figure 1). This function is first positive on $[0, \pi]$ subinterval and then it is negative on subinterval $[\pi, 2\pi]$. The fact that the function is first positive may be characterized in DC using the Boolean-valued function $sp(t) = 1_{\{x \mid 0 \leqslant sin(x)\}}(t)$, which returns 1 if $t \in \{x \mid 0 \leqslant sin(x)\}$ and 0 otherwise. In the first case, the value is 1, and in the second case, it equals 0; we identify *true* with 1 and *false* with 0. $sp(t)$ is called the characteristic function of the set $\{x \mid 0 \leqslant sin(x)\}$. To describe that the value of sin is negative, we can use the characteristic function $sn(t) = 1_{\{x \mid sin(x) < 0\}}$. This function may be obtained from the first one using logical negation $sn = \neg sp$, which swaps 1 and 0. It should be noted that DC does not allow direct access to sin and its integral, but only via Boolean-valued functions characterizing its values.

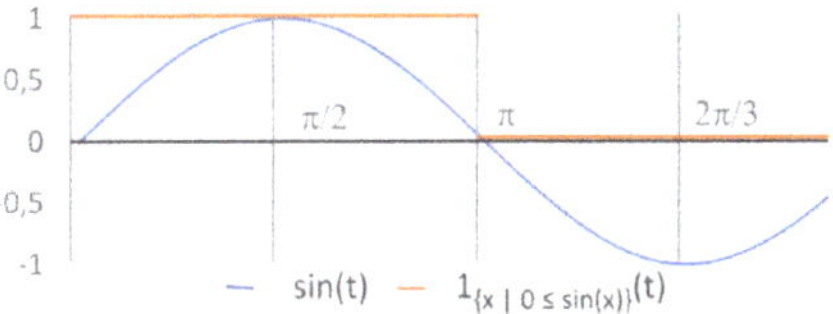

Figure 1. sin-function and the corresponding state function $1_{\{x \mid 0 \leqslant sin(x)\}}$.

Duration calculus enables one to express such properties as the duration of states—for example, the period of time over which sin remains positive, with the help of state function sp. DC formulas are evaluated always in respect to an underlying time interval. Length function ℓ returns the length of the underlying interval. It is defined using the integration operator $\int$ of DC: $\ell = \int 1$. The semantics of this operator is defined relatively to an underlying interval $[a, b]$—in this case, $\int 1 = \int_a^b 1\, dt$.

For a Boolean-valued function, the integral operator measures the time when the function is true. For intervals of length 2π, the length of subinterval where $sin(t)$ is positive is equal to the length of the subinterval in which it is negative. This fact may be expressed using the length function ℓ:

$$\ell = 2\pi \Rightarrow \int sp = \int sn \qquad (1)$$

In this formula, we use the integral $\int sp$, which, for a given interval $[a,b]$, returns the value $\int_a^b sp\ dt$, i.e., the value of the integral for the interval, and similarly for sn. The formula states that if the length of an interval equals to 2π, then the time when the *sp* is true equals to the time, the duration, when *sn* is true. We do not have to consider the fact that the function is equal 0 at the end of intervals, since the functions are characterized in terms of integrals and for subintervals of length 0, the integral value has no influence on the entire integral.

For interval $[0,2\pi]$, it holds that $\int sp = \pi$, since the duration of *sp* being true is π. In fact, this expression is true for every interval of the form $[0,x]$, where $\pi \leqslant x \leqslant 2\pi$, but it is false for intervals of the form $[x,2\pi]$. Similarly, we can express the fact that the duration of *sin* being negative is π: $\int sn = \pi$. This expression is true on every interval of the form $[y,2\pi]$ where $0 \leqslant y \leqslant \pi$. In both cases, we have to use the auxiliary functions sp, sn to characterize *sin* in terms of being positive or negative, respectively.

3.2. Temporal Modalities

In this subsection, we present three temporal modalities definable in DC: the chop, the after-modality, the sometime-modality, and the always-modality. The chop is denoted by $\frown$ (cf., e.g., [10]); it applies to two neighboring intervals. It allows one to express the fact that a property holds in the first interval and another property holds in the second interval. The last two modalities occur in all kinds of modal and temporal logics (cf., e.g., [2]), but they have a specific meaning in DC. The sometime-modality is denoted by $\Diamond$ and requires a property to hold for some time interval. The always-modality is denoted by $\Box$ and means that a property holds for all time periods.

Formally, for two formulas F, G, the formula $F \frown G$ means that for an interval $[a,b]$ in question, there exists a number x such that $a \leqslant x \leqslant b$, F holds for interval $[a,x]$ and G holds for interval $[x,b]$. For example, the fact that *sin* is initially positive and then negative (see the previous subsection) can be described in respect to two subintervals dividing the interval in question. We may express it by formula $(\int sp = \pi) \frown (\int sn = \pi)$. It is true on the interval $[0,2\pi]$ as the interval can be split into two subintervals $[0,\pi]$ and $[\pi,2\pi]$; thus, we chose $x = \pi$.

Note that the operator $\frown$ is associative; thus, the formula of the form $F \frown (G \frown H)$ is equivalent to formula $(F \frown G) \frown H$, since the formulas depend on splitting an underlying interval $[a,d]$ into three parts $[a,b],[b,c],[c,d]$ where the subformulas F, G, H have to hold, respectively. Consequently, we will drop the brackets when using chop.

Given a formula G and an interval $[a,b]$, the fact that $\Diamond\ G$ holds for interval $[a,d]$ means that it has a subinterval $[b,c]$, i.e., $a \leqslant b \leqslant c \leqslant d$, such that G holds in $[b,c]$. The may-modality may be expressed using the chop operator $true \frown G \frown true$, which means that we can divide the underlying interval into three subintervals and that G holds for the middle subinterval, whereas we do not require anything from its adjacent subintervals on the left- and right-hand sides. For example, we have the following property:

$$\pi < \ell \Rightarrow \Diamond\ 0 < \int sp = \ell \tag{2}$$

This formula says that if the length of the underlying interval is larger than π, then there is a subinterval of length larger than 0 where the property sp holds all the time, i.e., *sin* is positive.

The modal formula $\Box\ G$ means that for all subintervals of an underlying interval, the formula G holds. For example, for all subintervals of interval $[0,\pi]$, the function sp is true all the time: $\Box\ \int sp = \ell$. Modality $\Box$ can be expressed using the sometime-modality $\Diamond$: $\neg\Diamond\ \neg G$; the formula means that no subinterval exists where G does not hold. Vice versa, $\Diamond\ G$ can be expressed as $\neg\ \Box\ \neg G$. Thus, both modalities are dual. Both are definable in terms of the chop operator, but we will use them as syntactic sugar to facilitate readability of the formulas.

4. The Idea of Duration Calculus for Functions

In this section, we present the idea of duration calculus for functions (DC4F). The objective is to treat Riemann integrable functions with values of type real. Such functions can represent continuous as well as discrete time series. We informally present features of the proposed calculus and illustrate them with simple examples demonstrating its basic characteristics, the way it can be used, and the capabilities of the logic.

4.1. DC versus DC4F

DC allows one to treat Boolean-valued functions only and to reason about their durations. It does not allow to integrate general functions. In this subsection, we propose an extension of DC supporting all Riemann integrable functions. We present the basic features of the proposed logic. We also present examples that are specific to DC4F and, thus, cannot be expressed in DC.

In DC, a state function f, we can consider the time intervals, or periods, when this states holds. Such a case corresponds to considering the corresponding characteristic functions: $1_{\{y|0<f(y)\}}(t)$. The integrals of the characteristic functions correspond to the duration of the characterized state or property. For example, as explained in the previous section, the fact that sin is first negative and then positive can be expressed in DC only via the auxiliary characteristic functions sp and sn, but not directly. In DC, we can integrate function $sp(x)$ and obtain the duration over which it remains true, but we cannot integrate sin or any other function that is not Boolean. Thus, for example, we can express the property that for intervals of length 2π, sin is positive as equally long as it is negative $\ell = 2\pi \Rightarrow \int sp = \int sn$.

However, not all properties of integrals can be expressed in this way. For example, we cannot express properties such as $\ell = 2\pi \Rightarrow \int sin = 0$, i.e., that if the length of the underlying interval is 2π, then the integral is equal to 0. Thus, DC is not expressive enough to treat time series in general.

DC4F is a natural extension of DC as allows all integrable functions. For example, we can consider $\int sin$. In logical terms, it is a conservative extension of DC. Conservativity means that a formula of DC is a tautology in the sense of DC if, and only if, it is a tautology of DC4F; equivalently, it is satisfied in a DC model if, and only if, there is a DC4F extension of the model in which the formula is satisfied. In general, DC can be expressed in DC4F by restricting the set of integrable functions so that it contains Boolean-valued functions only.

4.2. Monotonicity

We formulate the definition of monotonicity in terms of intervals and integrals. The property that a function is monotone over an interval can be expressed in DC4F by the requirement that for two neighboring subintervals of the same length, the integral over the left one has smaller or equal value to the integral over the right one. It can be expressed by formula M of the form $\int f = x \wedge \ell = y ^\frown \int f = z \wedge \ell = w \wedge y = w \Rightarrow x \leqslant z$. This property must hold for all subintervals of this kind; thus, we have the formula: $\Box\, M$. Figure 2 shows a monotonically increasing function f.

Figure 2. Monotone function $f(t)$.

Alternatively, we can define the monotonicity property in the following way:

$$\forall_{x,y}\ \Box\ \{(\ell > 0 \wedge \int f/\ell = x) ^\frown (\ell > 0 \wedge \int f/\ell = y) \Rightarrow x \leqslant y\} \tag{3}$$

This formula, though it may seem complicated, is rather simple. It states that for a given interval and for all its adjacent subintervals, the normalized value of the integral for the left subinterval is lower than or equal to the normalized value of the integral for the right subinterval. It is illustrated in Figure 2. Intervals $[b, c]$ and $[c, d]$ are adjacent. The values of the integrals for these intervals have to be normalized by their length, as their lengths are different. We will abbreviate this property of function f as $MIncreasing(f)$. Analogously, we can define the property of monotonic decrease, $MDecreasing(f)$. We use a capital letter to indicate that both functions have a functional argument instead of a time value.

Although length operator ℓ occurs in the characteristics of the first and of the second interval, it returns independent values. Due to transitivity of smaller or equal relation $\leqslant$, the monotonicity property holds for all subintervals, not only for the adjacent ones. Thus, $\int_a^b f/\ell \; dt \leqslant \int_b^c f/\ell \; dt \leqslant \int_c^d f/\ell \; dt$ (see Figure 2).

4.3. Limits and Amplitude

In this subsection, we show how to define lower limits of functions and their minimal amplitudes. The definitions are based on temporal modalities. Upper limits and maximal amplitudes can be defined analogously. We use those definitions below when dealing with volcanic Strombolian effects.

We start with the case when a function $g(t)$ is an upper approximation of a function $f(t)$ or, equivalently, function $f(t)$ is a lower approximation of function $g(t)$ (see Figure 3). This property can be expressed by demanding that for each subinterval of a given interval, the integral of f is smaller than or equal to the integral of g:

$$\Box \int f \leqslant \int g \tag{4}$$

The property that g increases c times faster than c can be expressed in DC4F in a simple way: $\Box \int cf \leqslant g$. It should be noted that we cannot directly express it in DC.

The diagram presented in Figure 4 shows function h, which is above value a. The property that, for a given interval, values of function h are above a certain limit a all the time can be expressed as follows:

$$\Box \; \ell a \leqslant \int h \tag{5}$$

The integral of a is equal to the value ℓa. Consequently, we demand that it remains smaller than the integral of h for all subintervals of the underlying interval. The fact that function h exceeds threshold y for some time can be expressed as follows:

$$true \frown (0 < \ell \wedge \Box \ell y \leqslant \int h) \frown true \tag{6}$$

This formula states that for some subinterval of a nonzero length, the value of the integral is not lower than the value of a multiplied by the length of the subinterval. The function shown in Figure 4 extends beyond line y for a time period. Equivalently, the property can be expressed as follows:

$$\Diamond \; (0 < \ell \wedge \Box \ell y \leqslant \int h) \tag{7}$$

The fact that an amplitude of a function is at least d can be expressed as follows:

$$\exists_{x,y}(\Diamond \; (0 < \ell \wedge \Box \int h \leqslant \ell x)) \wedge \Diamond \; (0 < \ell \wedge \Box \ell \, y \leqslant \int h)) \wedge d \leqslant y - x \tag{8}$$

Formula (8) says that there exist two subintervals of nonzero length. The integral of h for the first one is smaller than or equal to $\ell \, x$, as it was expressed by formula (7). Analogously, the integral of h for the second one is larger than or equal to $\ell \, y$. Formula (8) requires also that the difference $y - x$ is equal to at least d.

In Figure 4, the intervals correspond to the parts of diagrams of h located below the x and above the y line, respectively, are indicated by bold lines. Below, we abbreviate formula (8) by $Amplitude(h,d)$. Formula $PeakAbove(h,y)$ is a shorthand of formula (6) and expresses the fact that function h is above threshold y for some time period.

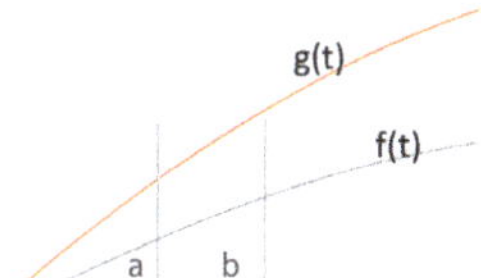

Figure 3. Functional limit.

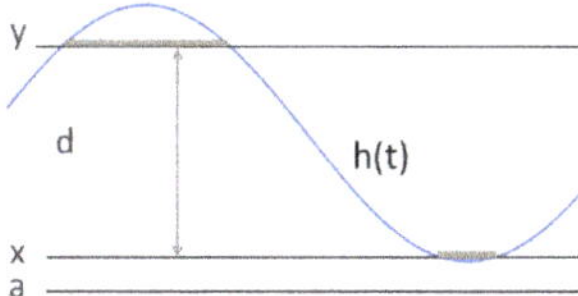

Figure 4. Constant limits and amplitude.

5. Syntax and Semantics of DC4F

In this section, we define the formal syntax of DC4F and describe its informal semantics. We list some basic properties of the proposed logic and show how to apply it in the multidimensional case, such as those involving two-dimensional images.

5.1. Syntax of DC4F

Every logic has its proper language, which has to be formally defined. In this subsection, we define the formal syntax of DC4F. This syntax is close to that of DC. The crucial difference consists in the fact that we allow all unary integrable functions, not only the Boolean-valued ones, and their integrals. Basically, there are constants, variables, and function symbols corresponding to the integrable functions, which can be added and multiplied by constants. There is also an integral operator. Formulas are either of the atomic form, when real numbers are compared, or are composed from other formulas.

To define the syntax formally, designated sets of variables, functions, and relation symbols are required. There are global variables, which remain unchanged. We assume that **VSym** is an unbound set of real-valued variables denoted by letters $x, y, z, \ldots$ We also assume that there is a set **CSym** $= \{0, 1, \ldots, true, false, \ldots\}$ of constant symbols. The set **FSym** contains unary functions symbols $f, g, h, \ldots$ corresponding to integrable functions. The set **FlSym** contains functional symbols $X, Y, Z, \ldots$ corresponding to functionals with arguments in the form of time intervals and values of type real $\mathbb{R}$ (cf. Appendix A).

The syntax of the logic comprises three categories of expressions: functions Fn containing integrable functions, terms $Real$ containing expressions of type $\mathbb{R}$, and formulas Fo containing logical formulas in the proper sense of the word. We assume in the definition below that $f \in$ **FSym**, $C \subset$ **CSym**, $x \subset$ **VSym**, $X \in$ **FlSym**.

- $Fn ::= f \,|\, C \,|\, C \cdot Fn \,|\, Fn \oplus Fn$, for $\oplus \in \{+, -, \vee\}$
- $Real ::= C \,|\, x \,|\, X \,|\, \int Fn \;\; |\, Real \oplus Real$, for $\oplus \in \{+, -, \cdot, /\}$
- $Fo ::= true \,|\, false \,|\, Real \oplus Real \,|\, \neg Fo \,|\, Fo \vee Fo \,|\, \exists_x Fo \,|\, Fo \frown Fo \,|\, Fo^*$, for $\oplus \in \{<, =\}$

The first category includes constants and functions of type real; the constants are treated here as functions with arity equal to 0. The functions can be multiplied by constants and added. There is also the maximum operator $\vee$; it denotes their maximum and is a reminiscent of DC (cf., e.g., [4]). In DC, there is an indefinable negation operator $\neg$ that applies to state functions (cf., e.g., [4]). It is characterized by the following property: $f(x) = 1$ if, and only if, $\neg f(x) = 0$. We can define the negation $\neg$ on integrable functions such that it coincides with negation on state functions: $\neg f =_{def} 1 - f(x) \vee f(x) - 1$. Consequently, if

$f(x) = 1$, then $\neg f(x) =_{def} 1 - 1 \lor 1 - 1 = 0$. If $f(x) = 0$, then $\neg f(x) = 1 - 0 \lor 0 - 1 = 1$, since $\lor$ is the maximum.

The second category includes constants, variables, and applications of the integral operator. There are also auxiliary operators X on intervals with values of type real. Complex terms can be constructed from simpler ones using arithmetic operations.

The third category comprises formulas as such. The atomic formulas include constants *true* and *false*; the atomic formulas are also formed from relational symbol applied to terms, such as $u < v$. Complex formulas are composed from other formulas by the application of logical operators: logical negation, alternative, existential quantification, concatenation, the chop operator, and iteration. Thus, if F and G are formulas, then the alternative $F \lor G$, the chop $F ^\frown G$, the iteration F^*, and the quantification $\exists_x F$ are formulas as well.

As an example of the above syntax, we can present formula $\int sin = 2 ^\frown (\int sin = -2)$, provided that *sin* belongs to **FSym**. The formula is obtained by the application of chop to two atomic subformulas. The first subformula is obtained by the application of equation $=$ to terms $\int sin$ and 2; the first term is obtained from function *sin* by application of the integral operator. The other subformula is obtained in a similar way.

It should be noted that relations $u \leqslant v$ and $u > v$ can be defined in terms of $<$ and $=$ using logical operators $\neg$ and $\lor$, e.g., in the first case, we have $u < v \lor u = v$. Similarly, the conjunction $F \land G$ is defined as $\neg(\neg F \lor \neg G)$. The implication $F \Rightarrow G$ is defined as $\neg F \lor G$. The general quantifier $\forall_x Fo$ is defined as $\neg \exists_x \neg Fo$. As mentioned in Section 3.2, modalities $\Box$, $\Diamond$ are defined using chop and negation. $\Box F$ is defined as $\neg(true ^\frown (\neg Fo ^\frown true))$ and $\Diamond F$ as $true ^\frown F ^\frown true$. The length operator ℓ is defined as $\int 1$ (see Section 3.1). We assume that $^\frown$ binds stronger than $\lor$ and $\land$. $\lor$ and $\land$ bind stronger than quantifiers $\exists$ and $\forall$. Quantifiers bind stronger than implication $\Rightarrow$. Implication binds stronger than modalities $\Diamond$ and $\Box$.

5.2. Informal Semantics of DC4F

In this subsection, we present the intuitive semantics of the previously defined syntax. The presentation is informal, but it can be easily formalized, as shown in the Appendix A. We discussed it for the three syntactic categories defined above: *Fn*, *Real* and *Fo*.

We assume here that the time domain $\mathbb{T}$ime is the set of non-negative real numbers $\mathbb{R}_+$. In the case of discrete time modeled by natural numbers $\mathbb{N}$, for every discrete function f, we define the corresponding step function g on $\mathbb{R}_+$ by extending the values of f to the corresponding unit interval: $g(t) = f(\lfloor t \rfloor)$, where $\lfloor t \rfloor$ is the largest integer $n \in \mathbb{N}$ such that $n \leqslant t$. Thus, we can treat discrete functions, or discrete time series, as if they were defined for continuous time.

Elements belonging to category Fn are interpreted as time series, and more precisely as Riemann integrable functions with the domain $\mathbb{R}_+$. The functions may be multiplied by constants, added, and subtracted; thus, the set corresponds to a linear space. The function $f(t) \lor g(t)$ returns the maximum of $f(t)$ and $g(t)$.

The category *Real* contains terms of type $\mathbb{R}$. The constants, e.g., 0 or 1, are interpreted as the corresponding real numbers. Variable symbols are evaluated by functions mapping them into the set $\mathbb{R}$. For $f \in Fn$ and an interval $[a, b]$, where $a, b \in \mathbb{R}_+$ and $a \leqslant b$, term of the form $\int f$ is interpreted as the integral $\int_a^b f(t)dt$. This operator is linear in respect to functions: $\int cf = c \int f$ and $\int f + g = \int f + \int g$.

A term t depends on variable $x \in$ **FSym** if t contains it. In general, the value of terms containing variables depend on the values of its variables. Valuations are mappings of variables into $\mathbb{R}$. We can, for example, define valuation val_1 as mapping of the form $x \mapsto 1, y \mapsto 2, z \mapsto 5$. Let term p be of the form $x + y + z + 2$. The value of p for val_1 is 10; it does not depend on an underlying interval. The value of a term depends on an underling interval if contains operator $\int$. In the case of term q of the form $(\int 1) + z$, the value depends on the underlying interval and variable z; for interval $I_1 = [0, 2]$ and val_1, the value is 7.

The category of formulas Fo plays a crucial role, as formulas are the proper objects of logical reasoning. In fact, formulas can be understood as Boolean-valued terms. The satisfaction of formulas, i.e., the fact whether formulas are true or false, is defined in respect to an underlying interval and a valuation of the variables they include. The values of these terms may be compared using $<$, $=$, given the underlying interval and the valuation. The result of this comparison is either true or false, depending on the value of the corresponding terms. For example, for interval I_1 and valuation val_1, formula $q < p$ is true but formula $q = p$ is false.

Let $I = [a, b]$ be an interval and val be a valuation. Formula $\neg F$ is satisfied in interval I for valuation val, if formula F is false in interval I for val. Similarly, the alternative $F \vee G$ is satisfied if F or G is satisfied in I for val.

In case of chop, $F \frown G$ is satisfied in interval I for valuation val if there exists $c \in [a, b]$ such that F is satisfied in $[a, c]$ for val and G is satisfied in $[c, b]$ for val. Iteration F^* of formula F is satisfied in I for val if there is an n such that we can split the interval into subintervals $[a, a_1], [a_1, a_2], \ldots, [a_{n-1}, b]$ such that F is satisfied in each subinterval for val (cf. [4] Section 4). An example of the iteration formula is presented below and its formal semantics are presented in the Appendix A.

Quantified formula $\exists_x F$ is satisfied in interval I for valuation val if there exists a valuation val' which differs from val only for variable x such that F is satisfied in I for val. The fact that val and val' differ only at x means that for every variable y different from variable x, it holds that $val(y) = val'(y)$. For example, $\exists_x x + y > 25$ is satisfied for valuation val_1 defined above; in fact, we can define val' as val_1 but $val'(x) = 30$.

5.3. Basic Properties

In this subsection, we list some basic properties of the logical operators, such as the linearity and monotonicity of the integral. We use them in an exemplary reasoning.

The DC4F calculus has the following logical properties:

1. $\int 0 = 0, \quad \int c = c \int 1 = c\ell$
2. $\int cf = c \int f, \quad \int f + g = \int f + \int g$
3. If $f(x) \leqslant g(x)$, for every $x \in \mathbb{R}_+$, then $\int f \leqslant \int g$
4. $\int f = x \frown \int f = y \Rightarrow \int f = x + y$
5. $F \Rightarrow F \frown (\ell = 0), \quad F \Rightarrow (\ell = 0) \frown F$
6. $F \Rightarrow true, \quad (F \Rightarrow G) \Rightarrow (F \wedge H \Rightarrow G \wedge H), \quad F \wedge G \Rightarrow G \wedge F,$
7. $(F \Rightarrow G) \Rightarrow (F \frown H \Rightarrow G \frown H), \quad (F \Rightarrow G) \Rightarrow (H \frown F \Rightarrow H \frown G)$
8. If the formulas F, $F \Rightarrow G$ hold in an interval I for a valuation val, then G holds in an interval I for a valuation val.

Point (1) concerns integration of constants. Point (2) expresses the fact that integrals are linear operators, i.e., they commute with multiplication by constants and are distributive in respect to addition. Point (3) expresses the monotonicity of integral operators. If the antecedent inequality holds for all non-negative reals, then the consequent inequality holds for all time intervals. Property (4) can be expressed as $\int_a^b f\,dt + \int_b^c f\,dt = \int_a^c f\,dt$. The above-mentioned points are specific to DC4F because it allows integrable functions, not only propositional ones as in the case of DC. The following points are common with DC (cf. [4], Section 2). Property (5) states that every interval can be split into an interval of length 0 and the rest. Point (6) specifies three exemplary propositional tautologies (cf., e.g., [27]) that we use later in an exemplary derivation. The first tautology says that truth is implied by any formula. The second tautology expresses the monotonicity of the conjunction operator. The third one expresses the commutativity of conjunction. It should be noted that all propositional tautologies hold in case of DC as well as DC4F. Point (7) specifies the monotonicity of the chop operator. Property (8) is the modus ponens reasoning rule. It says that if the antecedent of an implication is true and the implication is true, then the consequent of the implication is true as well. The rule is used in many kinds of logic (cf., e.g., [2,27]).

5.4. Multidimensional Case

The functions considered in DC and DC4F are unary, i.e., they are of the form $f(t)$, where t is the time parameter. In this subsection, we consider integrals over multidimensional sets and spaces. Such cases may occur when, for instance, images of an area are considered and measurements are performed. The development of values over time may concern specific areas or multidimensional spaces. In this case, the integrated function has a number of variables, apart from the time variable. We show that in fact such a multidimensional case can be reduced to the case of unary functions. In fact, we utilize such integrals in the examples presented in the following section.

An image may be represented by a number of pixels with two coordinates x, y. Thus, the area of an image may be considered as a set $A \subseteq \mathbb{R}^2$. In the multidimensional case, it has the form $A \subseteq \mathbb{R}^n$. If the images are used to measure certain values, then the measurement can be modeled by a function. The measurements may be time-dependent. Thus, we consider time-dependent integrable functions of the form $f(t, x_1, \ldots, x_n)$ and multidimensional sets of the form $\mathbb{R}_+ \times A$. For an interval I, the corresponding integral has the form: $\int_{I\times A} f \; d(t, x_1, \ldots x_n)$, where $d(t, x_1, \ldots, x_n)$ is a product measure on the space $\mathbb{R} \times \mathbb{R}^n$. Equivalently, we can present the integral using the characteristic function 1_A corresponding to set A, the function has value 1 for the elements of A and 0 for all other arguments: $\int 1_A f(t, x_1, \ldots, x_n) \; d(t, x_1, \ldots, x_n)$.

This integral can be presented also in the form $\int_I \int_A f \; d(x_1, \ldots, x_n)dt$. Thus, the integral $\int_{I\times A} f \; d(t, x_1, \ldots x_n)$ has the form $\int_I g(t) \, dt$, where $g(t) = \int_A f(t, x_1, \ldots, x_n) \; d(x_1, \ldots, x_n)$. Consequently, it can be presented as an integral of unary function $g(t)$ depending on the time parameter t only.

6. Applications and Validation

In the case of multisensor systems, the data generated may be of various types, such as temperature, pressure, or density measurements, videos, or sound, among others. It is hard to bring them into a uniform and consistent model. Thus, quite often, textual specifications and informal validations are used, which is inherently imprecise. Specification, interpretation, comparison, and reasoning about such heterogeneous data pose a nontrivial problem. In this section, we apply the DC4F calculus to express measurements and complex behaviors. The goal is to illustrate the way DC4F can be applied and to demonstrate its capabilities for specifying temporal series. We use examples from the area of volcanic activity monitoring by multisensor systems because of their complexity and, thus, the challenge that they pose. We aim to show that the extension of DC that we propose in this paper provides a uniform language, models, and a reasoning system to integrate, describe, and reason about such data. The examples are used to show the practical applicability of DC4F and to demonstrate how it can be used. However, it is not our goal to adequately describe those phenomena per se. It should be also pointed out that the units of measurement do not play an essential role for the illustration. Our concepts concern mathematical properties and, thus, are independent of the units used.

6.1. Volcano Monitoring

To demonstrate the applicability of DC4F, we looked at several examples of signal processing and the corresponding temporal series. It turned out that volcano monitoring is an interesting and challenging task due to the complexity of the process, various monitoring methods used, and the heterogeneity of the acquired data. In this subsection, we present a brief presentation of volcano monitoring processes, as they are instrumental in illustrating our ideas.

As pointed out by Corradini et al. in [18,28], volcano monitoring processes require the correlation of satellite data, provided that satellites are present over the specific target. Satellite data are heterogeneous in nature and relate to temperatures measured in different wavelengths, among other measurements. Several phenomena, such as temperature picks, gas explosions, and magma eruptions, need to be considered (cf., e.g., [19,20]).

Strombolian volcano eruptions are a widely studied topic. They are characterized by specific cyclical patterns. Therefore, their specification is a challenging task. A Strombolian volcano eruption is characterized by regular, relatively mild blasts. It consists of the ejection of fragments of solidified lava, material deposited by previous volcanic eruptions, and masses of molten rock to altitudes ranging from tens to hundreds of meters (cf., e.g., [18–20]). Eruptions of this type are named after the Stromboli volcano. They are commonly observed in volcanoes fed by low to moderate viscosity magma. The explosions are caused by the bursting of gas slugs, which rise to the surface faster than the surrounding magma. Quakes associated with such explosions occur at low depths (cf. e.g., [20]). Spectroscopic measurements performed on the Stromboli volcano were used to demonstrate that gas slugs originate from the volcano–crust interface at the depth of approximately 3 kilometers, which may promote slug coalescence. In the case of the Stromboli volcano, quantitative constraints for the depth of conglomerates of high-pressure gas bubbles inducing Strombolian activity can be identified on the basis of spectroscopic measurements of the magmatic gas phase driving the explosions.

6.2. Dealing with Estimates

Various parameters are considered when a volcano eruption is monitored. These parameters need to be measured and compared. In this subsection, we show that functional limits may be dealt with in CD4F.

Thermal satellite images of the Etna eruption that took place in 2018 were compared with ground-based geophysical data of summit craters using the Moderate Resolution Imaging Spectroradiometer (MODIS) (see [22]). The technology provides infrared images and makes the identification of hot spots including pixels possible. The thermal anomalies may be quantified in terms of volcanic radiative power (VRP), expressed in watts (see [22] and the references there). The VRP value is calculated as follows:

$$VRP_{pIX} = Ap_{IX} \times 18.7 \times (L_{4alert} - L_{4bk}) \tag{9}$$

where A_{pIX} is the pixel area of approximately 10^6 m^2, L_{4alert} is the recorded the middle infrared radiance (W $m^{-2}sr^{-1}m^{-1}$), and L_{4bk} are the background pixels. These values allow to estimate the average lava erupted volume, i.e., the time average discharge rate (TADR), through a unique parameter called radiant density c_{rad} (J m^3):

$$TADR = \frac{VRP}{c_{rad}} \tag{10}$$

This parameter represents capacity of the lava body to radiate heat (see [22] and the references there). It allows to estimate the erupted lava volume by assuming the appropriate parameter c_{rad} for the lava flow in question.

The fact that the actual lava flow volume $TADR$ is between limits $LApp$ and $UpApp$ (see Figure 5) can be specified by the following formula, which is analogous to Formula (5) from Section 4.3. This fact cannot be expressed in DC, since DC only allows to consider the duration properties on intervals but not the values of integrals:

$$\Box \int LApp \leqslant \int TADR \leqslant \int UpApp \tag{11}$$

The inequalities express the desired property in terms of intervals. The formula means that the inequalities hold for every subinterval of the given interval (see Section 3.2).

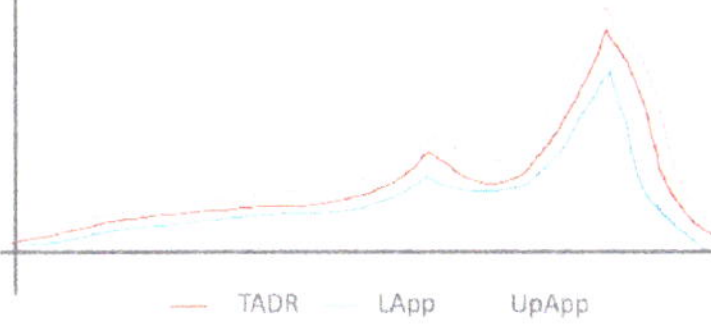

Figure 5. TADR with its lower and upper approximation.

6.3. Dealing with Two-Dimensional Spaces

There exist algorithms for detecting thermal anomalies in two-dimensional pictures related to volcano eruptions. In this subsection, we discuss the application of DC4F to such cases. We specify the units of measurement, but it should be noted that for the presentation, the units of measurement and exact numeric values obtained do not play a significant role in the presentation data, since in our framework we can deal with values of type $\mathbb{R}$ independently of their interpretation.

MODVOLC is a fully automated algorithm for the analysis of thermal satellite time-series data [1]. The algorithm is based on spectral wavelengths centred at 4 and 11–12 µm emitted by high-temperature volcanic sources and the surrounding Earth surface. In the case of a volcanic eruption, the radiance is measured from the corresponding pixel. One pixel corresponds to approximately 4 km. Images are collected every 15 min. In pixels that show a thermal anomaly, the radiance measured in the mid-infrared (4 µm) is significantly higher than in surrounding pixels where no thermal anomalies are depicted [1,21].

Figure 6 shows an excerpt of Figure 6 from [1]. It contains results obtained by the MODVOLC algorithm detecting thermal anomalies for different thresholds 2.0, 2.6, and 3.0, respectively, in case of the Anatahan volcano monitoring data. The following formula presents the Local Index of Change of the Environment $\otimes_v(x, y, t)$ representing the degree to which each pixel deviates from its normal behavior normalized by its variability (cf. [1,29]).

$$\otimes_v (x, y, t) = \frac{V(x, y, t) - V_{REF}(x, y)}{\sigma_v(x, y)} \tag{12}$$

An individual observation is modeled by the function $V(x, y, t)$ specifying the state of a pixel (x, y) at time t. Its normal behavior is expressed by function $V_{REF}(x, y)$ and its variability by $\sigma_v(x, y)$. One can then identify pixels (x, y) for which the function exceeds a given threshold $Tr \leqslant \otimes_v(x, y, t)$. The number of such pixels can be then computed for the entire area of interest A using its characteristic function 1_A, i.e., $(x, y) \in A \Leftrightarrow 1_A(x, y) = 1$ in the discrete case when single pixels are considered and each pixel has measure 1. A thermal anomaly is a pixel with a temperature measurement value above a defined threshold. The total number of thermal anomalies occurring at time t with values above threshold Tr is modeled by function $anomalies(Tr, t)$ defined by the following integral:

$$anomalies(Tr, t) = \int_A 1_{\{(x,y) \mid Tr \leqslant \otimes_v(x,y,t)\}}\, dxdy = \int_{\mathbb{R}\times\mathbb{R}} 1_A\, 1_{\{(x,y) \mid Tr \leqslant \otimes_v(x,y,t)\}}\, dxdy \tag{13}$$

Thus, the number of anomalies is the integral of the product of the characteristic function of set A, corresponding to the area of interest, and of the characteristic function identifying points with temperature above threshold Tr. In the continuous case, the value of the integral is equal to the measure of the set of thermal anomalies in the area of interest.

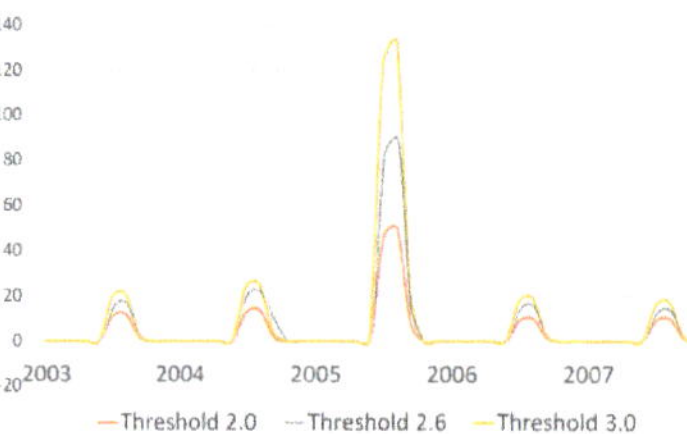

Figure 6. Number of thermal anomalies detected by the hybrid algorithm over MODVOLC using thresholds of 2.0, 2.6, and 3.0, respectively.

6.4. Specification of Strombolian Effects

Strombolian effects are characterized by cyclic periods of repetitive behaviors. Repetitive gas explosions and temperature peaks occurring in a periodic manner are one of such aspects. These explosions are caused by deep slag-driven explosive activity (cf., e.g., [20]). Those effects include also consecutive magma eruptions. In this subsection, we demonstrate that these kinds of effects can be specified in DC4F. We utilize the temporal properties formulated in the preceding section. The data that we use were obtained on 9 April 2002 during hours of passive and explosive degassing on the Stromboli volcano that was most active at the time [20]. The spectrometer was placed at a distance of 240 m.

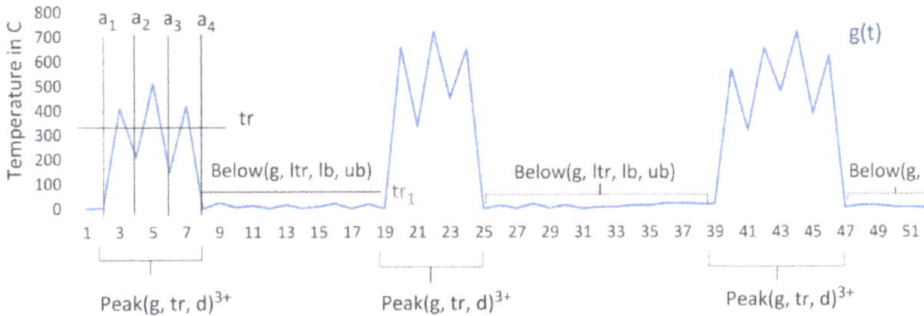

Figure 7. Periodic peaks in radiation corresponding to source temperature.

Repetitive temperature increases are shown in Figure 7. It shows the diagram presented in [20] (see Figure 1 and Figure S1 there) representing peaks in radiation corresponding to source temperature in Celsius. Figure 7 shows a function $g(t)$ with cyclic periods of consecutive value peaks and drops representing the level of radiation. The peaks can reach different maximal values, the drops can reach different minimal values, and their numbers are not fixed in a cycle. The time scale of Strombolian effects is usually hours and days (cf., e.g., [18,20,28]). In Figure 7 and below, we use only numeric time values, as the exact timescale does not play any role in the demonstration and may be changed in an arbitrary manner.

Each peak formation is characterized by a monotone increase of g followed by its monotone decrease. In consequence, a peak can be specified in DC4F by formula $MIncreasing(g); MDecreasing(g)$ (see Section 4.2). For example, Figure 7 shows a peak corresponding to interval $[a_1, a_2]$.

We can also require that the peak values be above a certain threshold tr and the amplitude be above the value of d. These requirements can be formulated as $PeakAbove(g, tr)$ and $Amplitude(g, d)$, respectively. The first formula states that function g reaches a value above threshold tr, and the second one states that its amplitude is at least equal to d (see Section 4.3). In Figure 7, the first peak is above 400 and the amplitude is above 100; of course, these parameters can be adjusted arbitrarily. The fact of reaching such a peak with the constraints concerning values of threshold and amplitude can be expressed by the formula $MIncreasing(g) \frown MDecreasing(g) \wedge PeakAbove(g, tr) \wedge Amplitude(g, d)$. This formula is the logical conjunction of the previous formulas; we abbreviate it as $Peak(g, tr, d)$. We use capital letters to indicate that functions have functional arguments, not only arguments of type $\mathbb{R}$.

Now, the peaks and the corresponding drops form a repetitive behavior. The left-hand side of Figure 7 shows three peaks, modeling COS bursts, which correspond to the intervals $[a_i, a_{i+1}]$, for $i = 1, \ldots, 3$. We can write the corresponding formula as $Peak(g, tr, d) \frown Peak(g, tr, d) \frown Peak(g, tr, d)$. It means that the interval may be split into three subintervals such that each satisfies the formula $Peak(g, tr, d)$. We abbreviate the formula as $Peak(g, tr, d)^3$.

In general, for a formula F, n repetitions of the form $F \frown \ldots \frown F$ are abbreviated as F^n. This formula means that the interval may be split into n subintervals in which F is satisfied. If the number of repetitions is arbitrary, then we write F^*. If it equals at least n, then we write F^{n+}, which is equivalent to $F^n \frown F^*$.

The peaks occur in cyclic series separated by periods of low volcanic activity. Let formula $Below(g, ltr, lb, ub)$ be the abbreviation of $(\Box \int g \leqslant ltr) \wedge lb \leqslant \ell \leqslant ub$. This means that g is always below threshold ltr in a given time interval and that the length of the interval is between time bounds lb and ub (cf. Section 4.3). We can compose the formulas again to specify the behavior shown in Figure 7; for $m = 3$ and $n = 3$, the formula has the form:

$$\{Peak(g, tr, d)^{m+} \frown Below(g, ltr, lb, ub)\}^{n} \tag{14}$$

This Strombolian activity consists of three cycles; each cycle is composed of at least three bursts followed by a period of a low volcanic activity. The three cycles of repetitive bursts are indicated in Figure 7 by brackets below the time axis. The periods of limited volcanic activity are indicated in a similar way.

Strombolian activities may be also detected using satellites equipped with thermal imaging cameras, which sometimes are the source of more reliable indicators than sensors on the ground (cf., e.g., [18,28]). However, if the satellites are not stationary or if visibility is reduced by clouds, then data from thermal imaging cameras cannot be used. We can specify the conditional visibility of thermal peaks by formula $\int visibility = \ell \Rightarrow Peak(f, tr', d')$. In this formula, $visibility$ is a Boolean-valued time-dependent function modeling visibility; it is true if there is visibility and false otherwise. Function f models temperature measurements at a certain frequency. Parameter tr' is the corresponding temperature threshold and parameter d' is the corresponding amplitude. The formula is conditional and reads as follows: if there is visibility, then a thermal peak is observed. We abbreviate this formula by $CondPeak(f, tr', d')$. Analogously, we can define $CondBelow(b, ltr', lb', ub')$ for the case in which the thermal radiance measurement is below a certain threshold. If the peaks in thermal anomalies coincide with the peaks in COS bursts, then we can transform formula (14), taking the thermal anomalies into account as well:

$$\begin{aligned} \Diamond \ \{CondPeak(g, tr, d)^{m+} \frown CondBelow(g, ltr, lb, ub))\}^{n} \wedge \\ \{CondPeak(f, tr', d')^{m+} \frown CondBelow(b, ltr', lb', ub'))\}^{n} \end{aligned} \tag{15}$$

The sometime-modality is used here to specify that we do not require anything for the time periods before and after the characteristic behavior. We may also specify here the minimum and maximum lengths of the periods of low activity, if needed.

Of course, the formulas become more and more complicated as we specify more kinds of behaviors and data, e.g., multisensor data, and constraints. However, the point is that the formulas may be specified in DC4F. Thus, we may precisely formulate fine hypotheses concerning complex behaviors. Furthermore, if implemented, the hypotheses may be verified in respect to existing data. They may also be used to automatically mine the available data, historic or incoming.

6.5. Reasoning in DC4F

In the previous subsections, we presented the capabilities of DC4F for the specification of periodic behaviors and the hypothesis framing. In this subsection, we present a relatively simple example of reasoning in DC4F utilizing the properties of DC4F defined in Section 5.3. The example is somewhat artificial, as we did not find proper examples in the literature, and we do not claim that the formulas truly express actual volcanic behavior. We simplify the proof, as its full version would be rather long. As in the previous section, we skip units of measurement as they do not play any role for the illustration.

Suppose that we know that a Strombolian effect is characterized by two properties occurring within 24 h: a series of COS bursts and an observation of thermal anomalies. The data is on COS burst is collected by sensors on the ground and the data on thermal anomalies are collected by a sensor placed on a satellite. The behavior is specified by functions $f_{COS}(t)$ and f_{anom} representing the emission level of COS and the number of observed anomalies, respectively. We use abbreviation $COSB$ for the formula obtained from formula (14) by substituting f_{COS} for g, 1 for n, and by setting other parameters somehow. We assume that $f_{anom}(t) = anomalies(t, Tr)$ for a certain threshold tr (cf. Section 6.3).

Let the Strombolian effect be characterized by at least 10 COS bursts and thermal anomalies of value 5 occurring within 24 h. Moreover, let it be an established fact that 12 COS bursts are always accompanied by thermal anomalies of value 7. The following formulas specify these two properties:

$$\ell \leqslant 24 \wedge \lozenge\ COSB^{10} \wedge 5 \leqslant \int f_{anom} \tag{16}$$

$$\ell \leqslant 24 \wedge \lozenge\ COSB^{12} \Rightarrow 7 \leqslant \int f_{anom} \tag{17}$$

Now, let the observation made by sensor on the ground be that within 24 h, there were 14 COS bursts, as specified by the following formula:

$$\ell \leqslant 24 \wedge \lozenge\ COSB^{14} \tag{18}$$

Suppose that the data on thermal anomalies are not available due to bad weather and, thus, the integral $\int f_{anom}$ cannot be computed. We will show that, despite that obstacle, the Strombolian behavior can be proved.

The formula $\lozenge\ COSB^{14}$ can be presented in the form $true \frown COSB^{12} \frown COSB^{2} \frown true$ (we apply here the definition of $\lozenge$ and unfold the exponent). The formulas $COSB^{2} \Rightarrow true$ and $true \frown true \Rightarrow true$ are tautologies of the predicate calculus (see point (6) in Section 5.3). Due to these two tautologies and the monotonicity of the chop operator (see point (7) in Section 5.3), we deduce the implication

$$true \frown COSB^{12} \frown COSB^{2} \frown true \Rightarrow true \frown COSB^{12} \frown true$$

Further, applying the monotonicity property of conjunction (see point (6) two times in Section 5.3) to the above implications, we derive the following implications:

$$\ell \leqslant 24 \wedge \lozenge\ COSB^{14} \Rightarrow \ell \leqslant 24 \wedge \lozenge\ COSB^{12},\ \ell \leqslant 24 \wedge \lozenge\ COSB^{12} \Rightarrow 7 \leqslant \int f_{anom}$$

Moreover, $7 \leqslant \int f_{anom} \Rightarrow 5 \leqslant \int f_{anom}$ and $\lozenge COSB^{12} \Rightarrow \lozenge COSB^{10}$. From the above implications, we derived the following implication:

$$\ell \leqslant 24 \wedge \lozenge\ COSB^{14} \Rightarrow Le \leqslant 24 \wedge \lozenge\ COSB^{10} \wedge 5 \leqslant \int f_{anom}$$

Antecedent of this implication is the observation (18). The consequent of this implication is the formula to be proved. The consequent follows from the antecedent and the implication by the application of modus ponens rule (see point (8) in Section 5.3). Thus, observation (18) and fact (17) imply that the Strombolian property holds.

6.6. Detecting Similar Behaviors

In this subsection, we consider the methods used in different areas, volcano monitoring included, for detecting similarities between functions and, in particular, between temporal series. Various similarity measures can be used in this case (cf., e.g., [23–26]). Different types of volcanic activity may be characterized not only in absolute terms, i.e., by general characteristics satisfied by all behaviors, but also relative to one another, or in relation to known examples of behaviors. We consider neither the application of specific similarity measures to specific cases, nor their adequacy, as we are not aiming to identify similarities in volcanic behavior. We are not in the position to judge which similarity measures would be most suitable for the Strombolian eruption patterns. We merely intend to stress that such measures and the corresponding algorithms can be used within the framework of DC4F.

Similarity measures may be formalized by functionals **FlSym**, which are interpreted as real-valued functions defined on intervals (see Section 5.1 and also the Appendix A). Similarity measures m can be seen as functionals that, for a given interval I and two functions f and g to be compared, return a real value being the measure of similarity. Thus, we can consider the application of a measure functional m to the two measured functions f and g and an interval of interest I: $m(f, g, I)$. Consequently, the result can be presented as a real-valued function on intervals $X_{f,g} \in \mathbf{FISym}$, which returns a real value r for the interval I (see Section 5.1 and also Appendix A). The measured level of similarity can be then used in framing descriptions and hypotheses concerning the time series—in particular, concerning volcanic eruptions.

Similarly, the standard Pearson correlation coefficient used in statistics may be integrated into CD4F as well. The correlation coefficient $\rho_{f,g}$ of two functions f, g is specified by integral $\int(f - \int f)(g - \int g) \,/\, (\int (f - \int f)^2 \int (g - \int g)^2)$. The integral returns for an interval I a value of type real, provided that the denominator is different from zero. Thus, coefficient $\rho_{f,g}$ depends on the interval in question, and as was the case in the example presented above, it may be represented as a real-valued function on intervals.

In general, some of the similarity functions can be defined directly in DC4F. Others, such as algorithms computing the similarity measures, can be defined simply by the interval mapping operators corresponding to functionals **FISym**. Thus, if there is an algorithm computing a similarity measure for intervals, then we can associate interval operator $X \in \mathbf{FISym}$ with the algorithm: $X([a, b]) = x$ if x is the value computed by the algorithm for interval $[a, b]$. The algorithm proposed in [19] can be treated along these lines.

7. Conclusions

In this paper, we proposed a duration calculus for integrable functions (DC4F). It is a natural, conservative extension of the well-known and widely used duration calculus. DC was used in numerous applications to specify various kinds of real-time and hybrid systems, to synthesize controllers, and so on. However, it allows to deal only with state functions, i.e., functions with Boolean values only. It did not aim to specify signals and measurements. The theoretical value of the proposed extension seems to be rather modest; however, its applications appear to be interesting. We are wondering why DC was not developed in this way from the beginning, as this would naturally broaden its application range and would be a consequent thing to do.

DC4F, like DC, is a type of interval logic providing a formal language, uniform mathematical models, and a reasoning system to deal with time series. Both calculi provide axioms and inference rules allowing to reason about properties of systems and to draw conclusions. However, they differ in expressivity, the scope of their applications, and, of course, maturity. Expressivity is achieved, in most cases, at the cost of complexity of the proof system and methods. This is the case of DC and DC4F as well. As the general form of DC is not decidable, i.e., there is no algorithm that decides whether a given formula is provable or not; the extended expressivity of DC4F relative to DC does not change much in that respect. DC, in contrary to DC4F, is aimed exclusively at the specification of durations of properties. The properties are specified by Boolean-valued functions. A great amount of research has been conducted about DC. Model-checking algorithms were created for its restricted versions and various tools were developed. It has been used to synthesize controllers. DC4F allows to integrate arbitrary Riemann integrable functions. Consequently, it provides a consistent general model for general form time series. The models can be handled using integral calculus, computer algebra, and so on. It allows one to specify several factors, parameters, and signals in one coherent language, to model them in a uniform manner, and to reason about them. Consequently, DC4F extends the ideas of DC to a qualitatively new area of applications.

We demonstrated that DC4F can be used to specify various temporal properties of time series, such as monotonicity, boundedness, and periodicity. We used it to specify different aspects of volcano monitoring activities, with a particular emphasis placed on Strombolian effects. DC4F proved to be a framework that is useful for integrating various types of data expressed in terms of time series and diagrams. Usually, in other papers, such types of data are informally related and reasoned about using informal textual descriptions in natural languages, such as English, and some mathematical formulas. This way of handling data is, of course, imprecise and error-prone. Thus, DC4f provides a remedy for this problem in the form of a unifying logic.

DC4F can be used to formulate specifications and hypotheses, reason about them, and validate them. Informally speaking, the difference between validation and deduction is that validation is performed for a concrete data set, on a specific model, semantic deduction concerns all possible data sets and all concrete models, and syntactic deduction relies solely on the application of sound deduction rules. The proposed logic can be combined with other methods, such as algorithms, and in particular with trained neural networks and other methods used in artificial intelligence (cf., e.g., [19]).

Funding: This research received no external funding.

Institutional Review Board Statement: Not applicable.

Informed Consent Statement: Not applicable.

Conflicts of Interest: The authors declare no conflict of interest.

Appendix A. Formal Semantics

To make the paper self-contained, we present in this section the formal semantics of the proposed DC extension. The semantics were already outlined in Section 5.2, but the formal definitions, thought a bit tedious, offer full mathematical precision.

These semantics are closely related to the original semantics of DC. The main difference is that we use general Riemann integrable functions instead of functions with Boolean values. DC4F generalizes the scope of integral operator, but it does not add new logical operators to DC. From the logic point of view, DC4F is a conservative extension of DC. Thus, a formula of DC is a tautology of DC4F if, and only if, it is a tautology of DC. We define the semantics of the formulas defined in Section 5.1. The definition has three parts corresponding to the three syntactic categories defined there: functions *Fn*, terms *Real*, and formulas *Fo*.

The time domain $\mathbb{T}$ime is the set $\mathbb{R}_+$ consisting of all non-negative real numbers. The discrete case, when the time domain is the set of natural numbers $\mathbb{N}$, can be dealt with as a special case of $\mathbb{R}_+$, as was shown in Section 5.2. In particular, if $\mathbb{T}$ime = $\mathbb{N}$, then we assume that the set of $\mathbb{IF}$un consists of functions that change their values at most at natural numbers and are constant elsewhere. Set $\mathbb{IF}$un contains some unary Riemann integrable functions on positive reals. The symbols from the set **FSym** are interpreted as functions from $\mathbb{IF}$un. For a function symbol $f \in$ **FSym**, the corresponding interpretation is an integrable function $I[f] \in \mathbb{IF}$un of one variable of type $\mathbb{R}$.

The semantics of *Fn* is defined in a standard way. Interpretation function I is the key here. Let $f, f_1, f_2 \in$ **FSym** and $c \in$ **CSym**. For the terms of category *Fn*, it is defined as follows:

- $I[0] = 0,\ I[1] = 1,\ I[true] = 1,\ I[c] \in \mathbb{R}$,
- $I[f] \in \mathbb{IF}$un
- $I[cf] = I[c]I[f]$,
- $I[g(f_1, f_2)] = I[g](I[f_1], I[f_2])$, for $g \in \{+, -, \vee\}$; here, the operations $+, -, \vee$ are interpreted as addition and subtraction, respectively.

The set $\mathbb{I}$nte consists of all intervals: $\mathbb{I}\text{nte} = \{[a,b] \mid a, b \in \mathbb{T}\text{ime} \wedge a \leqslant b\}$. The set $\mathbb{V}\text{al} = \{val \mid val : \textbf{VSym} \mapsto \mathbb{R}\}$ contains all valuations of variables. Let $\mathbb{F}$ls be a set containing functionals, i.e., functions that map time intervals to real numbers. The functional symbols **FlSym** (cf. Section 5.1) are interpreted as elements of $\mathbb{F}$ls. We assume that d is the usual measure on $\mathbb{R}$.

- $I[\int f](val, [a,b]) = \int_a^b f\,dx$
- $I[X](val, [a,b]) \in \mathbb{F}$ls, for $X \in$ **FlSym**

- $I[t_1 \oplus t_2](val,[a,b]) = I[\oplus](I[t_1](val,[a,b]), I[t_2](val,[a,b]))$, where t_1, t_2 are terms, $\oplus \in \{+,-,\vee,\diagup\}$, and $I[\oplus]$ are interpreted as the addition, subtraction, maximum, or division, respectively.

We say that two valuations val, val' differ at most on variable x if $val(y) = val'(x)$ for every variable y is different from x. To define the semantics terms, the interpretation function Int needs pairs consisting of a valuation function $val \in \mathbb{V}\mathrm{al}$ and an interval $[a,b] \in \mathbb{I}\mathrm{nte}$.

We define now the satisfaction relation $\models$ for the elements of category Fo. The relation has three arguments: a valuation, an interval, and a formula. It is defined as follows:

- $I(val,[a,b]) \models t_1 \oplus t_2$ iff the values $I[t_1](val,[a,b])$, $I[t_2](val,[a,b]))$ are in relation $I[\oplus]$, where t_1, t_2 are terms, $\oplus \in \{<,=\}$ and $I[\oplus]$ is the inequality or equality, respectively
- $I(val,[a,b]) \models \neg Fo$ iff $I(val,[a,b])Fo$ does not hold
- $I(val,[a,b]) \models Fo_1 \vee Fo_2$ iff $I(val,[a,b]) \models Fo_1$ or $I(val,[a,b]) \models Fo_2$
- $I(val,[a,b]) \models Fo_1 \frown Fo_2$ iff $I(val,[a,m]) \models Fo_1$ and $I(val,[m,b]) \models Fo_2$, for some $m \in [a,b]$
- $I(val,[a,b]) \models \exists_x Fo$ iff $I(val',[a,b]) \models Fo$ for some valuation val' which differs from valuation val at most on variable x
- $I[Fo^n](val,[a,b])$ if, and only if, either $n = 0$ and $a = b$ or $0 < n$ and there exist numbers $a_0,\ldots,a_n \in \mathbb{T}\mathrm{ime}$ such that $a = a_0 < a_1 < \cdots < a_n = b$ and $I[Fo](val,[a_i,a_{i+1}])$, for $i = 0,\ldots,n-1$

References

1. Koeppen, W.C.; Pilger, E.; Wright, R. Time series analysis of infrared satellite data for detecting thermal anomalies: A hybrid approach. *Bull. Volcanol.* **2011**, *73*, 577–593. [CrossRef]
2. Gabbay, D.M.; Guenthner, F. (Eds.) *Handbook of Philosophical Logic*; Springer: Berlin, Germany, 2001–2005.
3. Della Monica, D.; Goranko, V.; Montanari, A.; Sciavicco, G. Interval Temporal Logics: A Journey. *Bull. Eatcs* **2013**, *105*, 73–99.
4. Hansen, M.R.; Van Hung, D. A Theory of Duration Calculus with Application. in Chris George, Zhiming Liu, Jim Woodcock, Domain Modeling and the Duration Calculus. *Lect. Notes Comput. Sci.* **2007**, *4710*, 119–176.
5. Pering, T.D.; Ilanko, T.; Liu, E.J. Periodicity in Volcanic Gas Plumes: A Review and Analysis. *Geosciences* **2019**, *9*, 394. [CrossRef]
6. Castanedo, F. A Review of Data Fusion Techniques. *Sci. World J.* **2013**, *2013*, 704504. [CrossRef] [PubMed]
7. Bleiholder, J.; Naumann, F. Data fusion. *ACM Comput. Surv.* **2009**, *41*, 1–41. [CrossRef]
8. Liu, M.; Yang, W.; Zhu, X.; Chen, J.; Chen, X.; Yang, L.; Helmer, E.H. An Improved Flexible Spatiotemporal Data Fusion (IFSDAF) method for producing high spatiotemporal resolution normalized difference vegetation index time series. *Remote. Sens. Environ.* **2019**, *227*, 74–89. [CrossRef]
9. Meyer, R.; Faber, J.; Hoenicke, J.; Rybalchenko, A. Model checking Duration Calculus: A practical approach. *Formal Asp. Comput.* **2008**, *20*, 481–505. [CrossRef]
10. Hansen, M.R.; Brekling, A.W. On Tool Support for Duration Calculus on the Basis of Presburger Arithmetic. In Proceedings of the 2011 Eighteenth International Symposium on Temporal Representation and Reasoning, Lubeck, Germany, 12–14 September 2011; pp. 115–122. [CrossRef]
11. An J.; Zhan, N.; Li, X.; Zhang, M.; Yi, W. Model Checking Bounded Continuous-time Extended Linear Duration Invariants. In Proceedings of the 21st International Conference on Hybrid Systems: Computation and Control (part of CPS Week) (HSCC '18), HSCC 2018, Porto, Portugal, 11–13 April 2018; pp. 81–90. [CrossRef]
12. Larsen, K.G.; Rasmussen, J.I. Optimal conditional reachability for multipriced timed automata. *Theor. Comput. Sci.* **2008**, *390*, 197–213. [CrossRef]
13. Xiangjun, C.; Peng, D. Formalization Model and Safety Analyses of High Speed Train in CTCS-3 Control Mode. In Proceedings of the 2013 International Conference on Mechanical and Automation Engineering, Jiujang, China, 21–23 July 2013; pp. 161–167. [CrossRef]
14. Ramos, D.B.; e Silva, R.A.B.; Costa, I.C.; Colonese, E.M.; de Oliveira, J.M.P. Modeling conflicts resolution of Unmanned Aircraft System using a lightweight Duration Calculus. In Proceedings of the IEEE/AIAA 30th Digital Avionics Systems Conference, Seattle, WA, USA, 16–20 October 2011; pp. 5A6-1–5A6-9. [CrossRef]
15. Petnga, L.; Austin, M. Ontologies of Time and Time-based Reasoning for MBSE of Cyber-Physical Systems. *Procedia Comput. Sci.* **2013**, *16*, 403–412. ISSN 1877-0509. [CrossRef]
16. Dole, K.; Gupta, A.; Krishna, S.N. Robust Controller Synthesis for Duration Calculus. In *International Symposium on Automated Technology for Verification and Analysis*; Springer: Berlin/Heidelberg, Germany, 2020. [CrossRef]

17. Ody, H.; Fränzle, M.; Hansen, M.R. Discounted Duration Calculus. In *International Symposium on Formal Methods;* Springer: Berlin/Heidelberg, Germany, 2016; pp. 577–592. [CrossRef]
18. Corradini, S.; Montopoli, M.; Guerrieri, L.; Ricci, M.; Scollo, S.; Merucci, L.; Marzano, F.; Pugnaghi, S.; Prestifilippo, M.; Ventress, L.; et al. A Multi-Sensor Approach for Volcanic Ash Cloud Retrieval and Eruption Characterization: The 23 November 2013 Etna Lava Fountain. *Remote. Sens.* **2016**, *8*, 58. [CrossRef]
19. Dyea, B.C.; Morrab, G. Machine learning as a detection method of Strombolian eruptions in infrared images from Mount Erebus, Antarctica. *Phys. Earth Planet. Inter.* **2020**, *305*, 106508. [CrossRef]
20. Burton, M.; Allard, P.; Muré, F.; La Spina Magmatic, A. Gas Composition Reveals the Source Depth of Slug-Driven Strombolian Explosive Activity. *Science* **2007**, *317*, 227–230. [CrossRef] [PubMed]
21. Wright, R.; Flynn, L.; Garbeil, H.; Harris, A.; Pilger, E. MODVOLC: Near-real-time thermal monitoring of global volcanism. *J. Volcanol. Geotherm Res.* **2004**, *153*, 29–49. [CrossRef]
22. Laiolo, M.; Ripepe, M.; Cigolini, C.; Coppola, D.; Della Schiava, M.; Genco, R.; Innocenti, L.; Lacanna, G.; Marchetti, E.; Massimetti, F.; et al. Space- and Ground-Based Geophysical Data Tracking of Magma Migration in Shallow Feeding System of Mount Etna Volcano. *Remote. Sens.* **2019**, *11*, 1182. [CrossRef]
23. Vlachos, M.; Kollios, G.; Gunopulos, D. Discovering similar multidimensional trajectories. In Proceedings of the 18th International Conference on Data Engineering, San Jose, CA, USA, 26 February–1 March 2002; pp. 673–684. [CrossRef]
24. Raket, L.L.; Sommer, S.; Markussen, B. A nonlinear mixed-effects model for simultaneous smoothing and registration of functional data. *Pattern Recognit. Lett.* **2013**, *38*, 1–7. [CrossRef]
25. Olsen, N.L.; Markussen, B.; Raket, L. Simultaneous inference for misaligned multivariate functional data. *J. R. Stat. Soc. C* **2018**, *67*, 1147–1176. [CrossRef]
26. Koenig, L.; Lucero, J.; Perlman, E. Speech production variability in fricatives of children and adults: Results of functional data analysis. *J. Acoust. Soc. Am.* **2008**, *124*, 3158–3170. [CrossRef] [PubMed]
27. Barwise, J. (Ed.) *Handbook of Mathematical Logic;* Studies in Logic and Foundations of Mathematics; North Holland: Amsterdam, The Netherlands; New York, NY, USA; Oxford, UK, 1982.
28. Corradini, S.; Guerrieri, L.; Stelitano, D.; Salerno, G.; Scollo, S.; Merucci, L.; Prestifilippo, M.; Musacchio, M.; Silvestri, M.; Lombardo, V.; et al. Near Real-Time Monitoring of the Christmas 2018 Etna Eruption Using SEVIRI and Products Validation. *Remote. Sens.* **2020**, *12*, 1336. [CrossRef]
29. Pergola, N.; Marchese, F.; Tramutoli, V. Automated detection of thermal features of active volcanoes by means of infrared AVHRR records. *Remote. Sens. Env.* **2004**, *93*, 311–327. [CrossRef]

MDPI

Article

Modeling and Simulation of Very High Spatial Resolution UXOs and Landmines in a Hyperspectral Scene for UAV Survey

Milan Bajić, Jr. [1,*] and Milan Bajić [2]

1 Department of IT and Computer Sciences, Zagreb University of Applied Sciences, 10000 Zagreb, Croatia
2 Scientific Council HCR—Center for Testing, Development, and Training, 10000 Zagreb, Croatia; milan.bajic@ctro.hr
* Correspondence: mbajic@tvz.hr; Tel.: +385-99-266-8844

Abstract: This paper presents methods for the modeling and simulation of explosive target placement in terrain spectral images (i.e., real hyperspectral 90-channel VNIR data), considering unexploded ordnances, landmines, and improvised explosive devices. The models used for landmine detection operate at sub-pixel levels. The presented research uses very fine spatial resolutions, 0.945 × 0.945 mm for targets and 1.868 × 1.868 cm for the scene, where the number of target pixels ranges from 52 to 116. While previous research has used the mean spectral value of the target, it is omitted in this paper. The model considers the probability of detection and its confidence intervals, which are derived and used in the analysis of the considered explosive targets. The detection results are better when decreased target endmembers are used to match the scene resolution, rather than using endmembers at the full resolution of the target. Unmanned aerial vehicles, as carriers of snapshot hyperspectral cameras, enable flexible target resolution selection and good area coverage.

Keywords: explosive devices; hyperspectral data; simulation; Spectral Angle Mapping; UAV

Citation: Bajić, M., Jr.; Bajić, M. Modeling and Simulation of Very High Spatial Resolution UXOs and Landmines in a Hyperspectral Scene for UAV Survey. *Remote Sens.* **2021**, *13*, 837. https://doi.org/10.3390/rs13050837

Academic Editors: Piotr Kaniewski, Mateusz Pasternak and Stefano Mattoccia

Received: 10 January 2021
Accepted: 19 February 2021
Published: 24 February 2021

1. Introduction

1.1. Motivation

"Despite twenty-first-century technological advances by Western militaries for demining and the removal of improvised explosive devices, humanitarian demining relies mostly on mid-twentieth-century technology" [1].

Although we share this opinion—at least, regarding aerial survey technologies—we attempt to contribute to advancement by supporting the deployment of hyperspectral surveys by civilian users exposed to explosive device threats. We consider the following explosive devices: unexploded ordnances (UXOs), landmines (LMs), cluster munition (CM), improvised explosive devices (IEDs), homemade explosive (HME) devices, and explosive remnants of war (ERW). The civilian users that we consider are, among others: Single or group ground vehicles of humanitarian demining organizations, traveling from camp to the working area and returning, logistics convoys, medical, humanitarian aid, Red Cross, reconstruction, security forces, civilian VIP travelers, and others. The level of incidents and casualties for civilian vehicles and convoys dominate, when compared to military or security forces [2]. The focus of the technology reviewed and used in this work mostly considers the platforms, sensors, and software available to the civil sector.

There are several new aspects of these topics, and humanitarian mine action cannot be restricted only to the disposal of the landmines by humanitarian demining. Therefore, we briefly consider hazardous explosive threats, non-technical surveys (NTSs) [3], the technologies in use (based on the aerial survey), and advanced survey technologies under a high level of technical readiness. There are several promising sensor technologies, such as hyperspectral, non-linear junction detection (NLJD), LIDAR, longwave infrared, magnetometer, and ground-penetrating radar (GPR); however, in this article, we focus on passive hyperspectral data. This technology is specific, due to a lack of civilian (or public military) hyperspectral data regarding the considered explosive devices in a realistic and non-laboratory environment. A new positive fact is that hyperspectral imaging sensors used on unmanned aerial vehicles (UAVs) can provide pixel resolutions smaller than the explosive devices on the ground surface (very high spatial resolution), which is not practical when using aerial helicopter platforms.

We consider threats caused by several types of explosive devices: IEDs, UXOs, LMs, and ERW. Data about their spatial distribution are limited and, for IEDs, are typically classified. One possible solution is to predict the emplacement of explosive devices by simulation. Generally, several aspects should be considered in simulations: (1) terrain features, (2) explosive devices, (3) objects, and (4) methods of detection of explosive devices from data collected by sensors on remotely piloted aircraft systems.

Our research aims to derive modeling methods and to simulate explosive targets in a hyperspectral scene, through the use of real hyperspectral data of the considered types of explosive devices.

1.2. Possible Terrain Case

An exciting and valuable example of the situational and spatial behavior of IEDs has been presented in [4] and in a color video acquired from an unmanned aerial vehicle (UAV) [5]. The following photograph (Figure 1) shows a typical large-scale terrain and explosive targets in an arid region in Iraq. It is evident that this explosive hazard scene has a lack of vegetation, and the explosive devices are on the ground surface. The targets in such situations are "ideal" for hyperspectral survey and detection by UAV.

Figure 1. Remnants of a cluster munition in South Iraq; "Ideal" targets for hyperspectral detection from UAV [4].

Besides the need for detection and mapping of explosive devices, there exists needs and demands for the non-technical survey (NTS) of larger areas contaminated by various explosive objects. When one considers such areas for NTS, the most critical function of a survey system is the endurance (autonomy) of the UAV. The explosive hazard situation in many afflicted countries is similar to the above description, where the existing differences are mainly in the IED technique and application.

The United Nations peacekeeping forces are exposed to explosive threats; therefore, they have developed guidelines on IED threat mitigation [6]. The European Defense Agency realized the IED Detection Program from 2016 to 2019, in order to improve and to field-test IED detection capabilities to define future Route Clearance and Attack the Network capabilities [7].

1.3. The Civilian Aerial Survey Technologies for Explosive Threats

We have previously actively participated and contributed to the research, development, and operational deployment of a multisensory and multispectral non-technical survey (NTS) [3], mainly based on the detection of secondary indicators of mine presence (IMP) or indicators of the absence (IMA) of landmines in minefields [8,9].

The IMP and IMA depend on the situation, war history, terrain, climate, and vegetation cover, and are specific for each set of mentioned influencing factors. IMP and IMA have been identified in Croatia, Bosnia, and Herzegovina, [9]; similar IMP and IMA could be expected in Ukraine. However, for countries in the Near East (e.g., Iraq, Afghanistan, and Syria) and North Africa (e.g., Libya), the IMP and IMA will be significantly different. Due to a lack of vegetation in the mentioned countries, there exist chances for the direct detection of targets—that is, the explosive devices (UXO, LM, IED, CM, ERW)—on the ground surface by passive electro-optical sensors (see example from Iraq, Figure 1). With active sensors, these targets also can be detected in the soil.

The first UAV for humanitarian mine action appeared in the EU project ARC [8]. In the last 5 to 10 years, the application of UAVs with visible color sensors for humanitarian mine action has increased [4,5,9–12]. The statement made in Use of Aerial Imagery in Urban Survey & Use of UAVs in Mine Action—Lessons Learned from Six Countries, in simple words, generalizes the experience gained since 2019 regarding UAVs with advanced sensors:

> "No export restrictions. Platforms as small as possible. We want to operate equipment ourselves, not rely on external personnel. In the short term, detection capabilities are more important than the interface. Need to see real evidence of value before committing to field trials. Detection is only one stage of the clearance process. The combination of sensors and platforms must offer some advantage in terms of reduced false alarms or detection ability, not just the speed of coverage. Vegetation cover will be a major limiting factor in many places usually we cannot remove this in advance because of safety, cost, or environmental damage. The abilities of the sensor/data processing are what matters. Possible sensors: Thermal IR, Hyperspectral, Magnetometers?" [11].

Three essential facts have enabled the stepwise increase in survey efficiency by UAV: (A) The UAV industry (e.g., DJI) has produced very advanced systems which enable computer planning and automated airborne acquisition missions with several sensors. (B) The sensors industry has provided powerful devices matched to UAVs. (C) The software industry has provided tools for processing recordings collected by UAVs, producing the highest quality outcomes. Yet, only color visible cameras have been used in civilian humanitarian domain operations (as of May 2020). The (A), (B), and (C) changes have been finalized in the last several years; now, the average trained deminer-surveyor can use an UAV for their survey tasks, including preparing and planning field missions, pre-processing data into images, and processing these images into valuable and high-quality products for humanitarian mine action, in a short time. There has been an excellent example of the application of the UAV-based survey technology at the level of an entire country [12].

One exciting and challenging possibility is to detect buried explosive devices through spectral changes and derived indicators of the soil surface, as well as plant spectral changes, if exposed to the influence of landmines and explosives. Even if not applicable to arid surfaces, the hyperspectral assessment of plant spectral stress due to landmines and explosives has given promising initial results. Several research projects have been based on this assumption; one of the first was in 1997 [13]. In this project, a hyperspectral imager, named "casi", was used for the detection of buried landmines and blocks of explosive, reporting the probability of detection (in the range from 55% to 94%) and a false alarm rate (from 0.17 m^{-2} to 0.52 m^{-2}). Another research direction was to detect the difference of spectra of plants inside of a minefield, compared to the spectra of plants outside the minefield; that is, in areas that were clean of explosive and agricultural fertilizers. In [14–16], airplane platforms have been used, whereas [17] used different hyperspectral sensors onboard a UAV, helicopter, and ground-based vehicle [18]. This research direction considers changes in vegetation (plants, bushes, and trees) spectra after exposure to contamination with explosives [19,20]. Although the described domain is impressive, our feeling is that its operational potential is not high enough yet.

1.4. The Direct Detection of Explosive Targets and Detection of Their Secondary Indicators

Current UAV-based operational surveys for the detection of UXO, LM, IED, CM, and ERW only use (visible) color sensors [4,10,11]. Here, we comment on several other sensor technologies in the development or testing phase; although, we thoroughly only consider hyperspectral technology in later sections.

Longwave infrared (LWIR) or thermal infrared (TIR) sensors are now available as dual sensor units, together with a visible color camera (produced by FLIR and DJI) for application onboard UAVs. The dual-sensor delivers TIR and visible color images, either separated or overlapped. TIR sensors are also now available in a version which is optimized for UAV surveys [21,22]. One interesting solution for a long endurance survey is a tethered UAV [23].

For the active detection, via UAV, of targets in the soil or behind obstacles, non-linear junction detectors or harmonic radar (NLJD) can be used [24,25]. A ground-penetrating radar is under development for the detection of buried targets (GPR) from UAV [26–30].

The magnetometer on an UAV enables the automatic survey and detection of ferromagnetic UXO targets [31,32].

In the following chapters, we present the research results regarding the hyperspectral data of explosive targets. The first possible reference for the hyperspectral detection of landmines is from 1997 [13]. According to the best of our knowledge, there are no available/accessible hyperspectral data of explosive targets (UXO, LM, IED, CM, ERW) collected by UAV or, at least, collected by ground-based hyperspectral sensors in the considered afflicted countries. The exceptions are the cases where a limited amount of hyperspectral data of UXO, LM, and minefields were collected in the European environment, in minefields and exploded ammunition depots, by helicopter-, UAV-, and ground-based acquisition systems [9,17,18,33]. Some data collected by fixed-wing plane are available for Africa [14,15] and in Germany [16]. Ground-based hyperspectral data collection of landmines has also been carried out in Lebanon [34–36].

1.5. Hyperspectral Sensors and Platforms

1.5.1. Sensors

For this research, we used two hyperspectral imaging sensors. The first was a Specim ImSpector V9 (see Figure 2), a hyperspectral prism-grating-prism imaging spectrograph [37] which has a spectral range of 430–900 nm, a spectral resolution of 7 nm, sampling 5 nm, and 95 channels (product specifications). The second was a Cubert UHD-185 snapshot camera (see Figure 3), with a spectral range of 450–950 nm, spectral resolution of 8 nm, 125 channels, sampling at 4 nm, and spatial resolution of 1000×1000 pixels for panchromatic or 50×50 pixels for spectral (product specifications). The third sensor was a point measuring FieldSpec3 Spectroradiometer, from Analytical Spectral Devices, Inc., Boulder, CO, USA (ASD), ranging from 350 to 2500 nm, resolution from 3 to 10 nm, and 512 channels (product specifications); which was used for point measurements of targets and materials

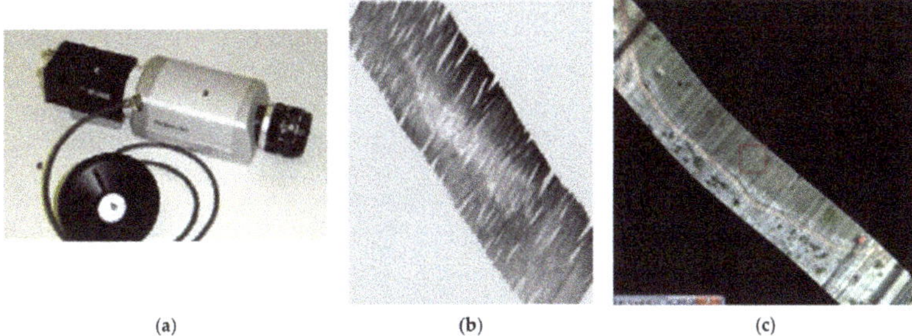
(a) (b) (c)

Figure 2. Hyperspectral push-broom sensor provides real hyperspectral data, while complex processing is needed to produce a calibrated hyperspectral cube: (**a**) Hyperspectral line scanner; (**b**) data are collected in scan lines, which need parametric geocoding; and (**c**) hyperspectral cube needed calibration.

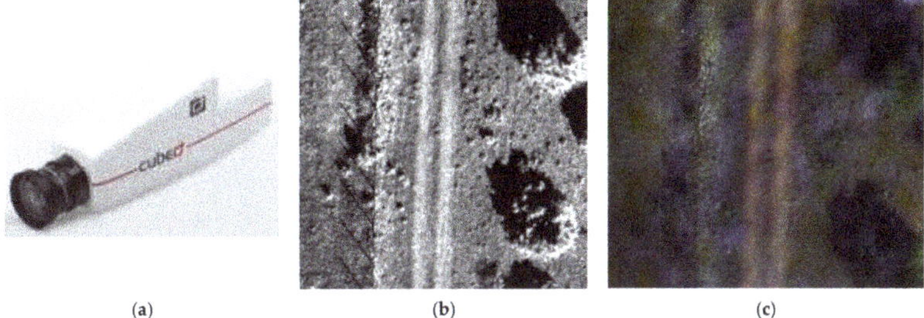

(a) (b) (c)

Figure 3. (**a**) Hyperspectral snapshot camera UHD-185; (**b**) panchromatic image of the scene (1000 × 1000 pixels); and (**c**) hyperspectral cube obtained by sharpening real hyperspectral recording of 50 × 50 pixels with 1000 × 1000 panchromatic pixels.

1.5.2. Ground-Based and Aerial Platforms

We have previously used the hyperspectral line scanner V9 to study minefields, landmines, and unexploded ordnances, initially through a ground-based mechanical scanner on a gantry [9,18,38]. We have used the V9 on several aerial platforms, such as the helicopters Mi-8c [39] and Bell-206. Note that the V9 was also used onboard the helicopter Mi-8 for detection of ship-sourced oil pollution on the sea [40].

Since 2012, we have applied the true spectral scanner V9 on UAVs, along with the pan-sharpening snapshot imaging scanner UHD-185 [17,41]. The helicopter Mi-8 platform was skipped in this research, as the spatial resolution was too low for the sake of target detection (due to blurring). Besides this primary purpose, the same mechanical scanner with V9 has been applied for archeological research [42] and in vineyards [43].

1.5.3. Portable Carry-on and Handheld Hyperspectral Cameras

Portable carry-on (or handheld) hyperspectral cameras are novel technological devices, one appearing around 2015–2018 and subsequently disappearing (Headwall Hyperspec® SNAPSHOT VNIR, Headwall Photonics, Inc., Bolton, MA, USA), while the second appeared in 2020 (Specim IQ). The Hyperspec® SNAPSHOT VNIR Sensor can quickly render a high-resolution hyperspectral scene at distances of 1.5 km in the VNIR spectral range (380–1000 nm; Headwall 2014). This makes it an excellent sensor for military hyperspectral reconnaissance. In 2017, the authors asked Headwall to offer this sensor; however, the answer was that it is not in production.

Specim IQ is a portable carry-on hyperspectral camera that contains the features needed for hyperspectral data capturing, data processing, and visualization of results. It has a wavelength band of 400–1000 nm, 204 spectral bands, an image resolution of

512 × 512 pixels, spectral resolution of 7 nm, and 12-bit data output. A full field of view (FOV) is 31 × 31 degrees; at 1 m, it covers 0.55 × 0.55 m. It is equipped with WiFi, GPS, and a 32 GB SD memory card. This camera can serve as an excellent tool for collecting hyperspectral data about explosive targets, landmines, unexploded ordnances, cluster munitions, improvised explosive devices, and neighborhood terrain.

2. Materials and Methods

After the war in 1991–1995, Croatia had become contaminated with minefields, scattered landmines, and unexploded ordnances, cluster munitions, and other explosive remnants of war. When we proposed to apply airborne multi-sensor minefield detection in 2001 [44], the reaction was prompt and productive [45,46]. Our interest in the detection of unexploded ordnances (UXOs) was initiated after an unplanned explosion in 2011 at the ammunition depot at Padjene, Croatia, and the survey of UXOs was included in the project TIRAMISU [18,38,41]. Fifteen different kinds of scattered UXO samples have been measured by V9 in imaging mode. Hyperspectral cubes have been produced and Johnson parameters calculated for each type [47]. The UXO samples appear in different conditions (e.g., intact, damaged, burned, covered by rust, covered by soil, original paint), orientation, and on similar soil types. This set of true hyperspectral cubes is our source for further research on UXOs. While radiance data were collected, we converted them into reflectance using the atmospheric correction QUAC (Quick Atmospheric Correction, ENVI).

We measured several other UXOs with the imaging hyperspectral sensor V9; this is also a set of true hyperspectral data, Figures 4–9. The mean values of measured reflectance of UXOs are shown at Figure 10. The conditions of measuring were controlled, Figures 11 and 12. Several landmines and one plastic object were measured using the point measuring spectroradiometer ASD. This data set provides only one value of reflectance for each wavelength, Figures 13–17. Both sets are used in the current article.

(a)

(b)

Figure 4. (**a**) Spectralon (on the right side) and UXO targets (artillery shell, bullet, and mortar mine); (**b**) The geometry of the measured data of Figure 4a are corrected by interpolation, with the new GRD being 0.945 mm.

(**a**) (**b**)

Figure 5. Artillery shell (**a**) photography by handheld camera; (**b**) color-visualized hyperspectral cube (red = 650 nm, green = 550 nm, blue = 450 nm). The ground resolving distance (GRD) is 0.945 mm.

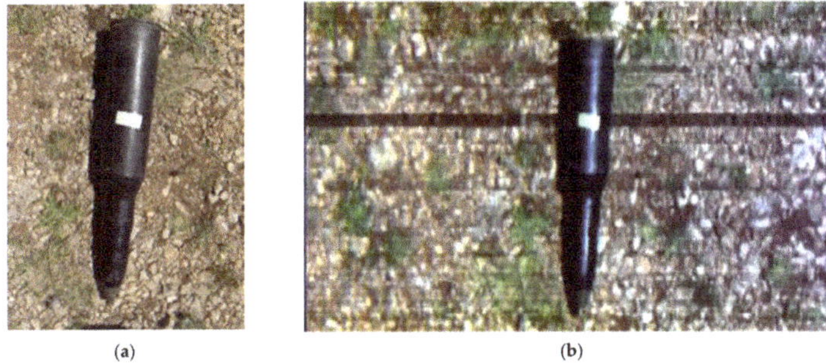

(**a**) (**b**)

Figure 6. Bullet: (**a**) photography; and (**b**) color-visualized hyperspectral cube (red = 650 nm, green = 550 nm, blue = 450 nm). The ground resolving distance (GRD) is 0.945 mm.

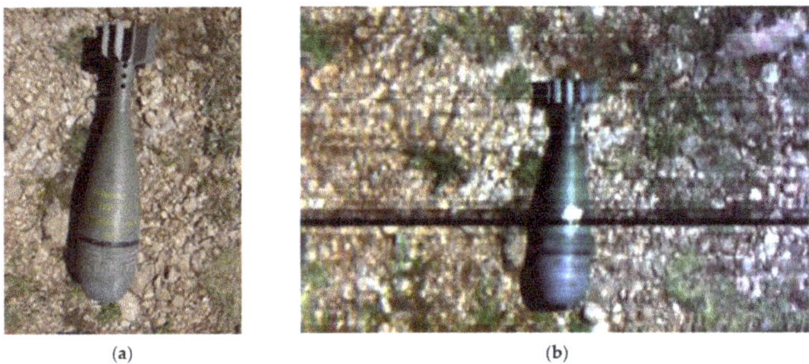

(**a**) (**b**)

Figure 7. Mortar mine: (**a**) photography; and (**b**) color-visualized hyperspectral cube (red = 650 nm, green = 550 nm, blue = 450 nm). The ground resolving distance (GRD) is 0.945 mm.

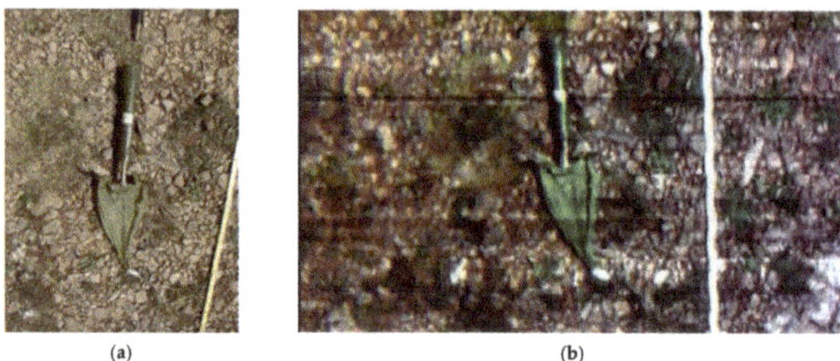

Figure 8. Cluster munition: (**a**) photography; and (**b**) color-visualized hyperspectral cube (red = 650 nm, green = 550 nm, blue = 450 nm). The ground resolving distance (GRD) is 0.945 mm.

Figure 9. UXO: (**a**) photography; and (**b**) color-visualized hyperspectral cube (red = 650 nm, green = 550 nm, blue = 450 nm). The ground resolving distance (GRD) is 0.945 mm.

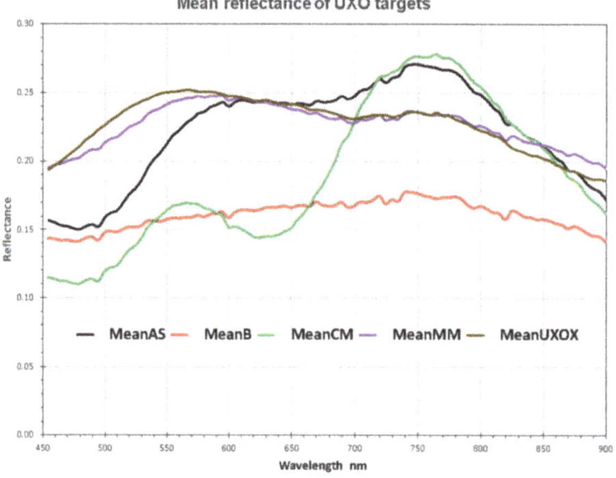

Figure 10. Mean spectra of five UXO targets. Legend: AS—artillery shell, B—bullet, CM—cluster munition, MM—mortar mine, and UXO_X—UXO of unknown type.

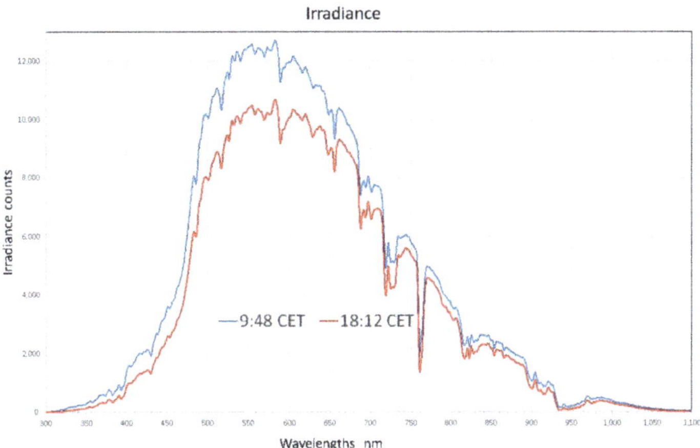

Figure 11. Irradiance counts were measured in the morning and the afternoon, using the UHD-185 imaging sensor onboard the UAV. This is the raw, uncalibrated Sun irradiance. The deep minimum of the irradiance (e.g., ~760 nm) must be corrected by interpolating irradiances at lower and higher wavelengths.

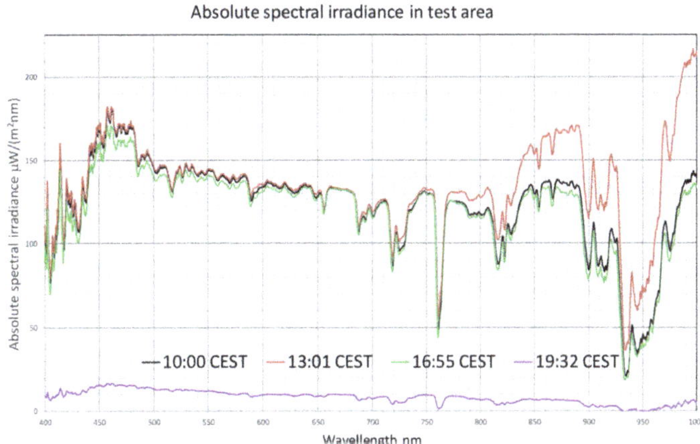

Figure 12. Absolute (calibrated) spectral irradiance, measured in the test area when hyperspectral data acquisition was carried out with the UAV.

The main obstacle was to provide hyperspectral cubes of the terrain ground surface, which should have pixel area smaller than the area of the considered UXO and landmines. The solution could be to use hyperspectral imaging of the minefields, which has been done using the V9 and UHD-185, onboard a Bell-206 helicopter or UAVs [17]. In the current article, we use only the terrain ground surface hyperspectral cubes collected by UHD-185 onboard a UAV.

2.1. The True Hyperspectral Data Cubes of UXO on the Ground

UXO samples have been measured by V9 in imaging mode, using the first, small version of the mechanic gantry [18]. The geometry of the acquisition mode is presented in Figure 4b.

The spectral radiance was measured, with vertical length A = 1.1 m, number of pixels M = 1164, and ground resolving distance of 0.945 mm. The horizontal length between Spectralon and the white-black-white panel is B = 2.0 m, the number of pixels is N = 556, and the ground resolving distance is 3.597 mm. The next step was geometric transformation and interpolation. For the interpolation, we tested the nearest neighbor, bilinear, and cubic methods and, as a result, decided to apply the nearest-neighbor method. The mean reflectance spectra values were below 0.280; see Figure 10.

The HR400 Spectrometer was used for irradiance measurements, both relative and absolute $\mu W/(m^2 nm)$ to the Sun in the periods when the hyperspectral missions were carried out; see Figures 11 and 12. Its role was to understand the dynamics of the absolute irradiance and to select times which are suitable for hyperspectral measurements.

2.2. Landmines and Plastic Objects, Whose Spectra Are Provided by Point-Like Measurements with ASD

The figures in the following section represent some of the targets that were measured by Point-Like Measurements with ASD. These sensor measures only one position where it is pointed. If we sample 10 or 15 points on target, they do not provide as much information variability as imaging sensor, covering entire target. When we created simulated targets by using ASD, we had available couple of points and simulation was not as realistic as from imaging sensor.

Figure 13. TMA-4 landmine.

Figure 14. VTMRP-6 landmine.

Figure 15. PMR-2A landmine.

Figure 16. Plastic bottle.

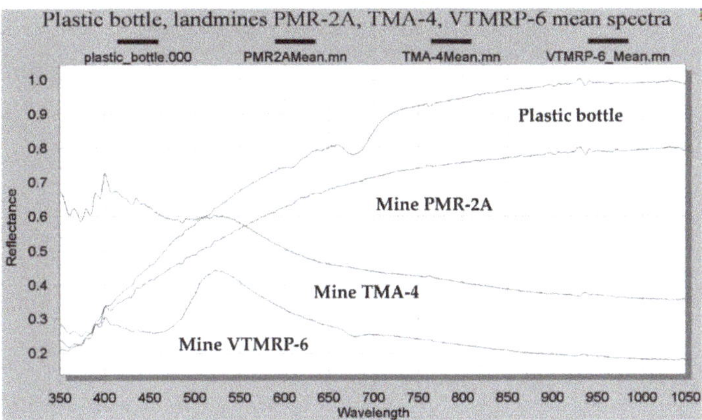

Figure 17. The reflectance of PMR-2A, TMA-4, and VTMRP-6 landmines, as well as that of a plastic bottle, measured by a point measuring unit ASD.

2.3. Hyperspectral Cubes of the Terrain Acquired by UHD-185

The hyperspectral cubes of terrain were acquired by the snapshot camera UHD-185, with 50 × 50 spectral pixels sharpened by 1000 × 1000 panchromatic pixels. The aerial platforms were UAVs at low altitude (Figures 18 and 19) and a Bell-206 helicopter at high altitude (Figure 20).

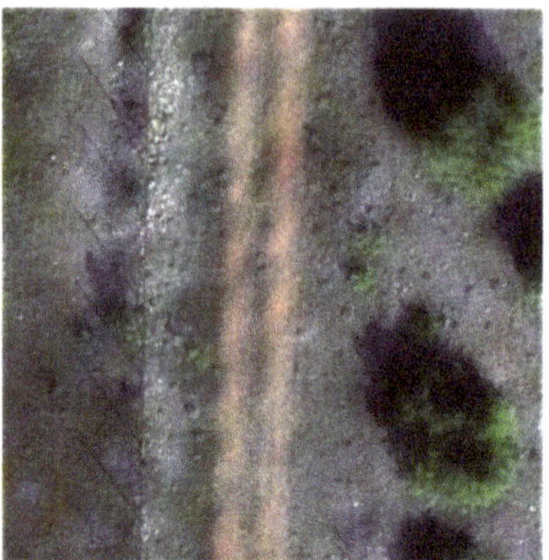

Figure 18. Hyperspectral scene 147. Dimensions 18.681 × 18.681 m, 1000 × 1000 pixels. Visualized with r = 650 nm, g = 550 nm, b = 460 nm.

Figure 19. Hyperspectral scene 227. Dimensions 18.681 × 18.681 m, 1000 × 1000 pixels. Visualized with r = 650 nm, g = 550 nm, b = 460 nm.

(a) (b)

Figure 20. Calibration area acquired from Bell-206 helicopter: (**a**) color-visualized hyperspectral cube (red = 650 nm, green = 550 nm, blue = 450 nm). Area ~ 72 × 72 m, GRD = 7.19 cm; and (**b**) Handheld oblique photography.

Note that the ground resolving distance (GRD) in Figures 18 and 19 of the pan-sharpened pixels is 1.868 cm, while the real spectral GRD was 37.36 cm. The consequences of pan-sharpening for spectral discrimination are generally qualitatively known, but we do not analyze them herein.

2.4. Simulation of the Spatial Distribution of the Explosive Objects

Data of the spatial distribution of threat-causing explosive objects are limited and, for IEDs, are classified. One possible solution is predicting their distribution by simulation, using the public sources considered in Section 1.3; we consider this in the current section. Once the spatial distribution is solved, the problem of how to implant the spectral data of targets in the hyperspectral data of terrain arises, which is considered in Section 2.5.

One older minefield simulation system [48] considers and models factors of airborne detection, including the type of background, time of day, swath width, number of steps, overlapping, minefield scenarios, false alarms, and landmine statistics.

Predicting the distribution of improvised explosive devices, in [49], had the purpose of examining how IED placement can be predicted using related historical data processed by artificial neural networks. Monte Carlo simulation and a logic-based examination of publicly available IED sources were performed, in order to simulate a population resembling the real world in relevant respects. Two cases were analyzed: flat terrain features and objects, and mountainous terrain features and objects [49].

The modeling and simulation of the detection of landmines and improvised explosive devices with multiple automatic target detection loops, as presented in [50], provided an example of a military approaches. In [51], the authors stated that a fully automatic target recognition process still fails to satisfy the operational requirements of minefield detection. This necessitates human interaction for verification and decision-making. It has been found that the operator would not be able to handle the number of segments to process effectively when the percentage of minefield segments in ground truth is more than 1% and when the false alarm rate for non-minefield segments is more than 1.5%.

From several promising detection technologies, we only consider passive hyperspectral data in this study. The crucial factor is the lack of civilian (or public military) data regarding explosive devices in a realistic, non-laboratory environment. The hyperspectral imaging sensors used on UAVs can provide pixels smaller than explosive devices on the ground surface, which simplifies the processing of collected hyperspectral data. The positive consequence is that the problems of target detection with sub-pixel dimensions are avoided. Two groups of hyperspectral target detection methods use only spectral information; not the size, shape, or texture of the target [52,53] (p. 066403-1). These are spectral matching detection algorithms and spectral anomaly detection algorithms. More information about both groups can be found in [54,55], and more about anomaly detection in [56], about deep learning classification in [57], and about the application of neural networks for landmine detection in [58]; we will not consider these further.

Consequently, we consider the modeling and simulation of explosive devices (targets) on a ground surface using their hyperspectral data obtained by hyperspectral measurement and their implanting in terrain hyperspectral cubes. We consider the effects of this process by assessing the outcomes of classification by the spectral angle mapping (SAM) method [59].

2.5. The Implanting Spectral Data of Explosive Targets in the Hyperspectral Scene of the Terrain

The analysis [60] by Basener et al., verified that "the utility of a hyperspectral image for target detection can be measured by synthetically implanting target spectra in the image and applying detection algorithms." Our aim is to implant spectral data of explosive targets, which was done in the following way: We implanted the true hyperspectral data of UXOs and landmines, measured with a ground resolving distance GRD_{UXO} = 0.945 mm, into hyperspectral scenes of terrain surface ($GRD_{terrain}$ = 18.68 mm), after spatial transformation and processing. Note that the ratio of the ground resolving distances, UXO/Terrain, is

0.05058 (or 5.058%). The dimensions of each UXO target are decreased and matched for implanting into pixels of the terrain hyperspectral data, such as in the example presented in Figure 21b.

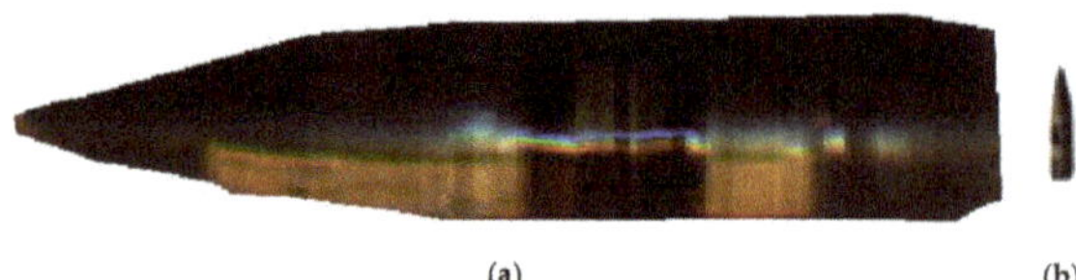

Figure 21. Artillery shell (a) extracted from its neighborhood (Figure 8b); (b) small targets, of decreased dimensions, in this figure are presented not in exact scale, in order to be recognizable. Small targets can be implanted with any orientation. This small target image is named nameR, where the suffix R means the decrease to 5.058%.

All operations were done with arrays (stacks) of images having MxNxL pixels, where M = number columns, N = number of rows, and L = number of wavelengths (channels). For processing, we used the ENVI and ImageJ software; see Table 1 and Figure 22. The format of spectral data was floating point 32 bits. The extracted targets contained all real hyperspectral data obtained by measurements, while the decreased targets contained from 0.23% to 0.27% of the data only; see Table 2. The following examples were used: hyperspectral scenes 147 and 227 of terrain (Figures 18 and 19; each 1000 × 1000 × 90 spectral data, 32-bit float); arrays with the same dimensions but zeroed data, named "blackboard"; and arrays of small targets (100 × 100 × 90 spectral data, 32-bit float), named AS-artillery shell, B-bullet, C-cluster munition, MM-mortar mine, UXOX-UXOX, TMA-4, VTMRP-6, PMR-2A, and Plastic bottle.

Table 1. Implanting spectral data of targets in the hyperspectral scene of the terrain.

No	Action	Description
1	Correcting the geometry of measured target data by NN interpolation.	Use raw measured data of the target Figure 4a. Corrected targets are shown in Figure 4b and Figures 5–9. ENVI
2	Extracting the target from its nearest environment	ENVI.
3	Decrease the extracted target (small target) to 0.05058 of its original dimensions.	ENVI: After decreasing dimension, export to the stack tifs. Match target pixels (0.945 mm) to pixels (18.681 cm) of terrain field (Figure 21b).
4	Export hyperspectral terrain field spectral data; 1000 × 1000 × 90, in 32-bit tiff stack. Co-ordinates can be pixel numbers or meters (if georeferenced).	Figure 18 (scene 147) and Figure 19 (scene 227). If desired scenes should be georeferenced, co-ordinates of pixels (in m) can be used. We recommend applying pixel co-ordinates and doing the georeferencing (if needed) on the simulation outcomes.
5	Implanting small targets in a blackboard stack (1000 × 1000 × 90), where blackboard pixels have value = 0; format 32 bits tiff.	Figure 22a. The targets are visible on the black background. ImageJ.
6	Inversing blackboard of step 5. Change targets area values to 1, and the values of the blackboard to 0, all in 32 bits floating-point format, in 90 channels.	This can be seen by inversing Figure 22b. ImageJ.
7	Implant areas of inversed small targets in blackboard stack of 1000 × 1000 × 90.	Figure 22b. ImageJ.
8	Locations of small targets into the scene of terrain	Figure 22c. Multiplying Figure 22b with the scene of the terrain in Figure 19. ImageJ.
9	Implanting small targets onto the scene of the terrain	Add blackboard array (Figure 22a) to the outcome of step 8. Result shown in Figure 22d.

The targets can be implanted in the hyperspectral data of the terrain in one or two of the following combinations: 1. Without interaction with its neighborhood—the whole area of the target is visible to the imaging hyperspectral sensor. 2. The area of the target is partially hidden, obscured, or covered by terrain. In the following text, we use the term obscured. 3. The spectrum of a target is mixed or overlaid by spectra of the terrain surface (e.g., partially by soil, sand, gravel, or vegetation). In the following text, we use the term overlaid.

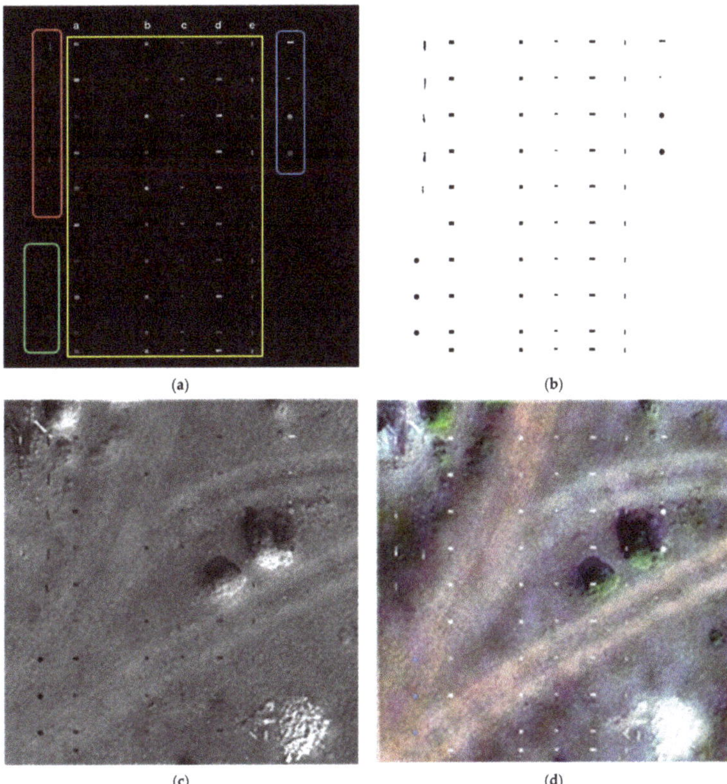

Figure 22. Main steps for implanting targets onto the hyperspectral scene of the terrain. Figure A9 is an example for terrain 147, Figure A10 is terrain 227: (**a**) Targets inserted on blackboard (1000 × 1000 pixels, 90 channels, floating point 32-bit, stack, tif). Red: AS, B, CN, MM, UXOX, Plastic bottle, mines PMR-2a, TMA-4, VTMRP-6; Yellow: False alarm objects; Green: Random uniform spectral values inside the minimum–maximum interval; (**b**) locations of targets: black—0, white—1, 90 channels, 32-bit; (**c**) hyperspectral terrain scene 227 (1000 × 1000 pixels, 90 channels, 32 bits, tif) multiplied by (**b**). One channel is shown; (**d**) Adding (**a**) to (**c**) in 90 channels, giving implanted targets on hyperspectral terrain scene 227. Color visualization (r = 650 nm, g = 550 nm, b = 460 nm).

Table 2. Spectral samples available in the measured targets and their decreased versions.

Target	Samples per Band in the Measured Target	Samples per Band in the Decreased Target	Percentage % of Implanted Spectral Samples	Target Area m^2
Artillery shell	45,661	108	0.2365	0.037690
Bullet	36,243	87	0.2400	0.030362
Cluster munition	24,653	63	0.2555	0.021986
Mortar mine	45,285	116	0.2562	0.040482
UXOX	19,251	52	0.2701	0.018147
Landmine PMR2A	7	1	14.2857	0.009500
Landmine TMA-4	8	1	12.5000	0.063340

2.5.1. Spectral Angle Mapping

From several spectral matching detection algorithms, we selected the Spectral Angle Mapping (SAM) algorithm, introduced in 1993 by F.A. Kruse et al., in The Spectral Image Processing System (SIPS) Interactive Visualization and Analysis of Imaging Spectrometer Data [58]. SAM allows for mapping of the spectral similarity of image spectra t_i to reference spectra r_i by calculating the angle, γ, between the two spectra, treating them as vectors in a space with dimensionality equal to the number of channels (L); see Figure 23 and Equation (1).

$$\gamma = \arccos\left[\sum_{i=1}^{L} t_i r_i / \left(\sum_{i=1}^{L} t2_i\right)^{1/2}\left(\sum_{i=1}^{L} r2_i\right)^{1/2}\right]. \quad (1)$$

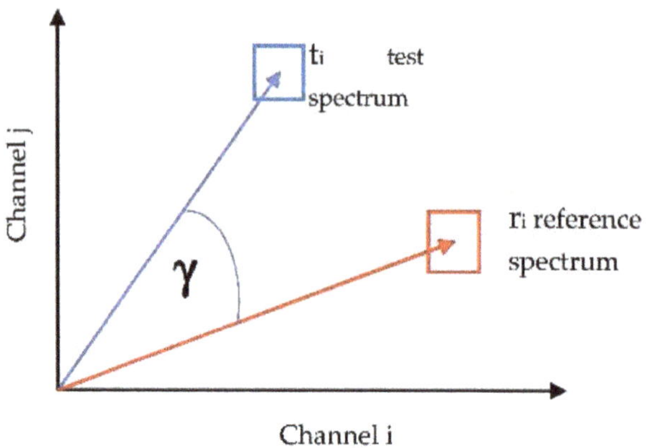

Figure 23. The reference spectrum r_i, ($i = 1, \ldots, L$), the test spectrum t_i ($i = 1, \ldots, L$), γ is the angle between them (in radians), and L is a number of channels [58] (p. 157).

This similarity measure is insensitive to gain factors, as the angle γ between the two vectors is invariant, concerning the lengths of the vectors. More information about SAM is available in many references (see, e.g., [56]).

The number of positive outcomes of SAM classification is a measure of detection success, which depends on the quality and quantity of spectral samples (endmembers) representing objects or materials and their areas; see Figure 24a, Table 2. The largest number of endmembers in the area belonged to Mortar mines (116; 0.040482 m^2), while the smallest belonged to UXOX (52; 0.018147 m^2). Figure 24a shows the mean SAM values of targets. Figure 24b shows the SAM values of 9 ASR targets obscured 25.7%. Obscured targets have larger dispersion and larger SAM angles. Figure 24c shows the SAM values of 10 ASR targets; their spectra are overlayed with 10% of terrain (scene 147). Targets overlayed with scene spectra have larger dispersion at smaller SAM angles. Similar behaviors appeared with other targets.

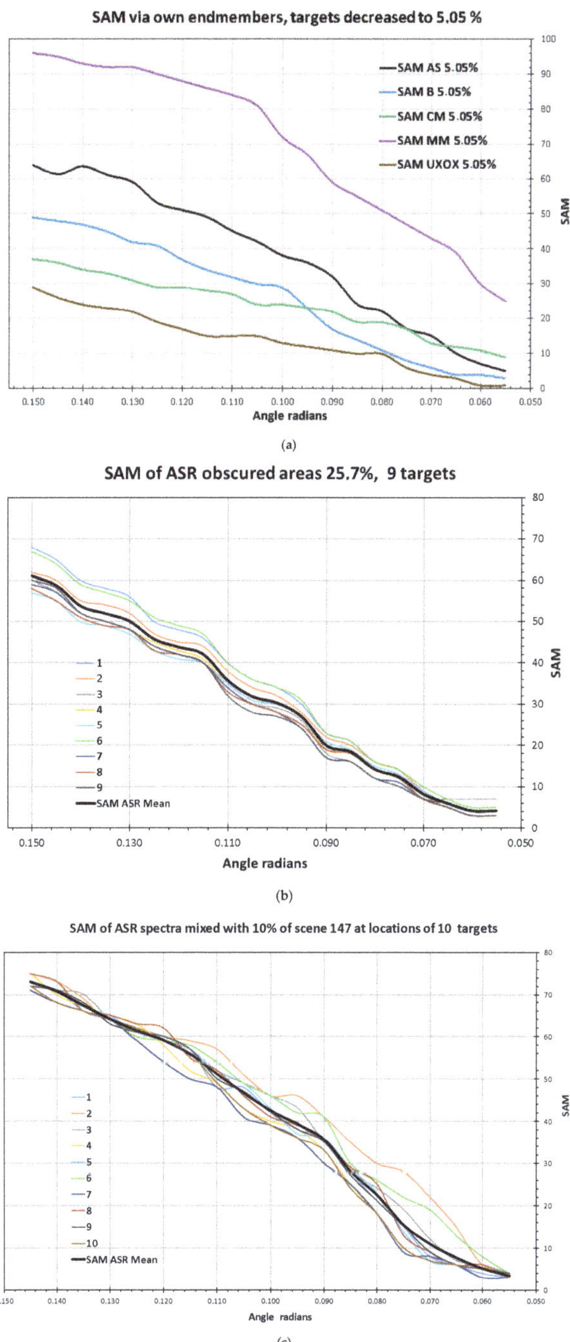

Figure 24. Results of spectral angle mapping processing: (**a**) SAM of targets calculated using their own endmembers; (**b**) SAM of 9 ASR targets with areas obscured by 25.7%; and (**c**) spectra of targets overlayed with 10% of scene 147 at locations of targets—SAM of 10 ASR targets.

2.5.2. Target Simulation Options

Our research aims to develop modeling and simulation of the spectral data of explosive targets, implanting them into spectral terrain scenes for civilian applications. Several approaches were analyzed or tested and considered:

1. The true spectral data of explosive targets, measured by a hyperspectral imaging scanner, and pixels matched to pixels of terrain scene spectra.
2. The average spectral data of explosive targets measured by a point measuring spectrometer. This kind of target's spectral data for land mines has only appeared in the literature.
3. Modeling the partial random obscuring of explosive targets on the ground surface.
4. Modeling the partial mixing spectra of the explosive targets and the background.
5. Simulation of random spectral data in the interval between the maximum and minimum of the spectral data of explosive targets, measured by a point measuring spectrometer. We tested the random generation of data using a uniform probability distribution and considered several other distributions.

In our research, we analyzed options 1, 2, 3, and 4, while 5 was the only one tested. The following conclusions were derived: The use of true spectral data of explosive targets, measured by a hyperspectral imaging scanner (case 1), and processed as described in Table 1 and Figure 22, gave reliable outcomes (Figure 25a,b). The average spectral data, (case 2) produced a high constant response (Figure 25c,d) and, so, should not be used. Histograms of spectral data comparing cases 1 and 2 (see Figure 26) provided additional evidence for this statement. Note that this kind of data has been used in several references, despite its weakness.

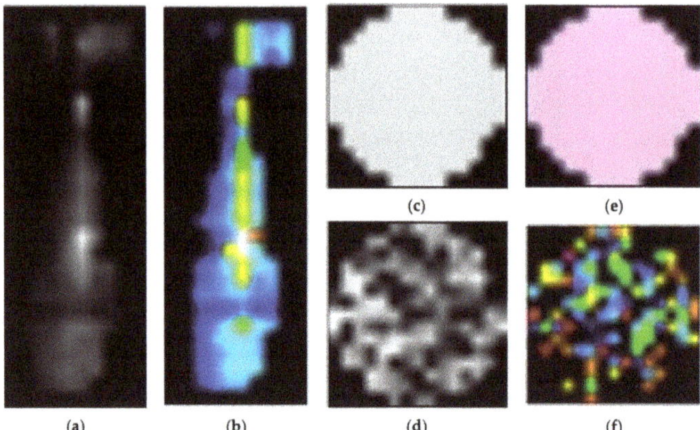

Figure 25. Examples of Case1: (**a**,**b**) target artillery shell; (**c**,**d**) Case 2: target mine TMA-4; (**e**,**f**) case 3: case target, mine TMA-4, spectral data randomly generated, with uniform distribution, in the interval from minimum to maximum.

The explosive targets in Figure 25 show their views at 550 nm, in the grayscale and artificial color lookup table. Note that case in Figure 26b has a stable constant view, which is not realistic in the natural environment. Figure 25e,f shows the same target's spectral data, generated by a random data generator with uniform distribution in the interval from minimum to maximum.

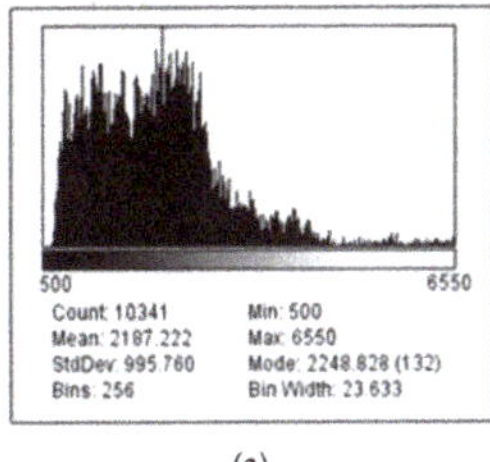

(**a**)

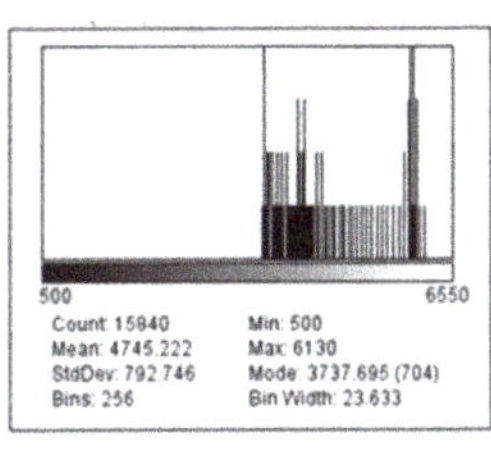

(**b**)

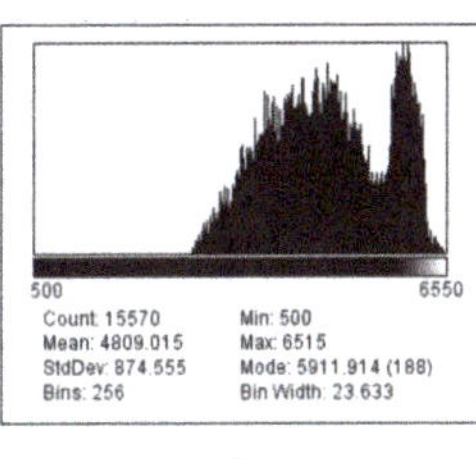

(**c**)

Figure 26. Histograms of spectral data of all channels: (**a**) Mortar mine; (**b**) TMA-4; and (**c**) randomly generated TMA-4 spectral data. Note that histograms (**a**) and (**c**) show significant variability of spectral values. Histogram (**b**) shows that constant values dominate the spectrum, and only several deviations appear.

2.5.3. Modeling the Obscured Spectra of the Explosive Target and the Overlayed Target's Spectra and the Spectra of Background

The general model for analysis of the effects of partially obscuring an explosive target and partially mixing its spectra with those of the neighboring terrain is [53]:

$$x = aS + bV, \tag{2}$$

where x is the spectrum of the observed pixel, S is the spectrum of the target, V is the spectrum of the background, $a \geq 0$ is the fraction of the considered pixel which is filled by the target, and $b \geq 0$ is the fraction of the considered pixel filled by the neighboring terrain. If the observed pixel is filled with the target ($a = 1, b = 0$), it is resolved or a full-pixel target. When part of the pixel is filled with the target ($a \neq 0, b \neq 0$), it is unresolved or a sub-pixel target. Although we mainly analyzed resolved (full-pixel) targets, we tested cases where part of the explosive target was randomly obscured (see Figure 27) and cases where its spectrum was overlaid with the spectrum of the background (i.e., $b > 0$).

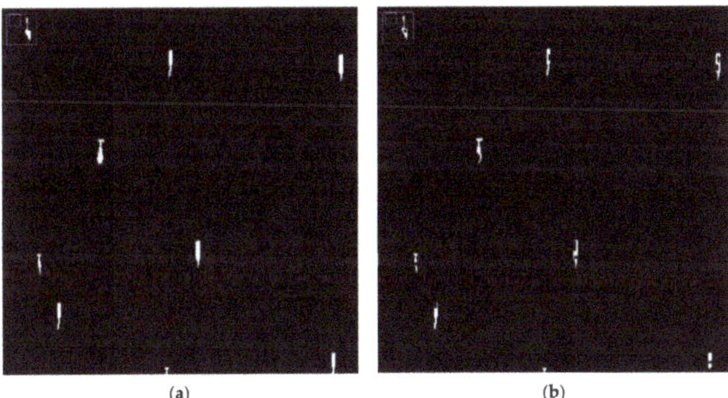

(**a**) (**b**)

Figure 27. Example of obscuring: (**a**) clear targets; and (**b**) areas of targets (obscured 25.7%).

The target and terrain (background) spectral combinations, defined by Equation (2), are indeed summations, although we often use the words "overlaid" and "mixed". Another combination is the partial obscuring of a target area by the terrain and its spectra (see Figure 27).

The obscuring 25.7% was applied to the areas of targets (ASR, BR, CMR, MMR, UXOXR) in scene 227 (Appendix C).

The mixing (overlaying) of the spectra in terrain scene 147 (Appendix C) with the spectra of the 10 targets (ASR, BR, CMR, MMR, UXOXR) was applied with $a = 1.0$, and $b = 0.10$, such that:

$$x = S + 0.10\ V. \tag{3}$$

2.6. Model of Target Detection

Our goal was to derive methods for modeling and simulating the explosive targets in a hyperspectral scene, using the real hyperspectral data of several types of explosive devices, where simulation should be suitable for application by civilians, which is narrower and less complex than the analysis of hyperspectral methods for target detection. Thus, we used, for the considered cases, the SAM algorithm as the detector, among several others (Cross-Correlation, Linear Unmixing, Matched Filtering). The outputs of SAM are a Spectral Angle raster, containing values of the spectral angle for each image cell, and a Class raster, in which cells are assigned to endmember classes based on the angle value set for the threshold value γ (see Figure 28).

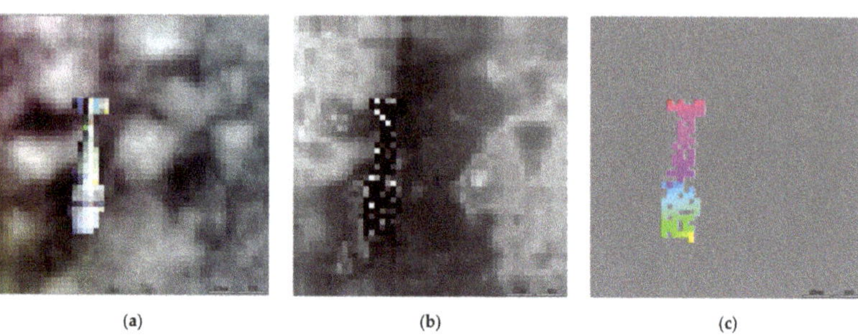

Figure 28. Spectral angle mapping (SAM) outputs for a decreased mortar mine MM: (**a**) The source of spectral endmembers, MM, in terrain 147 environment, visualized in color (r = 650 nm, g = 550 nm, b = 460 nm); (**b**) Spectral Angle raster, containing spectral angle values for each image cell, obtained with the angle threshold value of γ = 0.0174532925 radians; and (**c**) Class raster, in which cells are assigned to endmember classes based on the angle threshold value γ.

For our analysis, we used SAM Class raster values. The computing resources of SAM are proportional to $1/\gamma$ and, so, the smaller the value of γ, the larger the computing time. Thus, we selected γ as the independent variable.

The detection of a target was modeled as a Bernoulli experiment, where the binary random variable y took a value of y = 1 ("detected") with probability p and y = 0 ("not detected") with probability $1-p$ [60]. The parameter p was specific for each treatment and depends on the influence variables characterizing that treatment.

Let POD be the probability of target detection. If the number of opportunities to detect a target is n and the number of detections is y, the number of detections is binomially distributed with parameter p, where p = POD and q = $1-p$. The basic model for the analysis of mine detection POD confidence limits has been developed in [61], although we applied confidence limits—POD-lower and POD-upper—by Exact Confidence Interval using the Clopper–Pearson method [62]:

$$\text{PODupp} = 1 - \text{BetaInv}(\alpha/2, n - k, k + 1), \tag{4}$$

$$\text{PODlow} = 1 - \text{BetaInv}(1 - \alpha/2, n - k + 1, k), \tag{5}$$

where PODlow is the confidence interval lower limit, PODupp is the confidence interval upper limit, n is the number of trials, k is the number of successes in n trials, α is the percent chance to reject the true null hypothesis about detection incorrectly, and BetaInv has been defined in [63]. Usually, α = 0.05 (5%) and 1 − α is the 95% confidence.

The estimated false alarm rate (FAR) can be defined as the number of false alarms counted on an area divided by the size of that area (i.e., the average number of false alarms per square meter). The area calculated was the area of the terrain scene (147 or 227) minus the area of all detected targets. As we limited our concern only to models and simulations of explosive targets, the FAR was not considered.

3. Results

The results of our research were methods for modeling and simulating explosive targets in a real hyperspectral data scene. We considered improvised explosive devices (IED), unexploded ordnances (UXOs), and landmines. Spectral data of these objects are limited and, for IEDs, are classified. Our approach included spectral data of UXOs and landmines, measured by hyperspectral imaging sensors (line scanner and a snapshot camera) onboard a ground-based mechanic gentry, a helicopter, and a UAV. The spectral data of the terrain were acquired with a snapshot hyperspectral camera onboard a UAV and a Bell-206 helicopter. The measured spectral data of the explosive targets had a very fine spatial resolution of 0.945 × 0.945 mm, while the spectral data of the terrain had a resolution of 1.868 × 1.868 cm. The dimensions of each UXO target were, thus, decreased to 5.0588% (see Figure 21) and, after this step, they could be implanted into the pixels of hyperspectral data of the terrain (see Figures 18 and 19).

A key concept in our research is a combination of tests and analyses, in which several factors appear. The factors were UXO targets (artillery shell—AS, bullet—B, cluster munition—CM, mortar mine—MM, unexploded ordnances of unknown type—UXOX), landmines—PMR-2a, TMA-4, VTMRP-6—and plastic bottles) and the spectral angle mapping classifier (detector). The independent variable was the spectral angle (from 0.055 to 0.150 radians). A detector was tested with each UXO target in two situations: Spectra of targets overlaid with 10% of terrain spectra, or targets obscured by 25.7%. The overlaid and obscured UXO targets were implanted into the terrain hyperspectral cubes 147 and 227, which introduced additional variability; an example with terrain 227 is shown in Figure 22d. Figures 13–16, several targets had only one spectral value for each wavelength (see Table 2, Figure 25c,d), and were excluded from the following analysis. The histogram of the spectra in all channels of one considered UXO target showed rich variability, while the randomly simulated spectra were also very variable (see Figure 26c). In contrast to the discussed cases, where only the mean value per channel was known, spectral values were uniformly distributed in the majority of channels (see Figure 26b). We are aware that such cases appear often; therefore, we initially tested simulation with random spectral values, if besides mean values, the minimum and maximum values of the reflectance spectra were known (Figure 25e,f).

3.1. Probability of Target Detection POD, Confidence Intervals

The SAM classification outputs (an example is shown in Figure 28c) were used as the detector outputs. At the same time, the estimated probability of detection for a particular factor level combination is the ratio of the number of detected targets to the total number of opportunities to detect a target. The examples for ASR are shown in Figure 24b,c. While we assumed a binomial distribution for the number of correct positive indications, we also found the 95% confidence limits for the probability of detection, as indicated by relations in Equations (4) and (5).

The considered SAM class raster data models (Figure 24) of the explosive targets ASR, BR, CMR, MMR, and UXOXR were used, after normalizing each to its maximum value. For each target, the POD was derived, as well as the detection probability (see Figures 29 and 30 target 10% overlayed spectra; Figures 31 and 32 target obscured by 25.7%). As the POD and confidence interval data were non-monotonic, we applied a polynomial approximation (see Figures 30 and 32, as well as Appendices A and B).

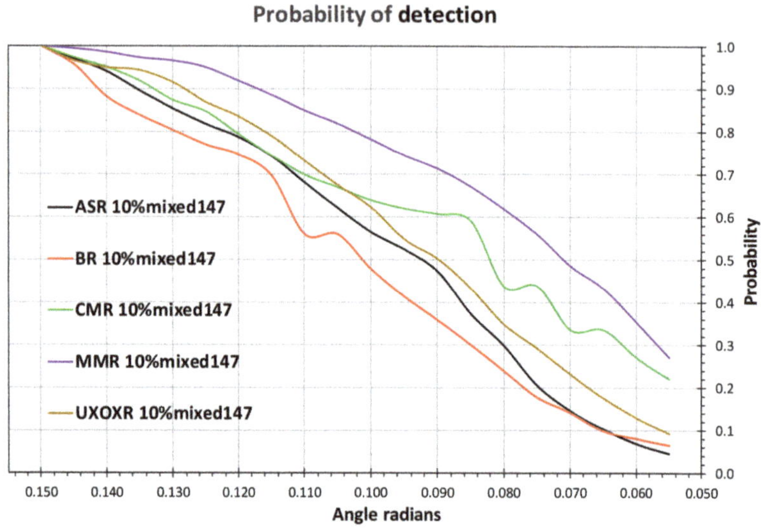

Figure 29. Probability of detection POD of ASR, BR, CMR, MMR, and UXOR targets, with their spectra overlaid with 10% of terrain 147 spectra.

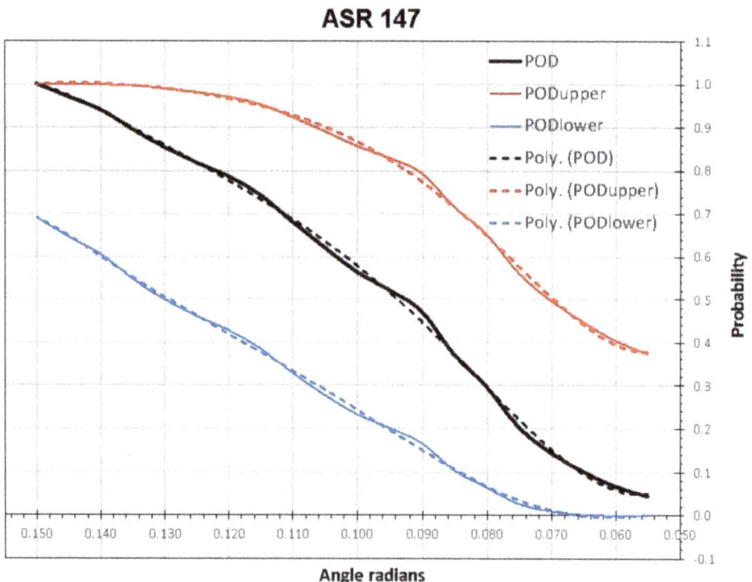

Figure 30. Probability of detection (POD), the confidence limits (PODupper, PODlower), and polynomial approximations (Poly) of ASR 147 overlaid spectra.

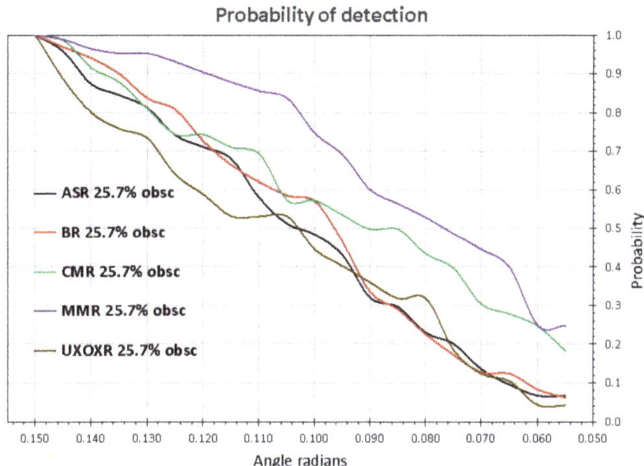

Figure 31. The probability of detection (POD) of ASR, BR, CMR, MMR, and UXOR in terrain scene 227; targets obscured by 25.7%.

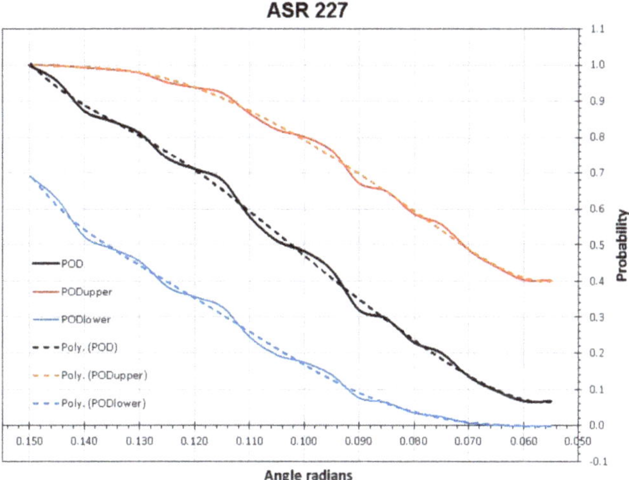

Figure 32. Probability of detection (POD), confidence limits (PODupper, PODlower), and polynomial approximations (Poly) of ASR 227, targets obscured by 25.7%.

The POD functions of targets with overlayed spectra (Figures 29 and 30, Appendix A) were smoother than the POD functions of obstructed targets (Figures 31 and 32, Appendix B).

3.2. Polynomial Approximations of POD, PODupper, and PODlower

The functions of the probability of detection (POD) and the associated confidence limits (PODupper and PODlower) were derived from empirical (measured) reflectance spectra. They are intended for use in civilian security applications, where they should be simulated and processed. Hence, we derived polynomial approximations for the considered targets (see Tables 3 and 4) Through approximation, we can avoid the need to read empirical POD, PODupper, and PODlower data, by using the corresponding functions.

Table 3. Polynomial approximation functions of targets overlayed with 10% of terrain 147.

	Polynomial Approximation
	ASR 147
PODupper	$y = 4E6x^6 - 3E6x^5 + 1E6x^4 - 166833x^3 + 14875x^2 - 663.1x + 11.849$
POD	$y = -4E6x^6 + 2E6x^5 - 240255x^4 - 662.15x^3 + 2877.7x^2 - 220.25x + 5.0015$
PODlower	$y = -9E6x^6 + 6E6x^5 - 1E6x^4 + 177823x^3 - 11922x^2 + 400.99x - 5.2922$
	BR 147
PODupper	$y = 6E6x^6 - 4E6x^5 + 940755x^4 - 131459x^3 + 10406x^2 - 429.58x + 7.4796$
POD	$y = 1E7x^6 - 7E6x^5 + 2E6x^4 - 199237x^3 + 14057x^2 - 526.59x + 8.1301$
PODlower	$y = 1E7x^6 - 6E6x^5 + 1E6x^4 - 152510x^3 + 9421.4x^2 - 306.4x + 4.1181$
	CMR 147
PODupper	$y = 8E6x^6 - 5E6x^5 + 1E6x^4 - 202551x^3 + 15760x^2 - 620.39x + 10.161$
POD	$y = 1E7x^6 - 7E6x^5 + 2E6x^4 - 303933x^3 + 24386x^2 - 990.91x + 16.095$
PODlower	$y = 8E6x^6 - 6E6x^5 + 2E6x^4 - 250338x^3 + 20267x^2 - 835.47x + 13.698$
	MMR 147
PODupper	$y = 806.5x^3 - 319.41x^2 + 42.148x - 0.8546$
POD	$y = 217.42x^3 - 141.66x^2 + 29.343x - 0.9423$
PODlower	$y = -403.37x^3 + 89.372x^2 + 2.0295x - 0.259$
	UXOXR 147
PODupper	$y = 19706x^4 - 8375.7x^3 + 1204.7x^2 - 61.222x + 1.3705$
POD	$y = 13569x^4 - 6690.5x^3 + 1123.3x^2 - 65.845x + 1.3051$
PODlower	$y = 1E7x^6 - 7E6x^5 + 2E6x^4 - 226501x^3 + 16046x^2 - 599.6x + 9.1702$

Table 4. Polynomial approximation functions of targets 25,7% obscured in scene 227.

Function	**Polynomial Approximation**
	ASR 227
PODupper	$y = 1E7x^6 - 7E6x^5 + 2E6x^4 - 242706x^3 + 18625x^2 - 743.66x + 12.321$
POD	$y = 1E7x^6 - 7E6x^5 + 2E6x^4 - 227551x^3 + 16594x^2 - 637.76x + 10.023$
PODlower	$y = 6E6x^6 - 3E6x^5 + 681375x^4 - 70842x^3 + 4006.8x^2 - 119.2x + 1.4781$
	BR 227
PODupper	$y = -3E7x^6 + 2E7x^5 - 5E6x^4 + 643587x^3 - 45100x^2 + 1633x - 23.579$
POD	$y = -5E7x^6 + 3E7x^5 - 7E6x^4 + 884396x^3 - 62088x^2 + 2252.8x - 33.03$
PODlower	$y = -3E7x^6 + 2E7x^5 - 4E6x^4 + 572692x^3 - 40514x^2 + 1480x - 21.823$
	CMR 227
PODupper	$y = 1E7x^6 - 7E6x^5 + 2E6x^4 - 222440x^3 + 16107x^2 - 590.99x + 9.0346$
POD	$y = -1E6x^6 + 302964x^5 + 37216x^4 - 19070x^3 + 2276.7x^2 - 103.6x + 1.7106$
PODlower	$y = -9E6x^6 + 5E6x^5 - 1E6x^4 + 145792x^3 - 9412.4x^2 + 316.94x - 4.3969$
	MMR 227
PODupper	$y = -1E6x^6 + 1E6x^5 - 346262x^4 + 57389x^3 - 5092.5x^2 + 241.73x - 4.1651$
POD	$y = -3E6x^6 + 3E6x^5 - 997759x^4 + 163684x^3 - 14079x^2 + 621.58x - 10.871$
PODlower	$y = -4E6x^6 + 4E6x^5 - 1E6x^4 + 188319x^3 - 15692x^2 + 663.92x - 11.22$
	UXOXR 227
PODupper	$y = 1E7x^6 - 1E7x^5 + 3E6x^4 - 427371x^3 + 34075x^2 - 1388.2x + 22.829$
POD	$y = 2E7x^6 - 1E7x^5 + 4E6x^4 - 523864x^3 + 40335x^2 - 1605.7x + 25.685$
PODlower	$y = 1E7x^6 - 8E6x^5 + 2E6x^4 - 269836x^3 + 19689x^2 - 755.96x + 11.847$

3.3. Simulation of Target Placement

The placement of targets in the terrain hyperspectral scene is defined in Figure 31, with futher examples given in Appendices A–C. We created three sets of fused scenes with targets. The first set contained targets, as described earlier. The second had a 10% overlay of spectral information from the position of target placement. The third set had an obscured, partially randomly hidden 25.7% area of targets (Figure 27). In the second case—where targets were overlaid with the scene—we were able to test whether and how different terrain would influence the outcomes. In the third case, we could see how the spectral footprint was changed by hiding randomly chosen different parts of 5 targets at 10 locations. The locations of targets in the scenes were picked to match as much variety as possible, and different positions were picked for each scene.

3.4. SAM Detection Endmembers and Results

We tested the detection results with endmembers from full-scale targets vs. targets decreased to 5.058% (to match scene resolution). Less accurate results were achieved with endmembers of the full-size targets and, so, we continued with the endmember collection containing all the pixels of the reduced-size targets. The use of true spectral data of explosive targets, measured by a hyperspectral imaging scanner and processed as described in Table 1 and Figure 22, led to reliable outcomes and is suitable for civilian security applications (see Figure 25a,b). The average spectral data produced a strong constant response (Figure 25c,d), and should not be used.

4. Discussion

The subjects tasked with explosive ordnance disposal and the disposal of improvised explosive devices are always exposed to explosive threats and, often, to ambushes. The civilian subjects are generally a single or group of the ground vehicles of a humanitarian demining organization, traveling from camp to the working area and returning, logistics convoys, medical, humanitarian aid, Red Cross, reconstruction convoy, security forces, civilian VIP travelers, or similar. The level of incidents and casualties for civilian vehicles and convoys dominate, when compared to military or security forces. Several survey technologies could be considered as a tool for analysis and decreasing the associated risk. These include hyperspectral, non-linear junction detection (NLJD), LIDAR, longwave infrared, magnetometer, and ground-penetrating radar (GPR) technologies. Several cited references have provided initial insights into these domains, although we focused solely on passive hyperspectral technology. We chose this specifically, due to the lack of data about the considered explosive devices in a realistic, non-laboratory environment. A positive fact is that the hyperspectral imaging sensors used onboard unmanned aerial vehicles (UAVs) can provide pixels smaller than the explosive devices on the ground surface (very high spatial resolution), which is not practical for piloted helicopter platforms. This is the new opportunity provided by hyperspectral sensors, allowing them to serve as an efficient tool for the detection of targets on the ground. In this study, we developed several solutions for modeling and simulating UXOs and landmines, which are suitable for application in civilian security. The conducted research relied on several assumptions; our work has found them to be generally valid.

The possibility of synthetically implanting target spectral images of explosive targets in a hyperspectral image was verified. The true spectral images of UXOs and landmines, measured by hyperspectral imaging sensors, as well as ground- and aerial (UAV and helicopter)-based imagery were fused with the spectra of terrain spectral images. The generalizability of these results is subject to certain limitations:

- The true spectral data of the UXO and landmines were measured by ground-based hyperspectral imaging sensors, with ground resolving distance (GRD) of 0.954 mm.
- The spectral data of the terrain—that is, of the minefields and their surroundings—were acquired by UAV with hyperspectral imaging sensors, with GRD of 18.68 mm.

- The best value of GRD target/terrain ratio was 0.05058 (or 5.058%) for the available explosive targets and terrain spectral images. Smaller values of this ratio cannot provide acceptable outcomes.

The goal of the current study was to derive modeling and simulation methods for implanting the spectral data of explosive targets into a hyperspectral terrain scene, not detection methods themselves. Thus, to quantify the success of the modeling and simulation, we considered several hyperspectral classification methods: spectral angle mapping (SAM), cross-correlation, and linear unmixing. SAM was ultimately selected and used in this research. The independent variable of the SAM method was the spectral angle γ, while the dependent variable was the value of the classification raster. The spectral samples (endmembers) representing the targets (UXOs, landmines) were provided by measured true spectral images of full-scale targets or decreased (reduced dimension) targets. The spectral images of the explosive targets were available in the following ranges, and the generalizability of these results is also subject to the following limitations:

The number of endmembers of UXO targets ranged from 19,251 to 45,661; while the number of endmembers of decreased (reduced) UXO targets ranged from 53 to 108. The area of UXO targets ranged from 0.018147 to 0.040482 m^2. The number of endmembers of landmines and plastic objects ranged from 1 to 8, while the number of endmembers of decreased (reduced) landmines and plastic objects was 1. The area of landmine targets and plastic objects ranged from 0.00950 to 0.066040 m^2.

The landmines and plastic object were excluded from further research in the study, while the available spectral endmember data was limited to one sample per wavelength.

Three types of interaction between targets and terrain were considered in the study:

- Without interaction with its neighborhood, such that the whole area of the target was visible to the imaging hyperspectral sensor.
- The area of the target was partially hidden or obscured or covered by terrain (for which, we used the term obscured).
- The spectrum of a target was mixed or overlaid by the spectra of terrain surface (e.g., partially by soil, sand, gravel, vegetation; for this, we used the term overlaid).

The five targets were analyzed on two terrain spectral data sets; therefore, further research is recommended, including statistically significant cases.

5. Conclusions

1. The motivation for our research into methods for modeling and simulating the implantation of spectral data of explosive targets into terrain spectral data was caused by the lack of civilian (or public military) hyperspectral data, regarding the considered explosive devices, in a realistic, non-laboratory environment. The lack of considered data can be compensated for by using the developed modeling and simulation methods.
2. The empirical research presented started with taking measurements using imaging hyperspectral sensors, line scanners, and snapshot cameras onboard a UAV and on a ground-based gantry, considering terrain, unexploded ordnances (UXO), and landmines on the ground surface.
3. The endmembers of explosive targets should be acquired with an imaging sensor having a very high spatial resolution. For artillery shells, bullets, cluster munitions, mortar mines, and small UXOs, we collected 19,251–45,661 spectral samples. For other types of UXO, these data will differ.
4. The implantation of targets into terrain spectra was done after decreasing the spatial dimensions of the targets and spatially matching their pixels to pixels of the terrain. In the considered cases, the spatial decrease was to 5.058% of the original dimension. The corresponding number of endmembers ranged from 52 to 108; for other types of UXO, this number will be different.

5. In this study, we demonstrated, for the first time, that larger values of spectral angle mapping classification outcomes are achieved if the endmembers are used from smaller (spatially decreased) explosive targets, and not from full-scale targets.
6. If the area of the target is partially hidden or obscured, or if the spectra of a target and terrain are mixed or overlaid, the variability of the SAM data has different behavior.
7. The SAM classifier was used as the detector, where its outputs were considered as a binary outcome of the Bernoulli statistical model, along with its confidence intervals.
8. Further research should analyze more terrain spectral images, a statistically relevant number of simulated explosive targets, and a variety of terrain–targets spectral influence.
9. The empirical and analytical findings of this study provide a new understanding of the hyperspectral behavior of UXOs and landmines in natural environments.

Author Contributions: Conceptualization, M.B.J. and M.B.; Data triage and review, M.B.; Processing, formal analysis M.B.J. and M.B.; Writing—Draft, M.B.J. and M.B.; Writing—Review and Editing, M.B.J. All authors have read and agreed to the published version of the manuscript.

Funding: This research received no external funding.

Institutional Review Board Statement: Not applicable.

Informed Consent Statement: Not applicable.

Data Availability Statement: Restrictions apply to the availability of these data. Data was obtained from HCR-CTRO and are available from the authors with the permission of HCR-CTRO, www.ctro.hr/en accessed on 19 February 2021.

Acknowledgments: The authors are grateful to colleagues who contributed to the development of the described hyperspectral aerial and ground-based survey system: H. Gold, T. Kičimbači, M. Krajnović, D. Vuletić, T. Ivelja, A. Brook, A. Krtalić, M. Tomić, I. Racetin, and G. Skelac. The research leading to these results was possible due to data available from the European Community's Seventh Framework Programme FP7/2012–2015, under grant agreement n° 284747, project TIRAMISU.

Conflicts of Interest: The authors declare no conflict of interest.

Appendix A. Target Spectra Overlaid with 10% of the Terrain 147 Spectra

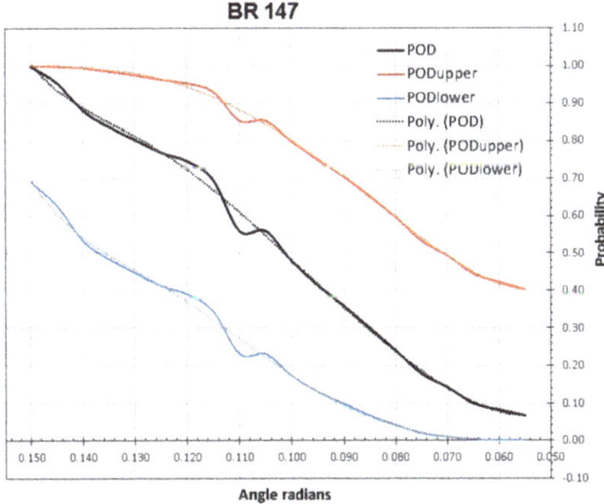

Figure A1. Probability of detection (POD), the confidence limits (PODupper, PODlower), and polynomial approximations (Poly) of BR 147.

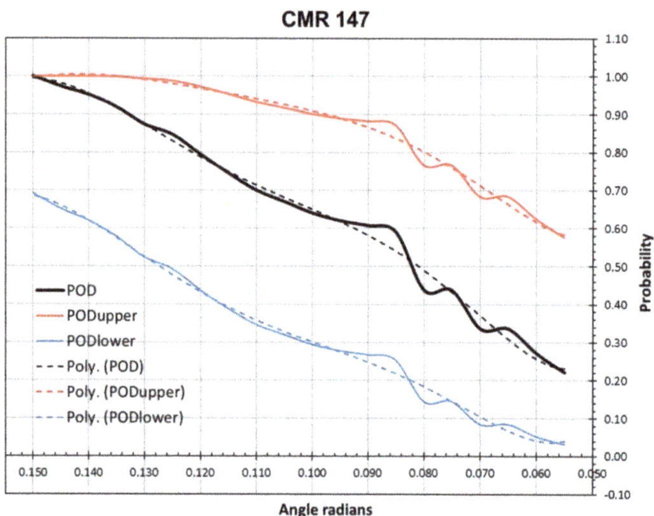

Figure A2. Probability of detection (POD), the confidence limits (PODupper, PODlower), and polynomial approximations (Poly) of CMR 147.

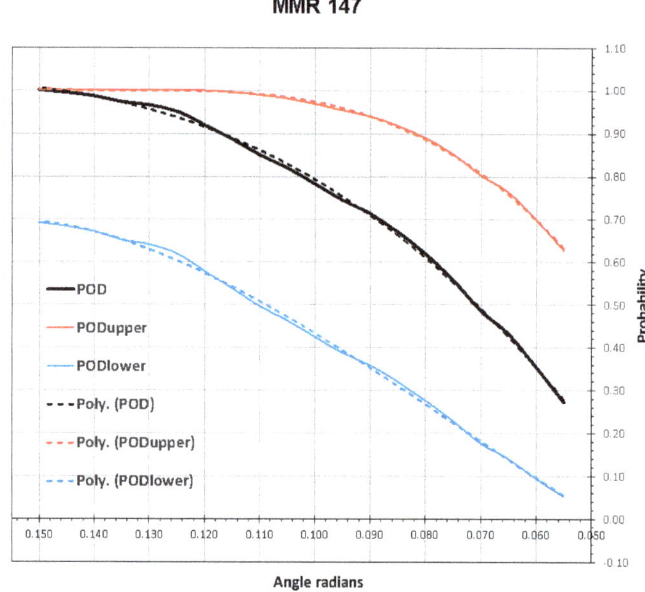

Figure A3. Probability of detection (POD), the confidence limits (PODupper, PODlower), and polynomial approximations (Poly) of MMR 147.

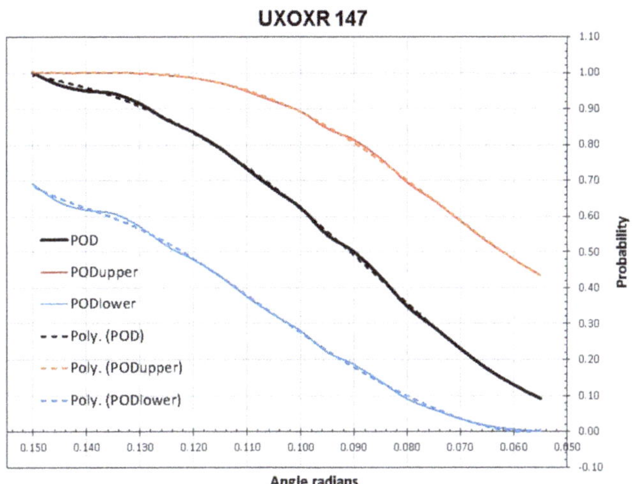

Figure A4. Probability of detection (POD), the confidence limits (PODupper, PODlower), and polynomial approximations (Poly) of UXOXR 147.

Appendix B. Targets Obstructed by 25.7% in Terrain 227

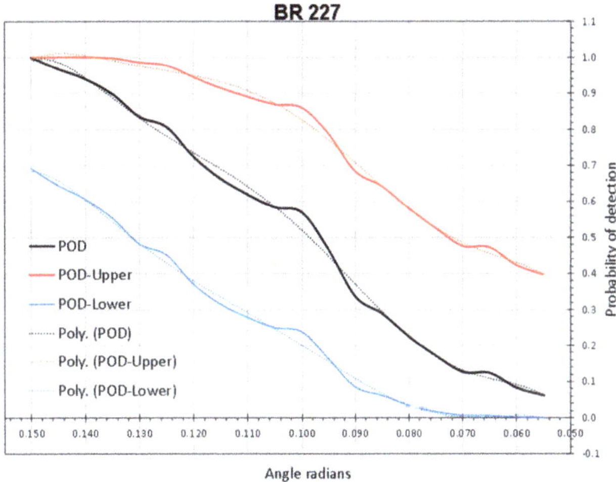

Figure A5. Probability of detection (POD), the confidence limits (PODupper, PODlower), and polynomial approximations (Poly) of BR 227.

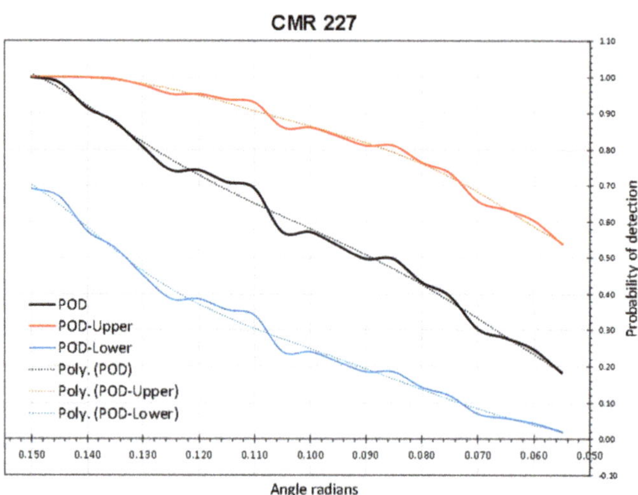

Figure A6. Probability of detection (POD), the confidence limits (PODupper, PODlower), and polynomial approximations (Poly) of CMR 227.

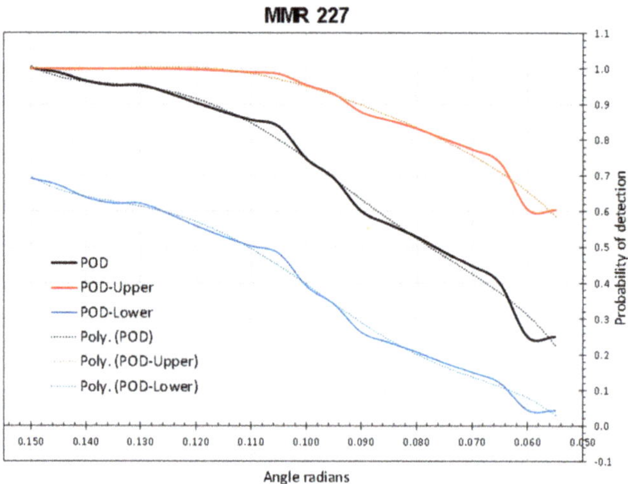

Figure A7. Probability of detection (POD), the confidence limits (PODupper, PODlower), and polynomial approximations (Poly) of MMR 227.

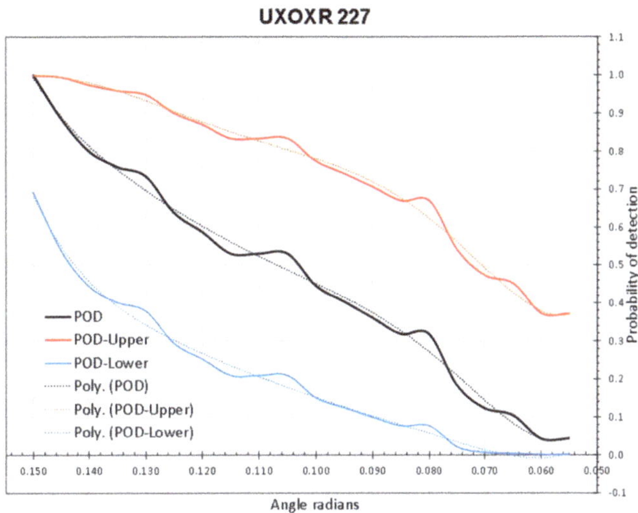

Figure A8. Probability of detection (POD), the confidence limits (PODupper, PODlower), and polynomial approximations (Poly) of UXOXR 227.

Appendix C. Terrain with Several Targets

Figure A9. Hyperspectral scene 147: Dimensions, 18.681 × 18.681 m; 1000 × 1000 pixels; color-visualized (r = 650 nm, g = 550 nm, b = 460 nm).

Figure A10. Hyperspectral scene 227: Dimensions, 18.681 × 18.681 m; 1000 × 1000 pixels; color-visualized (r = 650 nm, g = 550 nm, b = 460 nm).

References

1. Dorn, A.W. Eliminating hidden killers: How can technology help humanitarian demining? *Stab. Int. J. Secur. Dev.* **2019**, *8*, 5. [CrossRef]
2. Asylbek Kyzy, G.; Jung, Y.; Rapillard, P.; Hofmann, U. Geneva International Centre for Humanitarian Demining; SIPRI. In *Global Mapping and Analysis of Anti-Vehicle Mine Incidents in 2017*; Geneva International Centre for Humanitarian Demining (GICHD): Geneva, Switzerland, 2018.
3. Geneva International Centre for Humanitarian Demining. *IMAS 08.10. Non-Technical Survey 2009*; Geneva International Centre for Humanitarian Demining: Geneva, Switzerland, 2017.
4. Mats, H. Norwegian peoples IRAQ drone use and lessons learned. Norwegian People's Aid Workshop on Lessons Learned from the Use of Unmannedaerial Vehicles for the Identification and Assessment of Explosive Devices Threats, Presentation, Podgorica, Montenegro, 16–17 October 2019.
5. Mats, H. Video acquired by RPAS. In Norwegian People's Aid Workshop on Lessons Learned from the Use of Unmannedaerial Vehicles for the Identification and Assessment of Explosive Devices Threats, Podgorica, Montenegro, 16–17 October 2019.
6. United Nations: Department of Peacekeeping Operations, Department of Field Support. *Guidelines on Improvised Explosive Devices (IED) Threat Mitigation in Mission Settings*; United Nations, Department of Peacekeeping Operations, Department of Field Support: New York, NY, USA, 2018.
7. Kalbarczyk, M. Eda Ied Detection (IEDDET) Programme, Presentation. Available online: http://www.irsd.be/website/images/images/Activites/Colloques/presentation/2016-05-17/05-Mr-Marek-KALBARCZYK.pdf (accessed on 10 February 2020).
8. Shutte, K.; Sahli, H.; Schrottmayer, D.; Eisl, M.; Varas, F.J.; Bajic, M.; Uppsal, M.; den Breejen, E. ARC: A camcopter based minefield detection system. In Proceedings of the 5th International Airborne Remote Sensing Conference, San Francisco, CA, USA, 17–20 September 2001.
9. Toolbox Implementation for Removal of Anti-Personnel Mines, Sub-Munitions and UXO—TIRAMISU, EU FP7 Project 2012–2015, Grant Agreement Number 284747. Available online: http://www.fp7-tiramisu.eu/ (accessed on 9 January 2021).
10. Fardoulis, J. Drones in HMA lessons from the field 2019. In Proceedings of the 7th Mine Action Technology Workshop, GCIHD, Basel, Switzerland, 7–8 November 2019.
11. Nevard, M.; Mansel, R.; Torbet, N. Use of aerial imagery in urban survey & use of RPASs in mine Action—Lessons learned from six countries. In Proceedings of the 7th Mine Action Technology Workshop, GCIHD, Basel, Switzerland, 7–8 November 2019.
12. Lisica, D. Evaluation on use of UAVs in-country assessment of suspected hazardous areas in Bosnia and Herzegovina 2019. In Proceedings of the Norwegian People's Aid Workshop on Lessons Learned from the Use of Unmanned Aerial Vehicles for the Identification and Assessment of Explosive Devices Threats, Podgorica, Montenegro, 16–17 October 2019.

13. McFee, J.E.; Ripley, H.T. Detection of buried land mines using a casi hyperspectral imager. In *Detection and Remediation Technologies for Mines and Minelike Targets II*; International Society for Optics and Photonics: San Diego, CA, USA, 1997; Volume 3079, pp. 738–749.
14. Yoresh, A.B. Mine detection by air photography. In Proceedings of the 7th International Symposium; Humanitarian Demining; HCR Center for testing, development and training, Šibenik, Croatia, 27–30 April 2010; Volume 27.
15. Yoresh, A.B. Identification of minefields by aerial photography 2010. In Proceedings of the Third Mine Action Technology Workshop, Geneva, Switzerland, 6–8 September 2010.
16. Geneva International Centre for Humanitarian Demining. *Geomine Demonstration Test 2012/2013*; Geneva International Centre for Humanitarian Demining: Geneva, Switzerland, 2013.
17. Bajic, M.; Ivelja, T.; Brook, A. Development of a hyperspectral non -technical survey of the minefields from the UAV and the helicopter. *J. Conv. Weapons Destr.* **2017**, *21*, 11.
18. Bajić, M.; Krajnović, M.; Brook, A.; Ivelja, T. Ground vehicle based system for hyperspectral measurement of minefields. In *Book of Papers—International Symposium "Mine Action"*; HCR Center for Testing, Development and Training: Zagreb, Croatia, 2014; p. 13.
19. Manley, P.V.; Sagan, V.; Fritschi, F.B.; Burken, J.G. Remote sensing of explosives-induced stress in plants: Hyperspectral imaging analysis for remote detection of unexploded threats. *Remote Sens.* **2019**, *11*, 1827. [CrossRef]
20. Smit, R.; Schmitz, P.; du Plooy, N.; Cooper, A.; van Dyk, T.; Smit, E.; Ramaloko, P. The influence of explosives on plants using in-situ hyperspectral data, Presentation. In Proceedings of the 5th UNMAS/GICHD Bi-Annual Technology Workshop, Pretoria, South Africa, 18–20 June 2014.
21. Berg, A. *Detection and Tracking in Thermal Infrared Imagery*; Linköping University Electronic Press: Linköping, Sweden, 2016.
22. Nikulin, A.; de Smet, T.; Baur, J.; Frazer, W.; Abramowitz, J. Detection and identification of remnant PFM-1 'Butterfly Mines' with a UAV-Based thermal-imaging protocol. *Remote Sens.* **2018**, *10*, 1672. [CrossRef]
23. Smirnov, A.; Nikulin, A. Long-Range and tethered UAVs in UXO detection 2019, long-range and tethered UAVs in UXO detection, presentation. In Proceedings of the 7th Mine Action Technology Workshop, GCIHD, Basel, Switzerland, 7–8 November 2019.
24. Bajic, M. Testing of detectors for non-explosive components of the IED, the non-linear junction, and the control line-wire 2017. In *"Synergistic Technologies to Defeat Improvised Threat," Proceedings of the 3rd C-IED Technology Workshop*; Madrid, Spain, 24–26 October 2017, C-IED Centre of Excellence: Madrid, Spain, 2017.
25. Bajić, M. Propagation model of harmonic radar for detection of nonlinear contacts of improvised explosive device. *Polytech. Des.* **2017**, 210–218. [CrossRef]
26. Šipoš, D.; Gleich, D.; Malajner, M. Stepped frequency radar for landmine detection attached to hexacopter. Presentation and paper. In *Book of Papers, Proceedings of the 16th International Symposium MINE ACTION*; Dubrovnik, Croatia, 10 April 2019, HCR Center for testing, development and training: Dubrovnik, Croatia, 2019.
27. Šipoš, D.; Gleich, D. A Lightweight and low-power UAV-Borne ground penetrating radar design for landmine detection. *Sensors* **2020**, *20*, 2234. [CrossRef] [PubMed]
28. Mayr, W. FindMine UAV im humanitären einsatz presentation, Urs Endress Foundation. In Proceedings of the 7th Mine Action Technology Workshop, GCIHD, Basel, Switzerland, 7–8 November 2019.
29. Fasano, G.; Renga, A.; Vetrella, A.R.; Ludeno, G.; Catapano, I.; Soldovieri, F. Proof of concept of Micro-UAV-Based radar imaging. In Proceedings of the 2017 International Conference on Unmanned Aircraft Systems (ICUAS), Miami, FL, USA, 13–16 June 2017; pp. 1316–1323.
30. Targett, K. Amulet UAS with GPR. In Proceedings of the 7th Mine Action Technology Workshop, GCIHD, Basel, Switzerland, 7–8 November 2019.
31. Guldin, D. Development and Tests of a UXO Survey Drone System 2019. In Book of Papers, Proceedings of the 16th International Symposium MINE ACTION, Dubrovnik, Croatia, 10 April 2019; HCR Center for testing, development and training: Dubrovnik, Croatia, 2019.
32. Guldin, D. Development and Tests of a UXO Survey Drone System 2019. In Proceedings of the 7th Mine Action Technology Workshop, GCIHD, Basel, Switzerland, 7–8 November 2019.
33. Krtalić, A.; Bajić, M.; Ivelja, T.; Racetin, I. The AIDSS Module for data acquisition in crisis situations and environmental protection. *Sensors* **2020**, *20*, 1267. [CrossRef] [PubMed]
34. Makki, I.; Younes, R.; Francis, C.; Bianchi, T.; Zucchetti, M. A Survey of landmine detection using hyperspectral imaging. *ISPRS J. Photogramm. Remote Sens.* **2017**, *124*, 40–53. [CrossRef]
35. Makki, I. Hyperspectral Imaging for Landmine Detection. Ph.D. Thesis, Lebanese Univerity and Politecnico di Torino, Torino, Italy, 2017.
36. Makki, I.; Younes, R.; Khodor, M.; Khoder, J.; Francis, C.; Bianchi, T.; Rizk, P.; Zucchetti, M. RBF Neural network for landmine detection in H yperspectral imaging. In Proceedings of the 2018 7th European Workshop on Visual Information Processing (EUVIP), Tampere, Finland, 26–28 November 2018; pp. 1–6.
37. Aikio, M. *Hyperspectral Prism-Grating-Prism Imaging Spectrograph*; VTT Publications: Espoo, Finland, 2001.
38. Bajić, M.; Ivelja, T.; Krtalić, A.; Tomić, M.; Vuletić, D. The Multisensor and Hyper spectral survey of the UXO around the exploded ammunition depot, of the land mines test site vegetation. In Proceedings of the 10th International Symposium HUDEM, HCR Center for Testing, Development and Training. Šibenik, Croatia, 27–30 April 2013; Volume 9206, pp. 91–96.

39. Bajic, M.; Gold, H.; Pračić, Ž.; Vuletić, D. Airborne sampling of the reflectivity by the hyperspectral line scanner in a visible and near infrared wavelengths. In Proceedings of the 24th Symposium of the European Association of Remote Sensing Laboratories, Dubrovnik, Croatia, 15–27 May 2004; pp. 25–27.
40. Bajić, M. Airborne hyperspectral surveillance of the ship-based oil pollution in Croatian part of the Adriatic sea. *Geod. List* **2012**, *66*, 77–100.
41. Bajić, M.; Ivelja, T. Transfer of knowledge and technologies from mine action to counter improvised explosive devices (C-IED) domain. *Polytech. Des.* **2016**, *4*, 300–309.
42. Miljković, V.; Gajski, D. Adaptation of industrial hyperspectral line scanner for archaeological applications. *Int. Arch. Photogramm. Remote Sens. Spat. Inf. Sci.* **2016**, *5*, 343–345. [CrossRef]
43. Krtalić, A.; Miljković, V.; Gajski, D.; Racetin, I. Spatial distortion assessments of a low-cost laboratory and field hyperspectral imaging system. *Sensors* **2019**, *19*, 4267. [CrossRef] [PubMed]
44. Bajic, M. Impact of Mine polluted area characteristics on the suitability of the airborne multisensor mine field Detection—The case of Croatia. In Proceedings of the International Airborne Remote Sensing Conference and Exhibition, 4th/21st Canadian Symposium on Remote Sensing, Ottawa, ON, Canada, 21–24 June 1999.
45. Yvinec, Y.; Bajić, M.; Dietrich, B.; Bloch, I.; Vanhuysse, S.; Wolff, E.; Willekens, J. *Space and Airborne Mined Area Reduction Tools, SMART Project Final Report, V2.2, Classification: Public*; European Commission IST-2000-25044; European Commission: Brussels, Belgium, 2005.
46. Bajic, M.; Beckel, L.; Breejen, E.; Sahli, H.; Schrotmeier, D.; Upsal, M.; Varas, F.J. *Airborne Minefield Area Reduction-ARC*; European Commission Research Directorates General Project 2001–2003; IST-2000-25300; Information Society Technologies Programme: Brussels, Belgium, 2000.
47. Donohue, J. *Introductory Review of Target Discrimination Criteria*; Final Report, 16 April 1991–31 December 1991; Phillips Laboratory Air Force Systems, Command Hanscom Air Force Base: Wilmington, MA, USA, 1991.
48. Agarwal, S. Modeling and Performance Estimation for Airborne Minefield Detection System. Master's Thesis, University of Missouri-Rolla, Rolla, MO, USA, 2008.
49. Lerner, W.D. Predicting the Emplacement of Improvised Explosive Devices: An Innovative Solution. Ph.D. Thesis, Capitol College, Laurel, MD, USA, April 2013.
50. Johnson, D.; Ali, A. Modeling and simulation of landmine and improvised explosive device detection with multiple loops. *J. Def. Model. Simul. Appl. Methodol. Technol.* **2015**, *12*, 257–271. [CrossRef]
51. Rajagopal, A.; Agarwal, S.; Ramakrishnan, S. Simulation-Based performance modeling for war fighter in loop minefield detection system. In Proceedings of the Winter Simulation Conference, New York, NY, USA, 4–7 December 2005; pp. 1160–1169.
52. Manolakis, D.; Shaw, G. Detection algorithms for hyperspectral imaging applications. *IEEE Signal Process. Mag.* **2002**, *19*, 29–43. [CrossRef]
53. Manolakis, D.G. Taxonomy of detection algorithms for hyperspectral imaging applications. *Opt. Eng.* **2005**, *44*, 066403. [CrossRef]
54. Manolakis, D.; Marden, D.; Shaw, G.A. Hyperspectral image processing for automatic target detection applications. *Linc. Lab. J.* **2003**, *14*, 79–116.
55. Matteoli, S.; Diani, M.; Corsini, G. A tutorial overview of anomaly detection in hyperspectral images. *IEEE Aerosp. Electron. Syst. Mag.* **2010**, *25*, 5–28. [CrossRef]
56. Wang, L.; Zhao, C. *Hyperspectral Image Processing*; Springer: Berlin, Germany, 2015; ISBN 978-7-118-08646-1.
57. Paoletti, M.E.; Haut, J.M.; Plaza, J.; Plaza, A. Deep learning classifiers for hyperspectral imaging: A review. *ISPRS J. Photogramm. Remote Sens.* **2019**, *158*, 279–317. [CrossRef]
58. Kruse, F.A.; Lefkoff, A.B.; Boardman, J.W.; Heidebrecht, K.B.; Shapiro, A.T.; Barloon, P.J.; Goetz, A.F.H. The Spectral image processing system (SIPS)—Interactive visualization and analysis of imaging spectrometer data. *Remote Sens. Environ.* **1993**, *44*, 145–163. [CrossRef]
59. Basener, W.F.; Nance, E.; Kerekes, J. The target implant method for predicting target difficulty and detector performance in hyperspectral imagery. In Proceedings of the Algorithms and Technologies for Multispectral, Hyperspectral, and Ultraspectral Imagery XVII, SPIE, Orlando, FL, USA, 13 May 2011; p. 80481H.
60. Evans, M.; Hastings, N.; Peacock, B. *Statistical Distributions*, 3rd ed.; Wiley: Hoboken, NJ, USA, 2000; pp. 31–33.
61. Simonson, K.M. *Statistical Considerations in Designing Tests of Mine Detection Systems: I-Measures Related to the Probability of Detection*; Sandia National Laboratories: Albuquerque, NM, USA; Livermore, CA, USA, 1998.
62. Vollset, S.E. Confidence intervals for a binomial proportion. *Stat. Med.* **1993**, *12*, 809–824. [CrossRef] [PubMed]
63. Abramowitz, J. MATLAB. Available online: https://www.mathworks.com/help/stats/betainv.html (accessed on 9 January 2021).

Article

The Analysis and Modelling of the Quality of Information Acquired from Weather Station Sensors

Marek Stawowy [1], Wiktor Olchowik [2], Adam Rosiński [1,*] and Tadeusz Dąbrowski [2]

[1] Faculty of Transport, Warsaw University of Technology, Koszykowa 75, 00-661 Warsaw, Poland; marek.stawowy@pw.edu.pl
[2] Faculty of Electronic, Military University of Technology, gen. S. Kaliskiego 2, 00-661 Warsaw, Poland; wiktor.olchowik@wat.edu.pl (W.O.); tadeusz.dabrowski@wat.edu.pl (T.D.)
* Correspondence: adro@wt.pw.edu.pl

Abstract: This article explores the quality of information acquired from weather station sensors. A review of literature in this field concludes that most publications concern the analysis of data acquired from weather station sensors and their characteristic properties, estimating the missing values from the data, and assessing the quality of weather information. Despite the large collection of studies devoted to these issues, there is no comprehensive approach that would consider the modelling of information uncertainty. Therefore, the article presents a proprietary method of analysing and modelling the uncertainty of the weather station sensors' information quality. For this purpose, the structure of a real meteorological station and the measurement data obtained from it were analysed. Next, an information quality model was developed using the certainty factor (CF) of hypothesis calculation. The developed method was verified on an exemplary real meteorological station. It was found that this method enables the improvement of the quality of information obtained and processed in a multi-sensor system. This becomes practical when the influence of individual measurement system elements on the information quality reaching the recipient is determined. An example is furnished by a demonstration of the usage of two sensors to improve the information quality.

Keywords: information quality; weather station; sensors; modelling

Citation: Stawowy, M.; Olchowik, W.; Rosiński, A.; Dąbrowski, T. The Analysis and Modelling of the Quality of Information Acquired from Weather Station Sensors. *Remote Sens.* **2021**, *13*, 693. https://doi.org/10.3390/rs13040693

Academic Editor: Stefano Mattoccia

Received: 22 January 2021
Accepted: 11 February 2021
Published: 14 February 2021

1. Introduction

Information collected from weather station sensors is currently employed in many economy fields, e.g., agriculture, transport [1], and tourism. Based on the received information, it is possible to take rational actions to implement the specific activity in particular areas. This applies particularly to systems classified as critical national infrastructure. Many publications describe and analyse the acquired sensor data and their characteristic properties and estimate missing data in the original meteorological information. Some studies also present research on the quality of information obtained from these sensors. However, no approach takes into account uncertainty estimation of the information quality. When applied to the process of estimating the quality of information obtained from meteorological station sensors, uncertainty modelling allows one to increase the forecasted data reliability.

Analysing the state of knowledge in the field discussed in this article but also delving into the achievements of the scientific community, the following areas can be distinguished:

- publications describing meteorological stations, applied sensors, and construction solutions [1–4],
- publications on analyses of data obtained from sensors and their correctness,
- publications on the quality of information obtained from sensors used in meteorological stations,
- publications on the estimation of missing data in meteorological information,
- publications on the quality assessment of weather information.

The listed main research areas directly related to the subject of this article are analysed in detail below.

The study described in [5] describes issues related to the adoption of wireless sensor networks to assess air quality. The authors have rightly noticed that, having data from individual sensors on temperature, humidity, carbon monoxide (CO), and carbon dioxide (CO_2), it is possible to estimate air quality and decide about the occurrence of an emergency in the warning system. For this purpose, they implemented the classification tree algorithm with regard to entropy and information enhancement. This approach has a practical application, but it does not consider some factors influencing the quality of the information received from individual sensors.

Additionally, in the area of transport (especially in autonomous vehicles and on motorways), the quality of information obtained from meteorological sensors is crucial [6]. Study in this area is presented in monograph [7]. Owing to this, it is possible to detect dangerous weather events and inform drivers about them immediately. Similar studies of stationary weather stations applied in intelligent transport systems (ITS) are presented in [8].

A similar approach in the analysis of the obtained data from meteorological stations was adopted by the authors in the study [9]. They applied decision tree algorithms, analysing precipitation and minimum and maximum temperatures separately. Thanks to the application of algorithms devised by the authors, it is possible to identify flawed sequences contained in meteorological sensors. Similar studies regarding air quality classification using specific algorithms and a decision tree are presented in publication [5]. Inquiry in this area concerns not only land meteorological stations but also marine ones [10].

It is likely to estimate the correctness of meteorological data by comparing them with data from neighbouring meteorological stations. Then, it is possible to determine the consistency of data relating to a given meteorological phenomenon in a specific area [11].

The study described in [12] presents studies aimed at determining the forecast using a hybrid computing network. This approach enables forecasting weather conditions with an insufficient number of meteorological stations.

The next research area, highlighted by the authors of this article, contains publications on the estimation of missing data in meteorological information. Scientifically interesting considerations are presented in the study [13]. It proposes to employ a method consisting in finding time intervals with similar rainfall patterns. Thanks to their analysis, it is possible to interpolate the missing data with better quality compared to the methods used so far.

A study [14] also describes work in this research area. The team of authors proposed models enabling temperature interpolation in a geographical system for agricultural purposes. The conducted analyses resulted in finding that the application of multi-line regression is most beneficial.

Authors adopt various approaches to assess the quality of weather information. One of them is the quality of the data stored in big data. As data from many weather stations equipped with many different sensors are most often (except in sparsely populated areas) available, it is possible to pre-process them in order to eliminate errors. This approach was presented in publication [15]. By pre-processing the data, a weather forecasting system that used data of better quality could be designed.

In order to improve the quality of weather information, a data fusion solution is also employed. In this way, it is possible to combine data from different sensors. This increases the reliability of the weather information. This approach was described in the article [16]. The authors analysed the applied solutions in the area of intelligent transport systems. They considered that, in the fusion of data from sensors, the most important is the application of: fuzzy technique, ranging technique, integrated technique, and clustering technique. Despite considering these techniques and analysing their advantages, these lack the possibility to model uncertainty in estimating information quality. Similar considerations in this area in the field of transport are presented in the study [17]. This is a very important issue in the aspect of current research and design of autonomous vehicles. It also seems essential to use modelling of the uncertainty of estimating information quality, because it is possible to increase the level of safety of the means of transport.

Methods of variational assimilation of measurement data from various observational systems, including imagery, can also be distinguished among scientific studies in weather information analysis. In publication [18], the authors proposed using a proprietary data assimilation algorithm, which they presented in detail in a mathematical notation. However, they did not take into account the information quality from individual sources.

Some scientific studies propose the use of neural networks for the analysis of weather information [19]. The study in article [20] posits the application of deep neural networks (DNN) with the object of estimating the amount of precipitation on the basis of radar, microwave, and infrared data. The conducted simulations confirm the validity of using DNN to improve the forecast of the amount of precipitation. However, it seems that, by applying uncertainty modelling of estimating information quality, it is possible to increase the accuracy of the forecasted data. Therefore, the authors of this article conducted scientific scrutiny in this direction.

The investigation of the status of the issue allows one to conclude that most of the studies concern the analysis of the correctness of the obtained data from sensors and weather forecasting with the application of various algorithms. To the best of our knowledge, no publications considered the quality of the information received from a meteorological station at the time of conducting the studies. Studies in this area together with the results are presented by the authors in this article.

2. Uncertainty Modelling Applied to Estimate the Quality of Information Obtained from Sensors of a Meteorological Station

The information quality estimation method uses the calculations of the certainty factor of the hypothesis. The applied CF modelling is based on dependent and independent connections. Such modelling makes it possible to estimate the impact of selected quality dimensions and their factors on the quality of information and to identify reliable measurements from several different data sources (e.g., data from different types of sensors).

2.1. Information Quality

There are many ways to describe information quality [21–23]. The best known are the descriptions in reports and publications related to Massachusetts Institute of Technology Information Quality Program (MITIQ) [24]. They developed, among other things, an information quality model based on sixteen dimensions. Ultimately, the MITIQ defined the dimensions of information quality, which are described as [24–27]:

1. Availability (D_{av})—a dimension that defines the possibility of using an information and communication technologies (ICT) element on demand, at a given time, and by an authorized process. This dimension is directly related to information security.
2. Appropriate amount of data (D_{aad})—a dimension that determines how much data are adequate to complete the task while indicating that the amount is sufficient and more data could reduce information quality.
3. Believability (D_{bel})—a dimension which determines the degree to which information reflects reality. It may also be related to the credibility of the information source itself.
4. Completeness (D_{com})—a dimension that determines whether the data are sufficient to perform a specific task.
5. Concise representation (D_{ccr})—a dimension that determines the degree to which data are represented.
6. Consistent representation (D_{csr})—a dimension that specifies to what extent data are represented in the same format.
7. Ease of manipulation (D_{eom})—a dimension that determines how easily these data can be processed when applied to other tasks.
8. Free of error (D_{foe})—the dimension that determines the extent to which the data are error-free.

9. Interpretability (D_{inter})—a dimension that defines the extent to which data are clear and represented in appropriate languages and symbols.
10. Objectivity (D_{obj})—the dimension which determines to what extent data are not subjective.
11. Relevancy (D_{relev})—a dimension that determines the usefulness of data in performing a specific task.
12. Reputation (D_{reput})—a dimension that determines the extent to which data are assessed in terms of its sources and content.
13. Security (D_{sec})—a dimension that determines the access limits to data to isolate them from unauthorized access.
14. Timeliness (D_{tim})—the dimension that determines the extent to which data are available on time to complete a task.
15. Understandability (D_{uns})—a dimension that determines the understandability of data.
16. Value-added (D_{vadd})—a dimension that determines the benefits of using data and whether they themselves are beneficial to the task.

Figure 1 shows all the above-mentioned dimensions affecting information quality. Each of the dimensions has a direct impact on information quality. Assuming that each value of the dimension (dimension factor) may vary in the range from 0 to 1, the dimension that does not affect the quality of information has the value of 1. The dimension that significantly reduces the quality has the value of 0. Taking the value range <0.1> allows calculating information quality by statistical methods (e.g., a probability of error Pe can be used as the free of error dimension coefficient = 1—Pe) but also adopting methods of estimating uncertainty, such as mathematical evidence based on the Dempster–Shafer theory or CF modelling [28–30].

Figure 1. Information quality components (own study based on [25,26]).

In general, information quality (IQ) consists of the above-mentioned dimensions. Thus, IQ can be described by the formula:

$$IQ = f(w_1, w_2, \ldots, w_m), \quad (1)$$

where:

- m—the number of dimensions, information quality components (equals 16 according to the number of the above dimensions),
- w—a variable that determines the impact of a given dimension (i.e., a value in the range <0.1>).

In the study below, modelling based on the certainty factor of hypothesis [31,32] was applied.

2.2. Modelling Certainty Factor of Hypothesis

As mentioned above, a convenient model for describing information quality may be modelling based on CF of the hypothesis. It is assumed that this factor's value is a direct value indicating the information quality related to the given hypothesis.

Accurate presentation requires describing formalisms [31,32]. The formal simplified description of the certainty factor is defined as:

$$\mathrm{CF(s)} = \mathrm{MB(s)} - \mathrm{MD(s)}, \quad (2)$$

where:

- CF—certainty factor,
- MB—knowledge mapping, i.e., measure of belief,
- MD—hypothesis based on some information.

One has to bear in mind that:

$$\mathrm{MB} \rightarrow \langle 0,1\rangle; \mathrm{MD} \rightarrow \langle 0,1\rangle; \mathrm{CF} \in \langle -1,1\rangle, \quad (3)$$

Interpretation of the measure of belief (MB) and the measure of disbelief (MD) to probability can be defined as:

$$\mathrm{CF(s)}\begin{cases} 1 & \mathrm{P(s)} = 1 \\ \mathrm{MB(s)} & \mathrm{P(s)} > \mathrm{P}(\neg \mathrm{s}) \\ 0 & \mathrm{P(s)} = \mathrm{P}(\neg \mathrm{s}) \\ -\mathrm{MD(s)} & \mathrm{P(s)} < \mathrm{P}(\neg \mathrm{s}) \\ -1 & \mathrm{P(s)} = 0 \end{cases}, \quad (4)$$

where:

- P—probability,
- s—hypothesis based on some information.

However, as mentioned, we do not aim at determining probability because our quality measure is to be related to the CF of final hypothesis of the model.

Since there are many varieties of CF modelling, the basic dependents used in this paper are described below [31,32].

2.2.1. Parallel Basic Model

The formula for calculating the transition according to Figure 2 between two parallel observations and the hypothesis are described as [30]:

$$\mathrm{CF(h,e1,e2)} = \begin{cases} \mathrm{CF(h,e1)} + \mathrm{CF(h,e2)} - \mathrm{CF(h,e1)} \cdot \mathrm{CF(h,e2)} & \text{if} \quad \mathrm{CF(h,e1)} \geq 0 \text{ and } \mathrm{CF(h,e2)} \geq 0 \\ \frac{\mathrm{CF(h,e1)} + \mathrm{CF(h,e2)}}{1 - \min(|\mathrm{CF(h,e1)}|;|\mathrm{CF(h,e2)}|)} & \text{if} \quad \mathrm{CF(h,e1)} \cdot \mathrm{CF(h,e2)} < 0 \\ \mathrm{CF(h,e1)} + \mathrm{CF(h,e2)} + \mathrm{CF(h,e1)} \cdot \mathrm{CF(h,e2)} & \text{if} \quad \mathrm{CF(h,e1)} < 0 \text{ and } \mathrm{CF(h,e2)} < 0 \end{cases}, \quad (5)$$

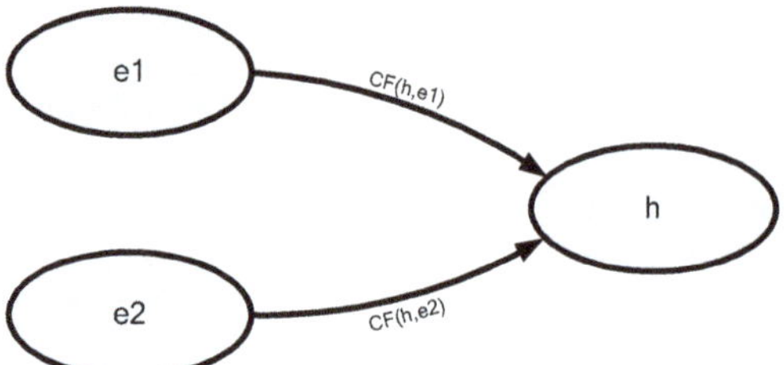

Figure 2. Parallel transitions between two observations and a hypothesis.

2.2.2. Serial Basic Model

In the case of a serial model for positive values (such appears in the modelling described later), according to Figure 3, the following dependent was used [31,32]:

$$\mathrm{CF(h,e1,e2)} = \begin{cases} \mathrm{CF(e2,e1)} \cdot \mathrm{CF(h,e2)} & \text{if} \quad \mathrm{CF(e2,e1)} > 0 \\ 0 & \text{if} \quad \mathrm{CF(e2,e1)} \leq 0 \end{cases}, \tag{6}$$

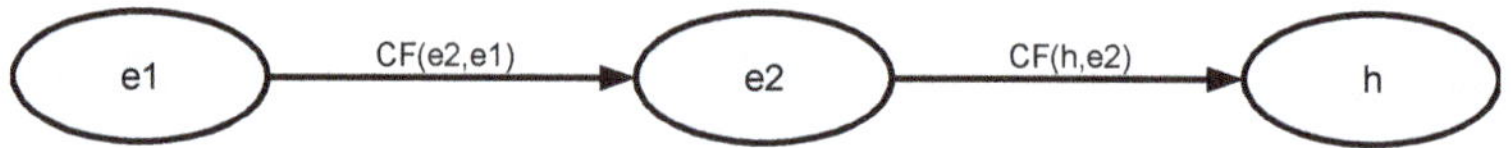

Figure 3. Serial transitions between two observations and a hypothesis.

Both connections, parallel and series, can be reduced to one connection, as shown in Figure 4. This property enables the simplification of calculations in the model proposed in the next chapter.

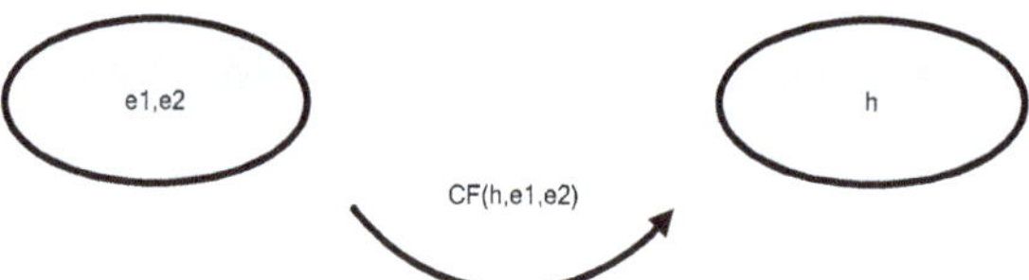

Figure 4. The result of the simplification based on formulas (5) or (6).

In the following considerations, the final hypothesis's certainty factor is the value of the information quality.

2.3. *Parallel-Serial Model of the Analysed Solution of the Meteorological Station*

In literature, many models are describing various states of information. The most developed ones can be found in [33], where they are called information processes. The following types of these information processes are listed below (Figure 5):

1. generating,
2. collecting,
3. storage,
4. processing [34–37],
5. transmitting [38,39],
6. sharing,
7. interpretation.

In Figure 5, the three information states are combined into one because they usually occur together. Such a presentation of information processes also makes it possible to slightly simplify the model, which does not affect the model's overall accuracy.

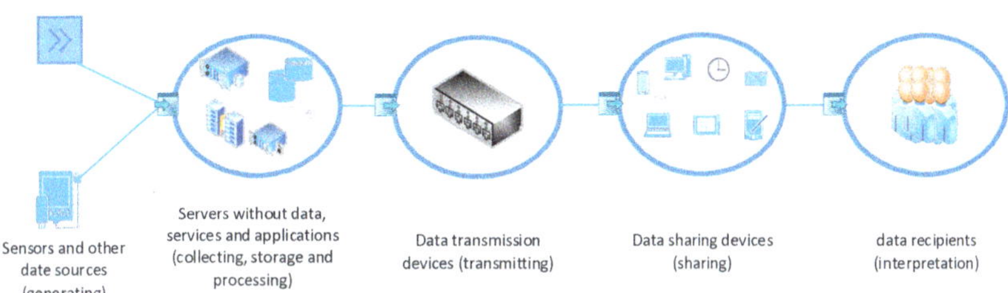

Figure 5. Diagram of a general information quality model of an information system [40].

A generalised information quality model can be presented as follows. Each of the previously mentioned information states can be a consecutive node of the information quality model and generally presented in Figure 6 [40].

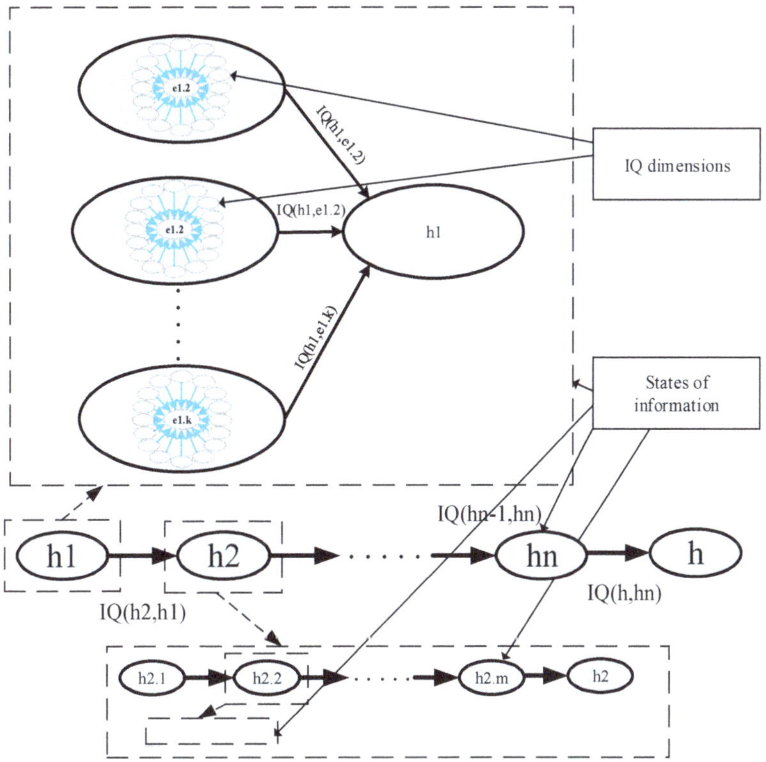

Figure 6. Diagram of the general information quality model of an information system [40].

In the presented case, the information quality model is limited to five information states, of which the fourth state contains three information processes, as shown in Figure 5. The general model consists of five hypotheses related to information states (Figure 6) and contains groups of factors, which influence measurement quality as below:

1. Dimensions related to the main data source. In this case, the data source is the weather station. The dimensions associated with this source influence the value of the indirect hypothesis h1. In the case of data source redundancy, the h1 hypothesis consists of many indirect hypotheses.

2. Dimensions related to collecting, storing, and processing of data. In this case, it is a computer system dedicated to performing specific tasks. The dimensions related to this state of information influence the value of the indirect hypothesis h2.
3. Dimensions related to data transmission. This group includes devices for data transport and transmission. Data transport factors influence the value of the indirect hypothesis h3.
4. Dimensions related to data sharing systems. This group includes imaging and sound devices transmitting data for interpretation as well as interfaces if the interpreter is a computer system, e.g., artificial intelligence (AI). The dimensions related to this state of information influence the value of the indirect hypothesis h4.
5. Dimensions related to data interpretation. This group includes people and—as in this case –computer systems, e.g., AI. The dimensions related to this state of information influence the value of the indirect hypothesis h5.

Each of the above points can be described with a full information quality model presented in Section 2.1. A schematic representation of such a model is shown in Figure 7.

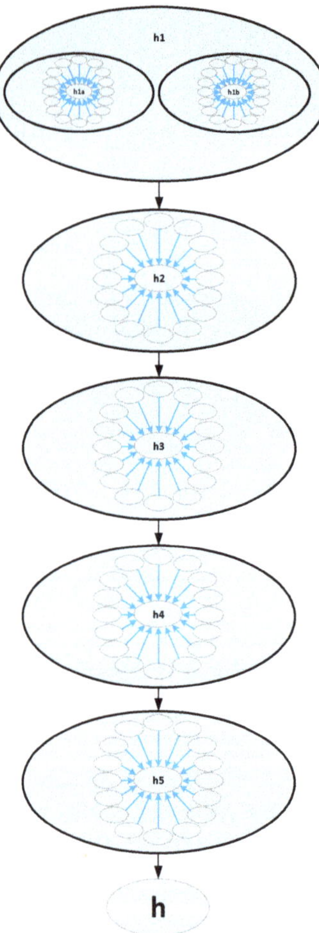

Figure 7. General model of information quality for weather stations.

In the case of the qualitative model, only those dimensions that significantly affect the result are of any interest. Thus, in the following description, only those factors that have such an influence are presented.

The final hypothesis is h—the data have been correctly interpreted. It consists of dependent indirect hypotheses (Figure 7):

- h1a—Basic data source provides valid data. Based on the observations of e1a.
- h1b—Auxiliary data source provides valid data. Based on observations from e1b.
- h1—The weather station delivers valid data. Based on observations of e1a and e1b.
- h2—Data collection, storage, and processing work properly. Based on the observations of e2.
- h3—Data transport systems work properly. Based on observations from e3.
- h4—Data sharing systems work properly and share data in the correct way. Based on the observations of e4.
- h5—Data are interpreted correctly. Based on observations e5.

Each of the indirect hypotheses created on the basis of observations results from observing factors in a given group. In order to simplify the final calculations, the following description includes only some of the events and the observations that may affect the quality of information.

The indirect hypothesis based on the observations e1a consists of independent observations:

- e1a.1—The detector is working properly.
- e1a.2—Detector failure.
- e1a.3—Lack of power.

The indirect hypothesis based on the observations e1b consists of independent observations:

- e1b.1—The detector is working properly.
- e1b.2—Detector failure.
- e1b.3—Lack of power.

The indirect hypothesis based on the observations e2 consists of independent observations:

- e2.1—Data collection, storage, and processing work properly.
- e2.2—Interruption of data transmission.
- e2.3—Data are not collected (e.g., lack of resources).
- e2.4—The data are not processed (e.g., insufficient capacity of the data processing system).

The indirect hypothesis based on the observations e3 consists of independent observations:

- e3.1—Data transport systems are working properly.
- e3.2—Link failure.
- e3.3—Power failure of network devices [41–43].

The indirect hypothesis based on the observations e4 consists of independent observations:

- e4.1—Data sharing systems are working properly and sharing data in the correct way.
- e4.2—Data transmission interruption [44,45].
- e4.3—Defective data sharing methods.

The indirect hypothesis based on the observations e5 consists of independent observations:

- e5.1—Data are interpreted correctly.
- e5.2—Badly trained staff (e.g., does not understand the message).
- e5.3—Incorrect data response of the interpreter.

Figure 8 shows a graph of the model of indirect hypotheses h1a. The hypotheses models h1b, h3, h4, and h5 are similar.

Figure 9 shows the graph of the model of indirect hypotheses h2.

Figure 10 shows the graph of the model of indirect hypotheses h1.

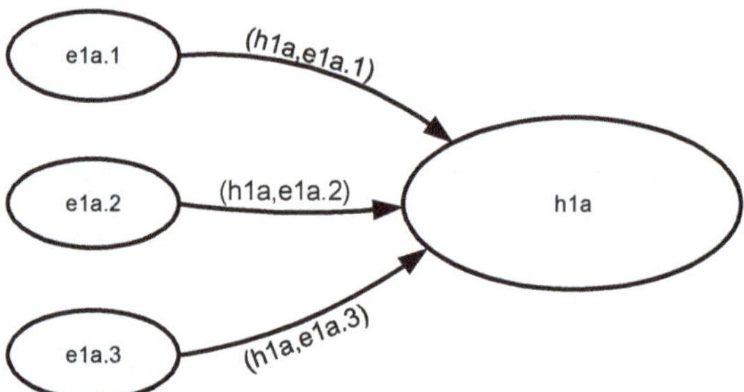

Figure 8. Model for the indirect hypothesis h1a, h1b, h3, h4, and h5.

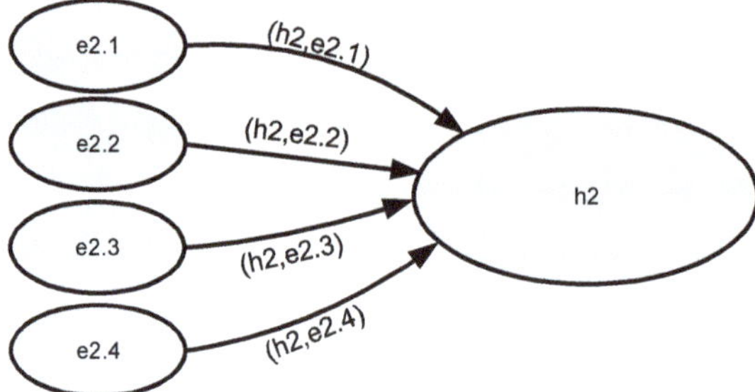

Figure 9. Model for the indirect hypothesis h2.

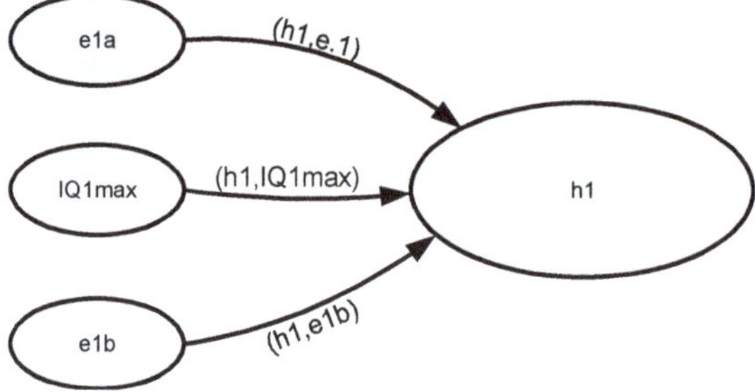

Figure 10. Model for the indirect hypothesis h1 (IQ1max—this is the maximum value that the hypothesis factor can reach [40]).

3. Method Verification and its Computer Exemplification

Sample calculations are presented below. Observation coefficients were estimated for the real measuring station shown in Figure 11 (observations e1a, e1b, and e2) and based on the authors'earlier publications [40,46]. The meteorological station is located in Poland in the northwest part of Warsaw on the premises of the Military University of Technology (geographical coordinates: 52°15′10.6″ N and 20°53′58.9″ E). The measurements were taken in May 2020. During the measurements, the following weather parameters were recorded: wind speed from 0 m/s to 12 m/s, temperature range from 3 °C to 25 °C, relative humidity from 35% to 90%.

The meteorological station includes the following sensors:

- Digital temperature and relative humidity sensor marked with the catalogue symbol SRH1A (abbreviation comes from the words: sensor, relative humidity) placed in an anti-radiation shield.

- Analogue temperature sensor with negative temperature coefficient (NTC) thermistor marked with the catalogue symbol ST1R (abbreviation comes from words: sensor, temperature) placed in an anti-radiation shield.
- Wind speed and direction sensor.
- Two independent solar radiation intensity sensors.

Figure 11. Meteorological station and its electronic module.

Additionally, the station includes a "Micropower" module which records and transmits data to the server. The station is situated on a two-meter-high aluminium mast. The digital temperature and relative humidity sensor marked with the catalogue symbol SRH1A is a measurement device which can operate both in external conditions and inside buildings. Its basic technical data [47] are shown in Table 1.

Table 1. Technical data of the digital temperature and relative humidity sensor.

Parameter	Relative Humidity (RH) Measurement	Temperature Measurement
Measurement range	0 ... 100%RH	−40 ... +70 °C
Accuracy at 25 °C	±1.8%RH (0 ... 90%RH) ±3.0%RH (>90%RH)	±0.3 °C (0 ... 70 °C), ±0.5 °C for the remaining values
Nonlinearity	<0.1%RH	-
Long-term stability	<0.25%RH/year	<0.02 °C/year
Measurement resolution	0.01%RH	0.01 °C

Analogue NTC temperature sensor with the catalogue symbol ST1R [48] is meant to measure air temperature. Its case is made of stainless steel which allows the use of the sensor in difficult atmospheric conditions. The basic parameters are as follows:

- Operation temperature range −50 ... +70 °C,
- Measurement accuracy ±0.5 °C,
- Measurement element 100 kΩ NTC,

- Sensor's dimensions ø6 × 60 mm,
- Level of security IP 67.

In the block diagram (Figure 12) of the meteorological station, the metrological data processing path for ambient temperature consists of blocks filled with background.

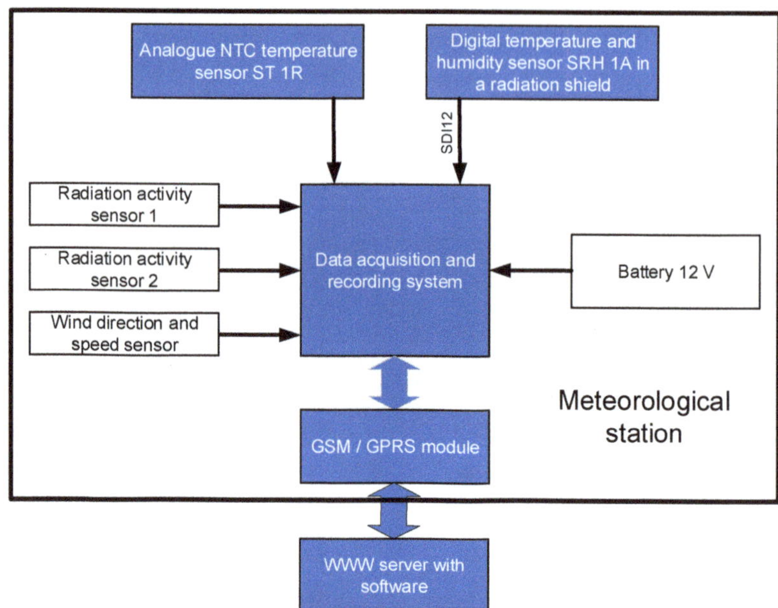

Figure 12. Block diagram of a meteorological station.

With reference to the diagram in Figure 5, the individual elements are related to the observations in accordance with the following list:

- e1a—these are observations related to an analogue temperature sensor with a cable connection,
- e1b—these are observations related to a digital temperature sensor with a cable connection,
- e2—these are observations related to the system for data acquisition and recording with an input expansion card and a memory card,
- e3—these are observations related to the digital cellular communication module.

The element related to e1a observations consists of an analogue NTC (negative temperature coefficient) temperature sensor with catalogue symbol ST1R, working properly in the range of 11–16 V supply voltage and a cable connection with a recorder. As described in the previous section of the article, the following characteristic observations were distinguished for this element:

- e1a.1—the sensors work correctly, the observation coefficient is 0.95,
- e1a.2—faulty analogue sensor or broken signal wire, observation coefficient is 0.02 based on observations, data from the manufacturer, and wiring reliability analysis,
- e1a.3—battery voltage supply below 11 V or interrupted power line, the observation coefficient is 0.04 based on observation of the facility exploitation.

The e1b element consists of a digital temperature sensor with catalogue symbol SRH1A that works correctly in the range of supply voltage 4–16 V and an interface for serial data transmission in the serial–digital interface, standard for microprocessor-based sensor (SDI-12 standard). As described in the previous chapter, characteristic observations were distinguished for this element:

- e1b.1—the sensor and the SDI-12 link work correctly, the observation factor is 0.99,
- e1b.2—faulty sensor or serial data transmission error, the observation coefficient is 0.01 determined on the basis of observations and data from the manufacturer,
- e1b.3—battery voltage supply below 4 V or interrupted power line, the observation coefficient is 0.002 based on observation of the facility exploitation.

Element 2 is a specialised recorder based on a single-chip micropower data logger microcontroller, requiring a supply voltage of 5–16 V and made in a technology that meets the IP67 standard of resistance to environmental factors. The recorder additionally includes an SDI-12 standard input expansion module and a memory card. Based on the observations, it was determined that the following events can occur in the e2 element:

- e2.1—the recorder is working correctly, the observation coefficient is 0.99,
- e2.2—faulty microcontroller or expansion modules, the observation factor is 0.005 determined on the basis of observations and data from the manufacturer,
- e2.3—data archiving not possible due to overflow or memory card fault, the observation factor is 0.004 determined on the basis of observations and data from the manufacturer
- e2.4—battery supply voltage below 5 V or power line interruption, the observation factor is 0.002 based on the observation of the facility exploitation.

The values in Table 2 were calculated on the basis of the annual observation time of the meteorological station, which is shown in Figure 11. The states of fitness and unfitness of individual elements included in the tested meteorological station were determined [49–52].

Table 2. Observation coefficients (hxx, exx.x).

	e1a	e1b	e2	e3	e4	e5
1.	0.95	0.99	0.99	0.892	0.865	0.781
2.	−0.02	−0.01	−0.005	−0.122	−0.152	−0.185
3.	−0.04	−0.002	−0.004	−0.03	−0.114	−0.251
4.			−0.002			

The value of the maximum coefficient of hypothesis (h1, IQ1max) was assumed at a level close to the value 1, namely 0.9999.

The coefficients of successive indirect hypotheses are determined using Equation (5).

$$\begin{aligned} \mathrm{CF}(\mathrm{h1a}, \mathrm{e1a.1}, \mathrm{e1a.2}) &= \frac{\mathrm{CF}(\mathrm{h1a,e1a.1}) + \mathrm{CF}(\mathrm{h1a,e1a.2})}{1-\min(|\mathrm{CF}(\mathrm{h1a,e1a.1})|;|\mathrm{CF}(\mathrm{h1a,e1a.2})|)} \\ &= \frac{0.95+(-0.02)}{1-\min(|0.95|;|(-0.02)|)} = \frac{0.83}{0.88} \cong 0.94898 \end{aligned}, \tag{7}$$

$$\begin{aligned} \mathrm{h1a} &= \mathrm{CF}(\mathrm{h1a}, \mathrm{e1a.1}, \mathrm{e1a.2}, \mathrm{e1a3}) = \frac{\mathrm{CF}(\mathrm{h1a,e1a.1,e1a.2}) + \mathrm{CF}(\mathrm{h1a,e1a.3})}{1-\min(|\mathrm{CF}(\mathrm{h1a,e1a.1,e1a.2})|;|\mathrm{CF}(\mathrm{h1a,e1a.3})|)} \\ &= \frac{0.94898+(-0.04)}{1-\min(|0.94898|;|(-0.04)|)} = \frac{0.9927}{0.96} \cong 0.94685 \end{aligned}, \tag{8}$$

h1b, h2, h3, h4, and h5 are calculated in a similar way and they amount to:

$$\begin{aligned} &\mathrm{h1b} \cong 0.98988 \\ &\mathrm{h2} \cong 0.98989 \\ &\mathrm{h3} \cong 0.87319 \\ &\mathrm{h4} \cong 0.82032 \\ &\mathrm{h5} \cong 0.64124 \end{aligned}$$

The next step is to determine the value of h1. Similarly, using Equation (5), the value of h1 is determined. h1a and h1b are replaced by values h1a′ = −1 + h1a and h1b′ = −1 + h1b.

$$\mathrm{CF}(\mathrm{h1a'}, \mathrm{IQ1max}) = \frac{\mathrm{CF}(\mathrm{h1}, \mathrm{IQ1max}) + \mathrm{CF}(\mathrm{h1a'}, \mathrm{IQ1max})}{1-\min(|\mathrm{CF}(\mathrm{h1}, \mathrm{IQ1max})|;|\mathrm{CF}(\mathrm{h1}, \mathrm{h1a'})|)} = \frac{0.9999+(-0.05315)}{1-\min(|0.9999|;|(-0.05315)|)} \cong 0.999894, \tag{9}$$

$$\begin{aligned} \mathrm{h1} &= \mathrm{CF(h1,h1a\prime,h2b\prime,IQ1max)} = \frac{\mathrm{CF(h1a\prime,IQ1max)+CF(h1b\prime,IQ1max)}}{1-\min(|\mathrm{CF(h1a\prime,IQ1max)}|;|\mathrm{CF(h1b\prime,IQ1max)}|)} \\ &= \frac{0.999894+(-0.01012)}{1-\min(|0.999894|;|(-0.01012)|)} \cong 0.999893 \end{aligned}, \tag{10}$$

Using equation 6, the final hypothesis coefficient can be determined as:

$$\mathrm{h} = \mathrm{h1} \cdot \mathrm{h2} \cdot \mathrm{h3} \cdot \mathrm{h4} \cdot \mathrm{h5} = 0.999893 \cdot 0.98989 \cdot 0.87319 \cdot 0.82032 \cdot 0.64124 \cong 0.45462, \tag{11}$$

4. Simulation and Results using Real Measurements

In order to present the influence of the observation coefficients on the indirect hypotheses and the final hypothesis, a series of simulations was performed. The results are presented below in the form of graphs. The simulations were run with a programme written (by the first author) for this purpose. The first graph in Figure 13 shows the effect of the observation coefficient values associated with the temperature sensors. The range of the coefficients e1a.1 and e1b.1 is from 0.5 to 0.99.

The next graph (Figure 14) shows the influence of the negative coefficients of observations on the analogue temperature sensor. The range of the coefficients e1a.2 and e1a.3 is from −0.09 to −0.01.

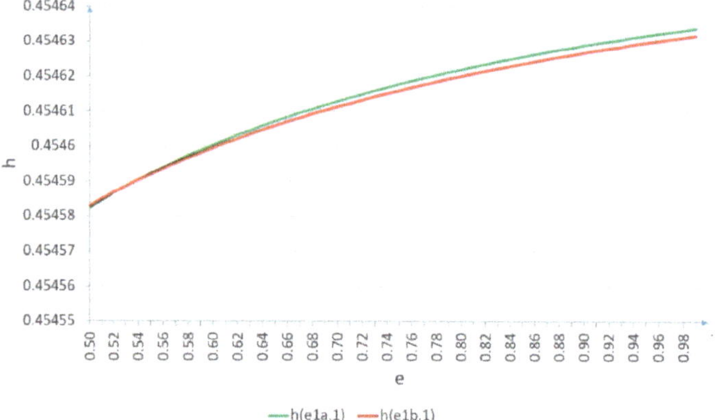

Figure 13. The result of the simulation of the h hypothesis value depending on the observation coefficients e1a.1 and e1b.1.

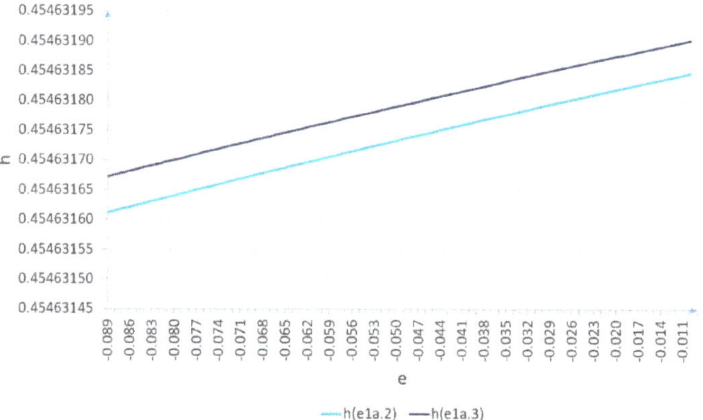

Figure 14. The result of the simulation of the h hypothesis value depending on the observation coefficients e1a.2 and e1a.3.

The next graph (Figure 15) shows the influence of the negative coefficients of observations related to the digital temperature sensor. The range of the coefficients e1b.2 and e1b.3 is from −0.099 to −0.001. The coefficient e1b.4 was omitted because function h (e1b.4) has the same values as function h (e1b.3).

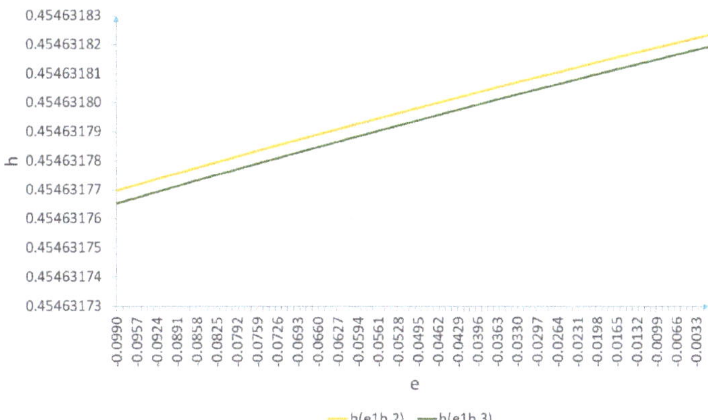

Figure 15. The result of the simulation of the h hypothesis value depending on the observation coefficients e1b.2 and e1b.3.

Figures 13–15 show some of the most important functions representing the impact of the selected and most important observations on the final hypothesis h (the data were correctly interpreted). Graphs of the presented functions show a tendency towards non-linearity. In the ideal model, they should aim asymptotically at the value, which for this model means absolute excellence of the system, as shown in Figure 16 as an idealised curve [40]. The graphs in Figures 13–15 also prove that each of the observation coefficients affects the final value of h, and the effect is non-linear.

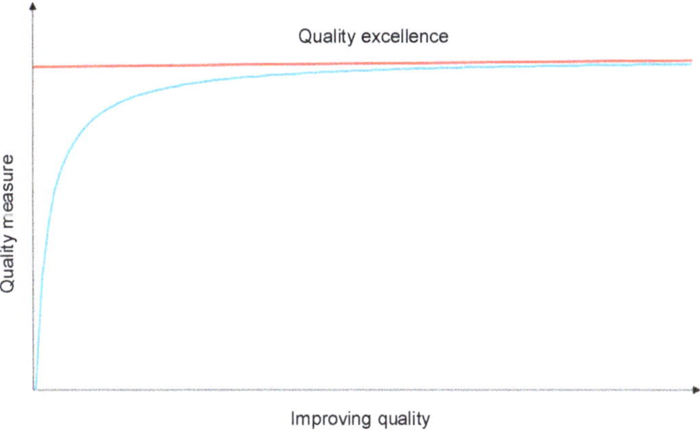

Figure 16. Illustration of the process of improving quality as a pursuit of excellence [40].

In practical terms, the presented simulation results make it possible to show whether the values calculated for the designed model are consistent with the assumptions.

5. Conclusions

The proprietary research presented in this article concerns the issue related to the quality analysis of information obtained from the weathers station's sensors. Currently, most scientific work is increasingly devoted to developing efficient and reliable sensors and weather station systems. A large body of studies also involves the analysis of data obtained from sensors of meteorological stations and their characteristic properties, estimating missing data in meteorological information and assessing the quality of weather information. This is a good research direction, but a broader perspective should also be adopted to assess the quality of information obtained from weather sensors. Such an approach is demonstrated in this article. The structure of a real meteorological station and the metrological data obtained from it were analysed. A set of factors influencing the indirect hypothesis was identified that constitute the final hypothesis (i.e., the data were correctly interpreted). The specific mathematical apparatus usage and the scrutiny carried out enabled the developing of an information quality model that uses calculations of the certainty factor (CF) of the hypothesis. The whole is a proprietary method of uncertainty modelling applied to estimate the quality of information obtained from meteorological station's sensors. The employment of the method allows, in practice, a more accurate defining of the value of information quality, taking into account many factors that determine it. In particular, it allows one to analyse the impact of individual information processing procedures on the quality of this information and the impact of quality dimensions and of redundancy on this quality. As a result, it becomes possible to identify those elements of the procedures of information acquisition and processing that negatively affect the quality of information.

The authors plan to continue their research with a model which includes a larger number of different sensors forming a meteorological station, with particular emphasis on the reliability and the exploitation dependencies between them.

Author Contributions: Conceptualization, M.S. and W.O.; methodology, M.S. and A.R.; software, M.S.; validation, W.O. and T.D.; formal analysis, M.S., A.R. and T.D.; investigation, M.S. and W.O.; resources, W.O. and T.D.; data curation, W.O.; writing—original draft preparation, M.S., W.O. and A.R.; writing—review and editing, A.R. and T.D.; visualization, M.S.; supervision, A.R. and T.D.; project administration, A.R. All authors have read and agreed to the published version of the manuscript.

Funding: This research received no external funding.

Data Availability Statement: The data presented in this study are available on request from the corresponding author.

Conflicts of Interest: The authors declare no conflict of interest.

References

1. Dorman, C.E. Early and Recent Observational Techniques for Fog. In *Marine Fog: Challenges and Advancements in Observations, Modeling, and Forecasting*; Koračin, D., Dorman, C., Eds.; Springer: Cham, Switzerland, 2017. [CrossRef]
2. Ilčev, S.D. Meteorological Ground Stations. In *Global Satellite Meteorological Observation (GSMO) Applications*; Springer: Cham, Switzerland, 2019. [CrossRef]
3. Olchowik, W. Simulation of systems with solar collectors in relation to the raw meteorological data. *Bull. Mil. Univ. Technol.* **2017**, *66*, 37–54. [CrossRef]
4. Sarkar, I.; Pal, B.; Datta, A.; Roy, S. Wi-Fi-Based Portable Weather Station for Monitoring Temperature, Relative Humidity, Pressure, Precipitation, Wind Speed, and Direction. In *Information and Communication Technology for Sustainable Development*; Tuba, M., Akashe, S., Joshi, A., Eds.; Springer: Singapore, 2020. [CrossRef]
5. Sugiarto, B.; Sustika, R. Data classification for air quality on wireless sensor network monitoring system using decision tree algorithm. In Proceedings of the 2nd International Conference on Science and Technology-Computer (ICST), Yogyakarta, Indonesia, 27–28 October 2016; pp. 172–176. [CrossRef]
6. Płanda, B.; Skorupski, J. Methods of air traffic management in the airport area including the environmental factor. *Int. J. Sustain. Transp.* **2017**, *11*, 295–307. [CrossRef]
7. Jaeger, A. *Weather Hazard. Warning Application in Car-to-X Communication*; Springer: Wiesbaden, Germany, 2016. [CrossRef]
8. Ryguła, A.; Brzozowski, K.; Konior, A. Utility of Information from Road Weather Stations in Intelligent Transport Systems Application. In *Tools of Transport Telematics*; TST 2015; Springer: Cham, Switzerland, 2015. [CrossRef]

9. Boulanger, J.; Aizpuru, J.; Leggieri, L.; Marino, M. A procedure for automated quality control and homogenization of historical daily temperature and precipitation data (APACH): Part 1: Quality control and application to the Argentine weather service stations. *Clim. Change* **2010**, *98*, 471–491. [CrossRef]
10. Li, X.; Zou, D.; Feng, W.; Xie, W.; Shi, L. Study of Quality Control Methods for Moored Buoys Observation Data. In Proceedings of the International Conference on Meteorology Observations (ICMO), Chengdu, China, 28–31 December 2019; pp. 1–4. [CrossRef]
11. Qu, S.; Feng, Y.; Li, T. Comparative Study on the Reliability of Weather Radar Intensity Data. In Proceedings of the International Conference on Meteorology Observations (ICMO), Chengdu, China, 28–31 December 2019; pp. 1–3. [CrossRef]
12. Vas, Á.; Tóth, L. Investigation of a Hybrid Sensor- and Computational Network for Numerical Weather Prediction Calculations. In *Distributed Computer and Communication Networks*; DCCN 2019; Vishnevskiy, V., Samouylov, K., Kozyrev, D., Eds.; Springer: Cham, Switzerland, 2019. [CrossRef]
13. Hema, N.; Kant, K. Reconstructing missing hourly real-time precipitation data using a novel intermittent sliding window period technique for automatic weather station data. *J. Meteorol Res.* **2017**, *31*, 774–790. [CrossRef]
14. Rosillon, D.; Huart, J.P.; Goossens, T.; Journée, M.; Planchon, V. The Agromet Project: A Virtual Weather Station Network for Agricultural Decision Support Systems in Wallonia, South of Belgium. In *Ad-Hoc, Mobile, and Wireless Networks*; ADHOC-NOW 2019; Palattella, M., Scanzio, S., Coleri Ergen, S., Eds.; Springer: Cham, Switzerland, 2019. [CrossRef]
15. Juneja, A.; Das, N. Big Data Quality Framework: Pre-Processing Data in Weather Monitoring Application. In Proceedings of the International Conference on Machine Learning, Big Data, Cloud and Parallel Computing (COMITCon), Faridabad, India, 14–16 February 2019; pp. 559–563. [CrossRef]
16. Sattar, F.; Karray, F.; Kamel, M.; Nassar, L.; Golestan, K. Recent Advances on Context-Awareness and Data/Information Fusion in ITS. *Int. J. Intell. Transp. Syst. Res.* **2016**, *14*, 1–19. [CrossRef]
17. Schubert, R.; Obst, M. The Role of Multisensor Environmental Perception for Automated Driving. In *Automated Driving*; Watzenig, D., Horn, M., Eds.; Springer: Cham, Switzerland, 2017; pp. 161–182. [CrossRef]
18. Penenko, V.V.; Tsvetova, E.A.; Penenko, A.V. Methods based on the joint use of models and observational data in the framework of variational approach to forecasting weather and atmospheric composition quality. *Russ. Meteorol. Hydrol.* **2015**, *40*, 365–373. [CrossRef]
19. Wei, C.-C.; Hsu, C.-C. Extreme Gradient Boosting Model for Rain Retrieval using Radar Reflectivity from Various Elevation Angles. *Remote Sens.* **2020**, *12*, 2203. [CrossRef]
20. Tang, G.; Long, D.; Behrangi, A.; Wang, C.; Hong, Y. Exploring deep neural networks to retrieve rain and snow in high latitudes using multisensor and reanalysis data. *Water Resour. Res.* **2018**, *54*, 8253–8278. [CrossRef]
21. International Organization for Standardization. *Data Quality—Part 8: Information and Data Quality: Concepts and Measuring*; ISO/IEC 8000-8:2015; ISO: Geneva, Switzerland, 2015.
22. International Organization for Standardization. *Quality Management Systems—Fundamentals and Vocabulary*; ISO/IEC 9000:2015; ISO: Geneva, Switzerland, 2015.
23. International Organization for Standardization. *Quality Management Systems—Requirements*; ISO/IEC 9001:2015; ISO: Geneva, Switzerland, 2015.
24. Massachusetts Institute of Technology Information Quality (MITIQ) Program. Available online: http://mitiq.mit.edu (accessed on 2 May 2020).
25. Fisher, C.; Lauria, E.; Chengalur-Smith, S.; Wang, R. *Introduction to Information Quality*; Authorhouse: Bloomington, IN, USA, 2011.
26. Wang, R.Y.; Pierce, E.M.; Madnick, S.; Fisher, C.W. (Eds.) *Information Quality. Advances in Management Information Systems*; M.E. Sharpe: Armonk, NY, USA, 2005.
27. Dempster, A.P. Upper and Lower Probabilities Inducted by a Multi-valued Mapping. *Ann. Math. Stat.* **1967**, *38*, 325–339. [CrossRef]
28. Krzykowska, K.; Krzykowski, M. Forecasting Parameters of Satellite Navigation Signal through Artificial Neural Networks for the Purpose of Civil Aviation. *Int. J. Aerosp. Eng.* **2019**, *1*, 1–11. [CrossRef]
29. Mazur, M. *Qualitative Information Theory*; Scientific and Technical Publishers: Warsaw, Poland, 1970.
30. Shafer, G. *A Mathematical Theory of Evidence*; Princeton University Press: Princeton, NY, USA, 1976.
31. Heckerman, D. The certainty-factor model. In *Encyclopedia of Artificial Intelligence*; Shapiro, S., Ed.; Wiley: New York, NY, USA, 1992; pp. 131–138.
32. Shortliffe, E.H.; Buchanan, B.G. *Rule-Based Expert Systems: The MYCIN Experiments of the Stanford Heuristic Programming Project*; Addison-Wesley Publishing Co. Inc.: Boston, MA, USA, 1984.
33. Oleński, J. *Economics of Information. The Basics*; Polish Economic Publishing House: Warsaw, Poland, 2001.
34. Rychlicki, M.; Kasprzyk, Z.; Rosiński, A. Analysis of Accuracy and Reliability of Different Types of GPS Receivers. *Sensors* **2020**, *20*, 6498. [CrossRef]
35. Jacyna, M.; Żak, J.; Gołębiowski, P. The EMITRANSYS model and the possibilities of its application for the analysis of the development of sustainable transport systems. *Combust. Engines* **2019**, *179*, 243–248. [CrossRef]
36. Jurczyk, A.; Szturc, J.; Otop, I.; Ośródka, K.; Struzik, P. Quality-Based Combination of Multi-Source Precipitation Data. *Remote Sens.* **2020**, *12*, 1709. [CrossRef]
37. Siergiejczyk, M.; Krzykowska, K.; Rosiński, A. Evaluation of the influence of atmospheric conditions on the quality of satellite signal. In *Marine Navigation*; Weintrit, A., Ed.; CRC Press/Balkema: London, UK, 2017; pp. 121–128. [CrossRef]

38. Bednarek, M.; Dąbrowski, T.; Olchowik, W. Selected practical aspects of communication diagnosis in the industrial network. *J. KONBiN* **2019**, *49*, 383–404. [CrossRef]
39. Dudek, E.; Kozłowski, M. Analysis of aeronautical information potential incompatibility—Case study. *J. KONBiN* **2017**, *41*, 59–82. [CrossRef]
40. Stawowy, M. *Method of Multilayer Modeling of Uncertainty in Estimating the Information Quality of ICT Systems in Transport*; Publishing House of Warsaw University of Technology: Warsaw, Poland, 2019.
41. Baggini, A. (Ed.) *Handbook of Power Quality*; John Wiley & Sons: Hoboken, NJ, USA, 2008. [CrossRef]
42. Watral, Z.; Michalski, A. Selected Problems of Power Sources for Wireless Sensors Networks. *IEEE Instrum. Meas. Mag.* **2013**, *16*, 37–43. [CrossRef]
43. Michalski, A.; Watral, Z.; Jakubowski, J. Energy Harvesting—A real possibility of alternative power supply to wireless sensor networks. In *Selected Aspects of the Use of "Energy Harvesting" Technology in Supplying Wireless Sensor Networks*; Military Academy of Technology: Warsaw, Poland, 2017; pp. 39–88.
44. Paś, J.; Rosiński, A.; Chrzan, M.; Białek, K. Reliability-Operational Analysis of the LED Lighting Module Including Electromagnetic Interference. *IEEE Trans. Electromagn. Compat.* **2020**, *62*, 2747–2758. [CrossRef]
45. Paś, J.; Rosiński, A.; Szulim, M.; Łukasiak, J. Modelling the Safety Levels of ICT Equipment Exposed to Strong Electromagnetic Pulses. In Proceedings of the 14th International Conference on Dependability of Computer Systems DepCoS-RELCOMEX 2019, Brunów, Poland, 1–5 July 2019; Zamojski, W., Mazurkiewicz, J., Sugier, J., Walkowiak, T., Kacprzyk, J., Eds.; Springer: Cham, Switzerland, 2020; pp. 393–401. [CrossRef]
46. Stawowy, M.; Perlicki, K.; Sumiła, M. Comparison of Uncertainty Multilevel Models to Ensure ITS Services. In Safety and Reliability: Theory and Applications. In Proceedings of the European Safety and Reliability Conference ESREL 2017, Portoroz, Slovenia, 18–22 June 2017; Cepin, M., Bris, R., Eds.; CRC Press/Balkema: London, UK, 2017; pp. 2647–2652. [CrossRef]
47. Humidity and Temperature Sensor SRH1A. Available online: http://www.pmecology.com/pl/wp-content/uploads/2019/09/RH_TEMP-Sensor-PM-Ecology_spec.pdf (accessed on 5 March 2020).
48. Temperature Sensor. Available online: https://www.pmecology.com/wp-content/uploads/2018/08/Temperature-sensor-ST1R-PM-Ecology_spec.pdf (accessed on 5 March 2020).
49. Będkowski, L.; Dąbrowski, T. *Basics of Maintenance, Vol. II Basic of Operational Reliability*; Military University of Technology: Warsaw, Poland, 2006.
50. Klimczak, T.; Paś, J. *Basics of Exploitation of Fire Alarm Systems in Transport Facilities*; Military University of Technology: Warsaw, Poland, 2020.
51. Duer, S.; Duer, R.; Mazuru, S. Determination of the expert knowledge base on the basis of a functional and diagnostic analysis of a technical object. *Rom. Assoc. Nonconv. Technol.* **2016**, *2*, 23–29.
52. Grabski, F. *Semi-Markov Processes: Applications in System Reliability and Maintenance*; Elsevier: Amsterdam, The Netherlands, 2015.

Article

Multi-Instance Inertial Navigation System for Radar Terrain Imaging

Michal Labowski * and Piotr Kaniewski

Faculty of Electronics, Military University of Technology, ul. gen. S. Kaliskiego 2, 00-908 Warsaw, Poland; piotr.kaniewski@wat.edu.pl

* Correspondence: michal.labowski@wat.edu.pl

Received: 24 September 2020; Accepted: 3 November 2020; Published: 6 November 2020

Abstract: Navigation systems used for the motion correction (MOCO) of radar terrain images have several limitations, including the maximum duration of the measurement session, the time duration of the synthetic aperture, and only focusing on minimizing long-term positioning errors of the radar host. To overcome these limitations, a novel, multi-instance inertial navigation system (MINS) has been proposed by the authors. In this approach, the classic inertial navigation system (INS), which works from the beginning to the end of the measurement session, was replaced by short INS instances. The initialization of each INS instance is performed using an INS/GPS system and is triggered by exceeding the positioning error of the currently operating instance. According to this procedure, both INS instances operate simultaneously. The parallel work of the instances is performed until the image line can be calculated using navigation data originating only from the new instance. The described mechanism aims to perform instance switching in a manner that does not disturb the initial phases of echo signals processed in a single aperture. The obtained results indicate that the proposed method improves the imaging quality compared to the methods using the classic INS or the INS/GPS system.

Keywords: INS; GPS; UAV; SAR

---

1. Introduction

Most navigation systems are aimed at minimizing slow-varying position and velocity errors of the object. These methods aim to keep the errors at a low level for as long as possible. However, there are certain applications that impose specific demands on navigation systems, e.g., synthetic aperture radar (SAR) terrain imagery.

Synthetic aperture radars allow for high-resolution terrain imaging with similar quality to optical methods. In contrast, however, this technique is independent of weather and lighting conditions. Currently, unmanned aerial vehicles (UAV) are commonly used as carriers of SAR systems. In side-looking aerial radar (SLAR), the direction of flight is called *azimuth* while the direction of observation is called *range* [1].

The advantages of SAR systems are accompanied by high requirements and limitations. To get focused and geometrically correct images, the UAV must move in a strictly defined manner, most often with uniform rectilinear motion, which ensures proper conditions for receiving and processing echo signals [2–9]. However, in real-world scenarios it is impossible to fulfill these requirements.

In [10], Fornaro distinguished three types of SAR-carrier trajectories: nominal, real, and measured. Due to measurement errors, the measured trajectory does not coincide with the real one. The discrepancy between the real and the nominal trajectories result from atmospheric disturbances and autopilot controls [3]. According to [6,11], the unfavorable effects of the atmosphere on the UAV lead to the range (radar to object) distance instabilities, nonuniform UAV motion in the azimuth direction, and changes in the UAV's attitude. Groundspeed variations can be compensated by adjusting the radar pulse repetition frequency (PRF) so the distance traveled by a UAV in one pulse period is constant [12]. The second method of groundspeed instability compensation consists in resampling the received echo signals [2,3]. The UAV attitude variation can be compensated by the radar antenna stabilization system [13]. As a result, the flightpath curvature has the most negative impact on the quality of radar images. According to Kirk [4] and Fornaro [3], UAV movement instabilities can be divided into low- and high-frequency errors. Low-frequency errors (with a period longer than a duration of the synthetic aperture) lead to a shift in the center frequency of the echo signal, which results in geometric distortions of the image. High-frequency errors (with a period shorter than the synthetic aperture) increase the amplitude of the side lobes, which blurs the image. Therefore, the compensation of these errors has the greatest impact on the quality of radar images [9]. As a result, in SAR systems it is necessary to compensate the influence of UAV motion instabilities on the received echo signal phase history. These procedures are called motion compensation (MOCO) [14]. MOCO algorithms can be divided into procedures using navigation equipment, as well as procedures based on the analysis of the echo signal (autofocus) [10,14].

In navigation procedures, the echo signal phase correction is performed based on the measured and calculated discrepancy between the nominal and actual trajectories. However, the navigation system is a source of additional errors, which could decrease the quality of SAR images. The analysis of the inertial measurement unit (IMU) and INS errors in the imaging process is presented in [15]. The paper [16] presents the SAR MOCO procedure using INS navigation corrections, which suggest that in the case of the limited duration of the measurement session (approx. 30 s), it is possible to significantly improve the image quality.

In the case of the navigation systems dedicated to cooperation with SAR radars, it is necessary to consider their accuracy in the short and long term. The INS system ensures high short-term accuracy, but in the long-term perspective an increase of the positioning errors is the main disadvantage of the inertial navigation [17]. This drawback can be compensated for using additional sensors (e.g., Doppler radar, GPS receiver), ensuring a high long-term accuracy [18]. The disadvantage of this approach, however, is high-frequency measurement errors, which transfer to the MOCO procedure (through a data fusion algorithm), causing errors in the initial phase of the echo signals.

In the article [19], Fuxiang and Zheng noticed that navigation systems proposed in the MOCO literature, and in particular the methods of INS and GPS data integration, are predominantly focused on reducing the long-term errors, which makes them more suitable for navigating than for SAR MOCO. For example, in [20] Gong and Fang presented the navigation system used in MOCO, consisting of INS and GPS with real time kinematic (RTK). They presented four strategies for integrating the measurement data: a Kalman filter, an extended Kalman filter (EKF), a combination of the unscented Kalman filter (UKF) with an EKF, and a model predictive filter (MPF) with an EKF. The authors focused on long-term error minimalization and did not consider the impact of the short-term errors on the radar images. Focusing on the navigation system, and disregarding its influence on the SAR subsystem, is a common practice in the literature [19,21,22].

In the work [19], Fuxiang and Zheng presented a proposal to solve the high-frequency navigation system errors caused by GPS. According to the authors, the period of the Kalman filter correction phase (the period of GPS data used in the filter) should be longer than the duration of a synthetic aperture. In such a system, the INS errors grow during the synthetic aperture, which, however, is much less disadvantageous than random changes introduced by the GPS. Unfortunately, this limits the duration of the synthetic aperture, which may be especially problematic in the case of UAVs moving with low speeds in the azimuthal direction. A short synthetic aperture decreases the spatial resolution of the radar image.

According to Kennedy's work [15], the INS used in a MOCO algorithm should not be reinitialized during the synthetic aperture. A proposed solution consists of the main INS (aircraft navigation system) and the auxiliary INS (IMU mounted on the radar antenna). A correction of the auxiliary INS (reinitialization) is performed after the synthesis of the aperture has been completed. This method is therefore analogous to that proposed by Fuxiang and Zheng [18]; they differ only in the source of the correction information (main INS or GPS receiver).

In [21], Fang and Gong presented a navigation system divided into two INS blocks using a common IMU. Errors of the main INS are estimated by a Predictive Iterated Kalman Filter (PIKF) by comparing results from the INS and GPS. As proposed by the authors, the main INS is only used during the approach to the radar scanning area. At the start of the measurement session, the output from the main INS is used to initialize the secondary INS (not corrected by the GPS), whose output is used in the SAR MOCO algorithm. The proposed mechanism has two goals. The first one is to reduce the operating time (and errors) of the secondary INS. The second goal is to reduce the influence of the GPS errors on the MOCO by isolating two INS blocks. The authors emphasize that the radar measurement session should last no longer than 15 s to keep INS errors at an acceptable level.

The second group of MOCO methods, called autofocus, consists in the analysis of the echo signals received by the radar sensor [14,23–33]. A great variety of autofocus methods exists due to their individual limitations, e.g., the Phase Gradient Autofocus (PGA) needs the presence of highly reflective objects in the imaged area [24], while in the Map-Drift method the area should be diversified (presence of edges, shadows, etc.) [32]. A majority of autofocus methods are iterative, which extends the time needed to obtain a result (image) [23]. For comparison, in the navigational MOCO procedures the image synthesis is performed once, and the corrections are computed independently in a dedicated subsystem. As a result, in navigational MOCO it is possible to reduce the computational effort (no iterations) and time consumption.

In summary, there are currently several navigational MOCO procedures using data from INS or INS/GPS systems. However, in many cases the navigation system is not optimized for the MOCO algorithm—these methods focus on the long-term accuracy of the system, ignoring the adverse impact of fast-changing errors contributed by the GPS receiver. In papers dealing with this problem, the authors have proposed methods limiting the duration of the synthetic aperture (shorter than the GPS update period) or limiting the duration of the measurement session (INS-only systems).

Based on this analysis, the authors propose a multi-instance INS (MINS) system, combining the advantages of the classic INS and INS/GPS systems. This paper is related to the works presented in [16,34–36]. An application of pure inertial navigation for SAR MOCO is presented in [16] and using INS/GPS for MOCO is described in [34]. A method of calculating position deviations from a theoretical, nominally rectilinear trajectory for a UAV-based SAR is explained in [36]. The MOCO method presented in this paper, however, is new and has not been presented before, both by the authors of this paper nor by other authors. The results presented in the further parts of this paper, based on real measurements obtained during UAV flights, show that the proposed MINS system is in many aspects superior with respect to the INS and the INS/GPS systems used for SAR MOCO applications.

The layout of the further parts of this paper is as follows: the idea of a multi-instance inertial navigation system is explained in Section 2, selected results of testing the algorithm with the use of real navigation data are presented in Section 3, whereas in Section 4 the influence of the MINS-based MOCO on SAR images is compared with the results obtained with other MOCO methods. Conclusions are provided in Section 5.

2. Materials and Methods

The proposed MOCO method, using the MINS system, is based on a combination of INS and INS/GPS systems described in [16,34,36]. The results presented in the mentioned papers show that each proposed solution improves the quality of SAR images, but in both cases the improvement is only partial and complementary. The INS system improves the image contrast (IC), entropy (E), spatial resolution (SR), peak to sidelobe ratio (PSLR), and integrated sidelobe ratio (ISLR), while the INS/GPS system better reduces the image geometric distortions.

The purpose of INS instances proposed in MINS is to keep INS errors limited in such a way that the corrections calculated by a navigation correction algorithm (NCA) [36] do not disturb the initial phases of radar echo signals in a single synthetic aperture. Therefore, in the proposed system, the INS error corrections are not performed during an aperture synthesis. During any single synthetic aperture, only the uncorrected INS system is used as a data source for an NCA, and a periodic INS correction, using the INS/GPS system, is conducted in a special manner that takes into account the specificity of the aperture synthesis algorithm.

The proposed method assumes a parallel work of the MINS and INS/GPS systems. A diagram illustrating the concept of such a system is shown in Figure 1.

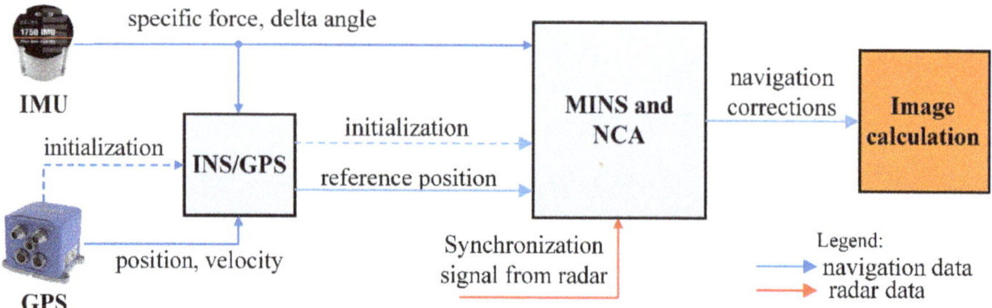

Figure 1. Data flow in multi-instance inertial navigation system (MINS)-based motion correction (MOCO).

The INS/GPS subsystem [34] determines the UAV's position, velocity, and attitude. Its output is then used to initialize the *i-th* INS instance, INS_i, in the MINS system. The integrated system also provides the position reference used to calculate MINS errors. Data obtained from the MINS are then used by the NCA to determine the navigation corrections for the SAR system. The INS/GPS system is initialized at the system launch and works until its power is off.

The detailed algorithm of MINS is presented in Figure 2. It has two main branches presented in Figure 2 using red and blue colored boxes and lines. The red branch is related to the overlapping INS instances, i.e., a situation when two INS instances work in parallel. The blue branch is executed when an INS overlap is finished or impossible to create and is related to a situation when only a single INS instance is working.

The SAR system synthesizes the image line for the central measurement within the synthetic aperture of length M_{SA}, where M_{SA} is the number of soundings that make up the synthetic aperture. Therefore, the IMU and radar data have to be buffered and the navigation algorithm's output is delayed by the duration of the aperture (usually less than 1 s). The switching of INS instances can be performed if the buffer contains enough IMU data to calculate navigation corrections for the new synthetic aperture. During the MINS system operation, the IMU measurements are used both by the INS/GPS and an *i–th* instance INS_i. Both systems calculate the position and velocity of the UAV's center of gravity (the origin of the body frame, *b-frame*). The results from the INS_i are then compared with the current INS/GPS estimates.

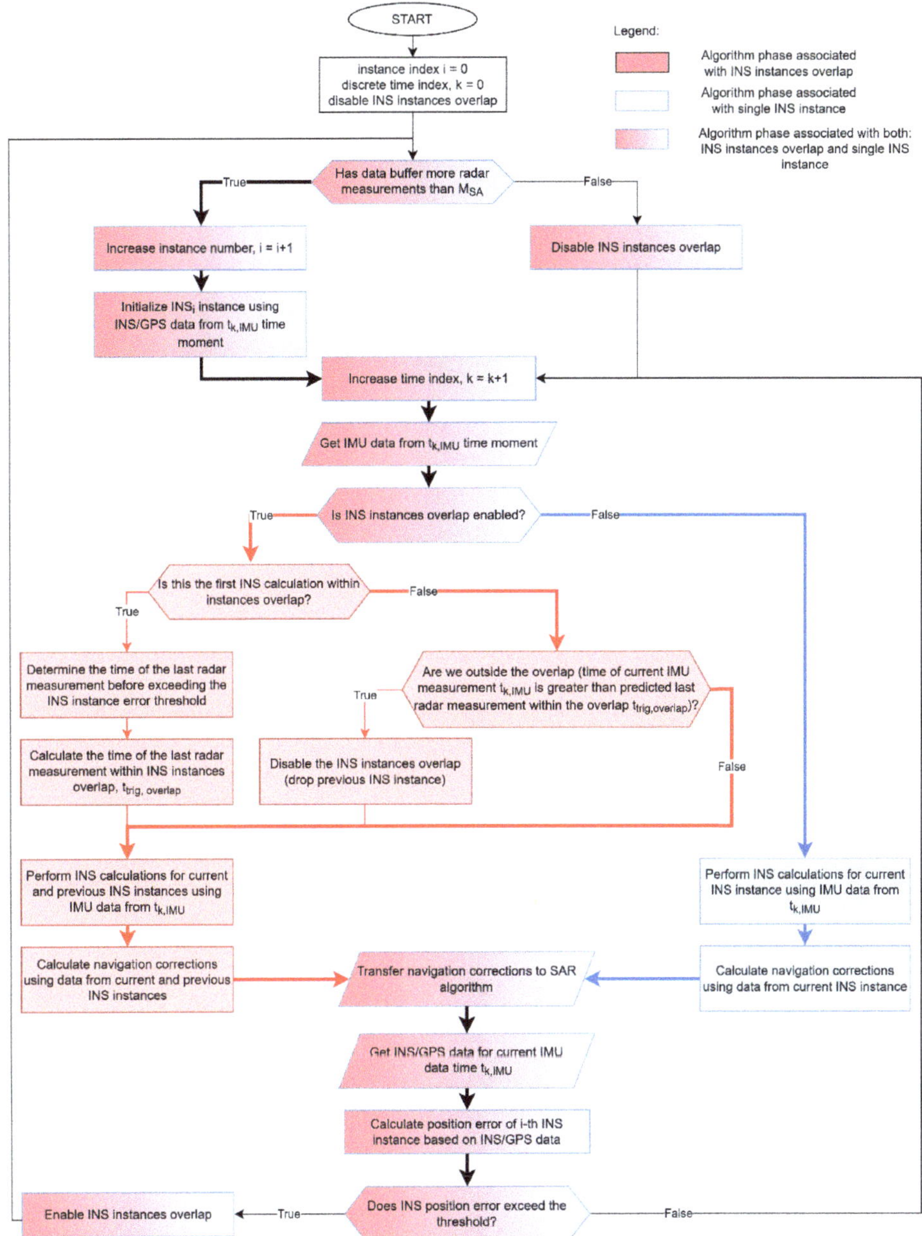

Figure 2. MINS algorithm.

When INS_i exceeds a position error threshold and data buffer contains enough IMU data, the instance number i is increased and the next INS instance is started (the red part of the algorithm becomes active). If there is not enough data in the buffer (e.g., due to the end of the measurement session), then no new INS instance is created and the current instance runs until all the buffered data have been processed. A new INS instance is initialized using the INS/GPS data determined for the moment when the position error threshold was exceeded. The INS_i position error, denoted as $||\Delta\mathbf{r}^n_{b,INS,INS/GPS}||$, is a norm of the vector describing the position difference of the *b-frame* origin expressed in the *n-frame* navigation system (North East Down, NED), determined by the *i-th* INS instance INS_i ($\mathbf{r}^n_{b,INS}$) and the INS/GPS system ($\mathbf{r}^n_{b,INS/GPS}$):

$$||\Delta\mathbf{r}^n_{b,INS,INS/GPS}|| = ||\Delta\mathbf{r}^n_{b,INS} - \Delta\mathbf{r}^n_{b,INS/GPS}|| = \left\| \begin{bmatrix} n \\ e \\ d \end{bmatrix}_{b,INS} - \begin{bmatrix} n \\ e \\ d \end{bmatrix}_{b,INS/GPS} \right\| \tag{1}$$

where *n*, *e*, *d* are position coordinates expressed in the *n-frame*. After exceeding the position error threshold $||\Delta\mathbf{r}^n_{tr}||$, i.e., when:

$$\left\|\Delta\mathbf{r}^n_{b,INS,INS/GPS}\right\| \geq ||\Delta\mathbf{r}^n_{tr}|| \tag{2}$$

the instances INS_{i-1} and INS_i run in parallel, as shown in Figure 3. The purpose of the simultaneous operation of both instances is to enable their switching in a way that does not occur within a single synthetic aperture, which could lead to an abrupt change of the calculated position. Switching the INS instance (the source for the NCA) is performed after the current aperture is synthesized and before moving to the next aperture, which ensures smooth navigation data used in each of them.

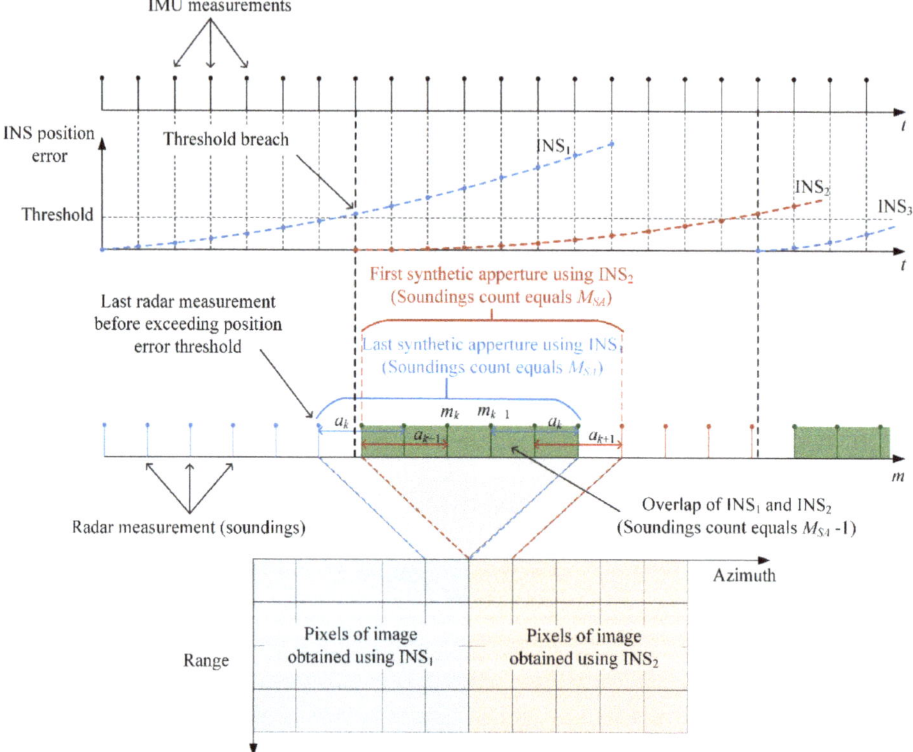

Figure 3. INS instances switching in MINS.

At the end of the overlap, the set of navigational corrections determined on the basis of the new INS_i instance is sufficiently large to allow for an aperture synthesis. As a result, after instance switching, the pixels that make up a given line of the image are created using the INS_i data, while the previous line was created using the INS_{i-1} instance. The length of the overlap can be expressed in terms of the number of radar soundings. The parallel work of instances begins at the time of the first IMU measurement after exceeding the INS error threshold, continues through $M_{SA} - 1$ radar soundings, and ends when the first IMU measurement occurs after processing the given number of soundings. In the radar system used in the further described experiments, M_{SA} is an odd number:

$$M_{SA} = 2a - 1 \tag{3}$$

where a is a rounded down number of soundings making up half of the synthetic aperture.

The navigational corrections are calculated using results of the INS instance. During the overlap, a pair of corrections are determined using INS_{i-1} and INS_i. The image synthesis algorithm, thanks to the knowledge about the length of the synthetic aperture shared with MINS, can detect the end of the correction data set and switch to the next source—the next INS instance.

In the presented algorithm, the INS instance initialization procedure can be interpreted as a form of INS error correction; however, the INS/GPS data are not used to correct the current INS instance (the errors of which continue to increase) but to initialize a new one. The initial INS errors correspond to small errors of the INS/GPS system. Thus, the proposed method combines the advantages of the INS and the INS/GPS systems. Thanks to the initialization of the instances, it is possible to keep INS errors at a low level, depending on the established error threshold, and at the same time the NCA uses the uncorrected INS results. According to the conclusions presented in [16] and [34], this should have a positive impact on the quality of the radar terrain images.

3. Results

The MINS system was tested using measurement data from the WATSAR radar system [37]. Before starting the navigation calculations, it was necessary to determine the value of $||\Delta \mathbf{r}^n_{tr}||$. If the value is too high, it leads to a shift between two image parts generated using INS data from the successive instances. The threshold value should also not be too low, as it would result in a short duration of INS instances, potentially shorter than the duration of the synthetic aperture, which in the WATSAR system was about 0.606 s. Taking into account the class of IMU used in the WASTAR system, the threshold was experimentally established and equaled 0.25 m.

After the threshold is exceeded, the errors of the INS_{i-1} instance still increase, but the time duration of this instance is limited to the time of the overlap. In a test flight (named here flight no 1), six INS instance switches were performed. Table 1 summarizes the duration of individual instances and the final values of position errors in relation to the INS/GPS trajectory.

Table 1. MINS results for flight no 1.

Instance Number	Duration [s]	Final Value of $||\Delta r^n_{b,INS,INS/GPS}||$ [m]
1	3.425	0.354
2	2.342	0.350
3	5.486	0.342
4	6.966	0.338
5	6.249	0.311
6	5.590	0.253
7	2.053	0.295

The time duration of instances is varied and ranges from 2.053 to 6.966 s. The reasons for this variation are initialization errors of INS instances and INS/GPS errors—GPS errors cause small step changes in $\Delta \mathbf{r}^n_{b,INS,INS/GPS}$. Figure 4 shows the position error of the first instance of flight no 1.

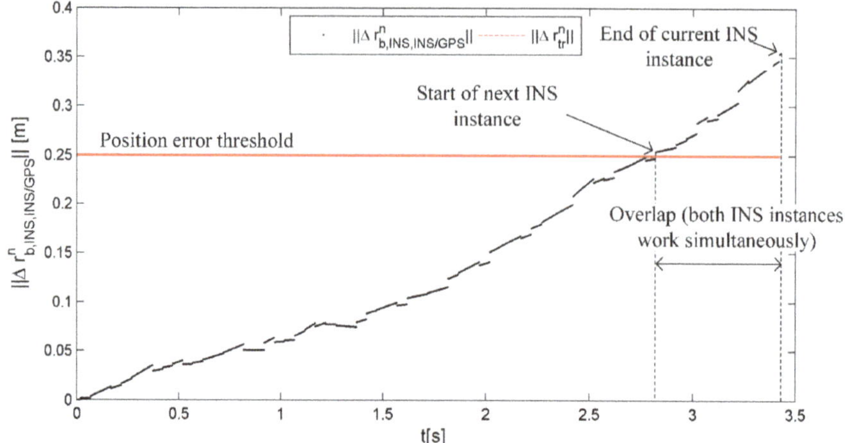

Figure 4. INS instance (no 1) position error in flight no 1.

A visualization of the UAV flight trajectory determined by the MINS system for flight no 1 is shown in Figure 5. The white markers indicate locations where the instances were initialized.

Figure 5. MINS trajectories for flight no 1.

A visual comparison of the position errors of the classic INS and the proposed MINS, determined with respect to the INS/GPS data, is shown in Figure 6, while Figures 7 and 8 show analogous results for the velocity and the velocity errors.

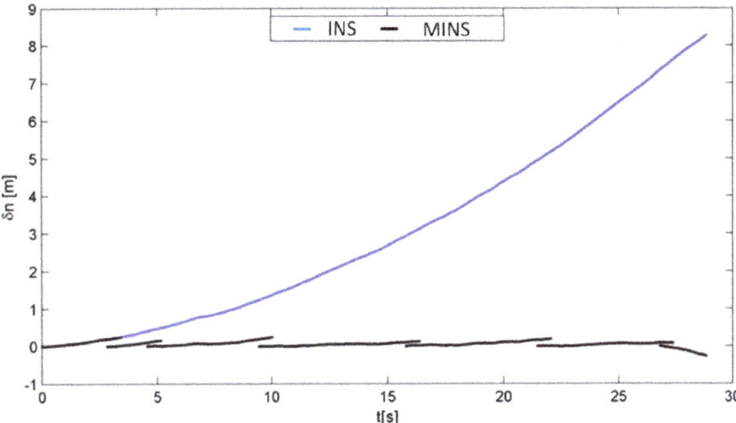

Figure 6. INS (blue) and MINS (black) position error with respect to the INS/GPS trajectory, North axis.

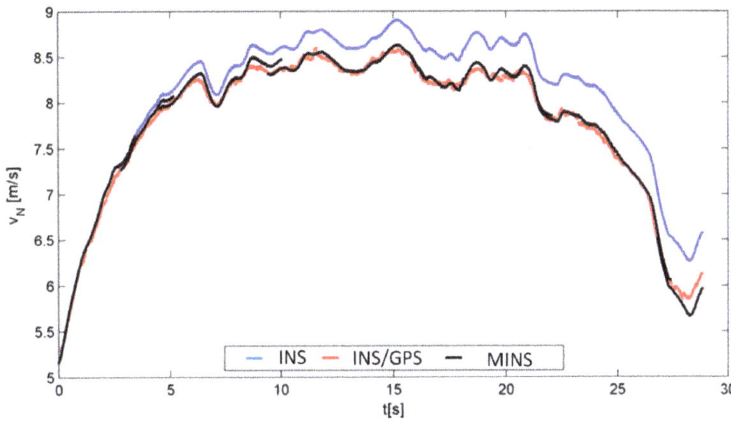

Figure 7. INS (blue), MINS (black), and INS/GPS (red) velocity, North axis.

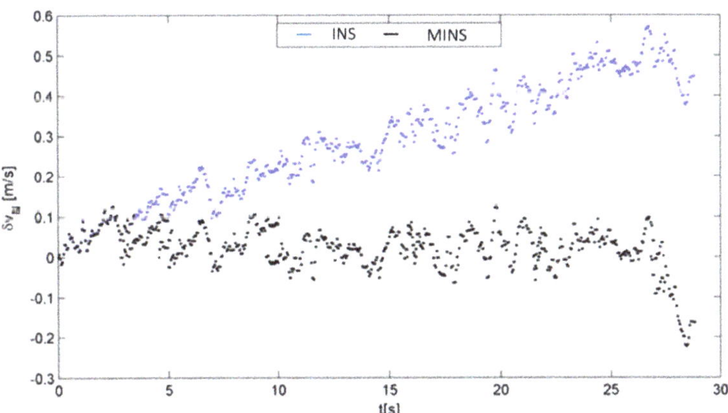

Figure 8. INS (blue) and MINS (black) velocity error with respect to the INS/GPS trajectory, North axis.

Thanks to the INS instance switching mechanism, the error in determining the navigation elements is kept at a low level, similar to the INS/GPS system, and the value of maximum error is related to the adopted threshold, which triggers a new instance. In the MINS system, the INS instances are initialized using INS/GPS data, thus the MINS trajectories originate on the INS/GPS trajectory. A random character of data presented in Figure 8 is caused by the error calculation mechanism, where from "smooth" INS data the "noisy" INS/GPS results were subtracted.

The MINS results are used by the NCA [36]. According to the procedure shown in Figure 2, INS overlaps lead to the navigation corrections overlap, which is presented in Figure 9.

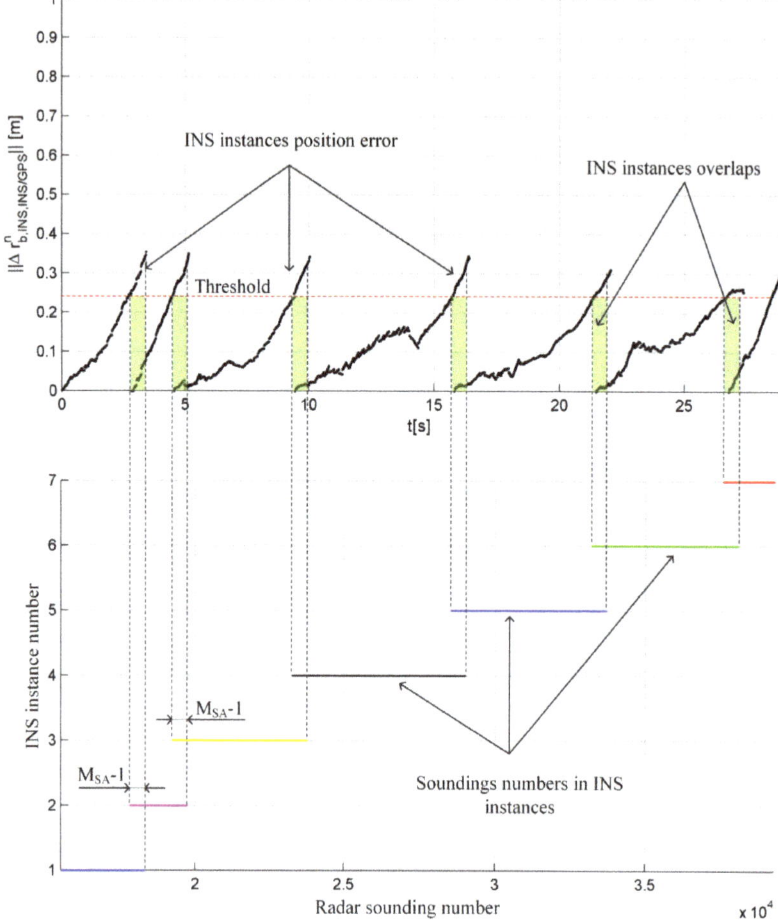

Figure 9. Navigation corrections data overlap.

4. Discussion

MINS-based MOCO results were compared with those obtained using other MOCO methods: INS [16] and INS/GPS [34]. Based on the corrections, the radar terrain images of the region presented in Figure 10 were calculated. The radar terrain image obtained without MOCO correction, for flight no 1, is shown in Figure 11. Figure 12 presents an image obtained using INS-based MOCO, Figure 13 shows an image calculated using INS/GPS-based MOCO, whereas Figure 14 shows an image obtained using the proposed MINS system.

Figure 10. Aerial photography of the imaged area.

Figure 11. Radar terrain image obtained without navigation correction.

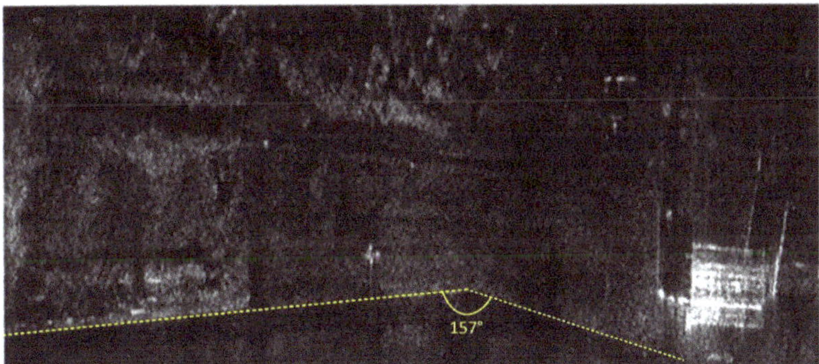

Figure 12. Radar terrain image obtained with the use of INS data; yellow line—the north edge of the taxiway determined using an INS-based image.

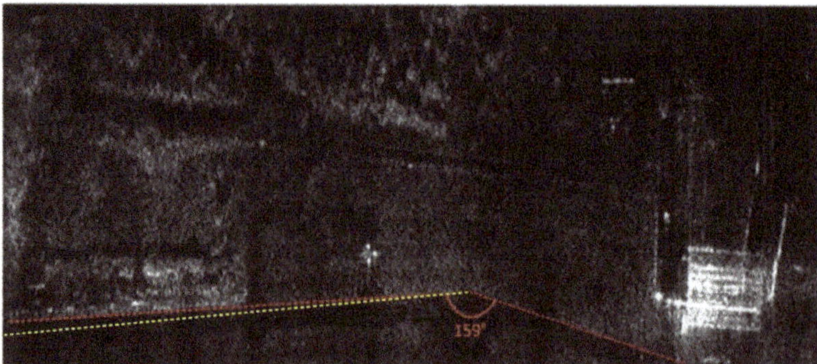

Figure 13. Radar terrain image obtained with the use of INS/GPS data; yellow line—the north edge of the taxiway determined using INS-based image, red line—the north edge of the taxiway determined using an INS/GPS-based image.

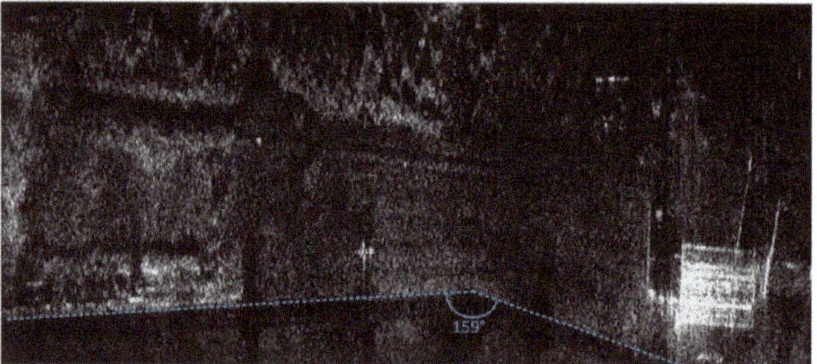

Figure 14. Radar terrain image obtained with the use of MINS data.

The total flight duration along the scanned area was approximately 29 s and was limited by the Visual Line of Sight (VLOS) UAV rules. Along the azimuth direction, the scanned swath had a length of 680 m. During imaging, the UAV moved along a semilinear trajectory. Its rectilinear shape was imposed by the radar aperture synthesis algorithm implemented in our system and described in [38]. In general, the applied SAR procedure allows for a nonrectilinear UAV motion as its displacements from an assumed linear trajectory are compensated using navigation corrections and the MOCO procedure. However, during large maneuvers (e.g., rapid turns) the spatial resolution of the image can be degraded. Therefore, our system is typically used for SAR imaging during flights along almost rectilinear trajectories. In order to present data representative for a normal operational use of the system, only straight flight trajectories were considered.

The geometric distortions of images were determined by measuring the angle between northern edges of the taxiway, whose true value is 159°. In Figure 12 (INS-based MOCO), the edge is marked in yellow, and the measured angle is 157°. The two-degree discrepancy is related to the slowly growing INS positioning errors with a final value of 11.7 m. In the case of INS/GPS-based MOCO, the taxiway edge is marked in red (Figure 13), while the measured angle is concise with the true one. In this image, the geometric distortions are reduced thanks to the INS error correction with the GPS receiver and the Kalman filter. In addition, in the case of the proposed MINS system, the taxiway angle has a proper value. In this system, the positioning error is kept low by the instance switching mechanism.

Vertical white lines, presented in Figure 15, mark the ranges of radar measurements with INS overlaps. As can be seen, these overlaps do not deteriorate the image quality or geometric conciseness of the image.

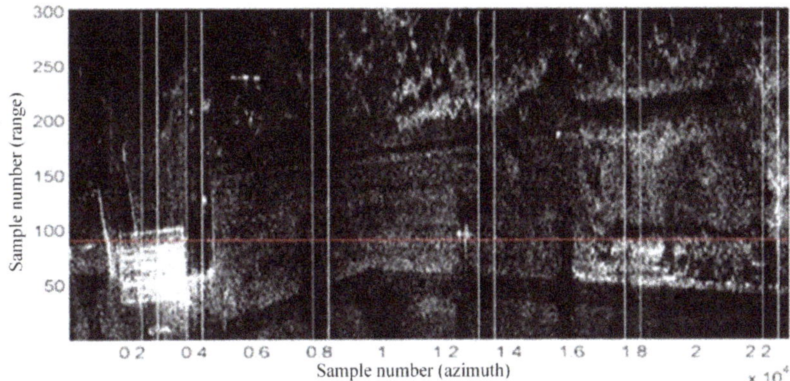

Figure 15. Radar terrain image obtained using MINS-based MOCO. Overlap bounds are marked white, whereas the red line is an image line chosen to determine SR, PSLR, and ISLR.

Radar terrain images obtained using the three considered MOCO methods (INS, INS/GPS, and MINS) were also compared using quality indicators such as image contrast (IC), entropy (E), spatial resolution (SR), PSLR, and ISLR [25,29,39–41]. The results are presented in Table 2.

Table 2. Parameters of the quality of the selected synthetic aperture radar (SAR) image.

Source of Corrections	Parameter of Image Quality				
	IC (↑)	E (↓)	SR [m] (↓)	PSLR [dB] (↓)	ISLR [dB] (↓)
none	4.06	14.43	0.913	−5.17	−5.85
INS	8.32	13.83	0.304	−8.45	−10.50
INS/GPS	7.56	13.94	0.119	−2.99	1.52
MINS	8.01	13.84	0.278	−8.57	−10.68

The image obtained using MINS-based MOCO has a better (higher) contrast than the image calculated without the navigation correction (IC increases from 4.06 to 8.01). This result is also better than the contrast of the INS/GPS-based image (IC = 7.56) and comparable to the contrast of the INS-based image (IC = 8.32). The improvement is also visible in the image entropy. Due to the usage of MINS-based MOCO, a reduction (improvement) of entropy was achieved in relation to the image without MOCO (E decreases from 14.43 to 13.84). The obtained result is also better than the entropy of the INS/GPS-based image (E = 13.94) and similar to the result obtained with a traditional INS (E = 13.83). The SR, PSRL, and ISLR parameters are determined based on point-like objects in the images. For this purpose, during preparations to the experiment a set of corner reflectors was placed in the central part of the imaged area and arranged in the shape of an arrow. In Figures 15 and 16, the horizontal red line marks the image line running through a corner reflector located in the lower arm of the arrow, for which SR, PSLR, and ISLR were calculated. The normalized amplitude of image pixels measured along this line is presented in Figure 17.

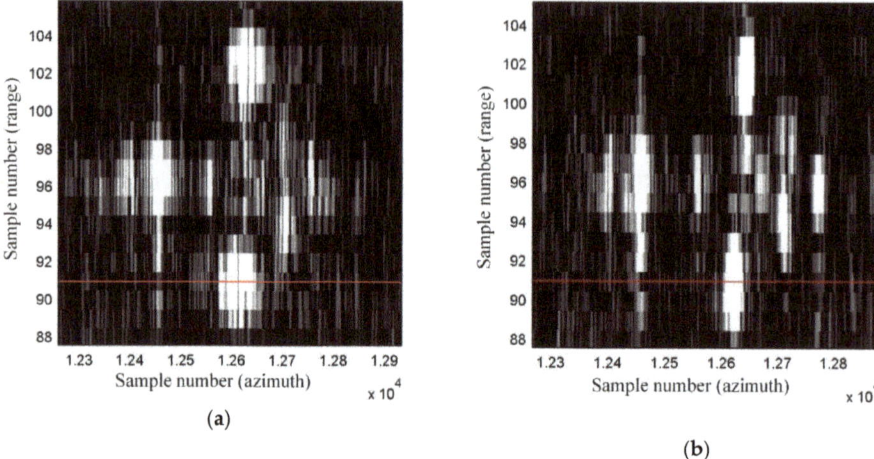

Figure 16. Image of the arrow made of corner reflectors: (**a**) with the INS/GPS-based MOCO, (**b**) with the MINS-based MOCO; red line—an analyzed row.

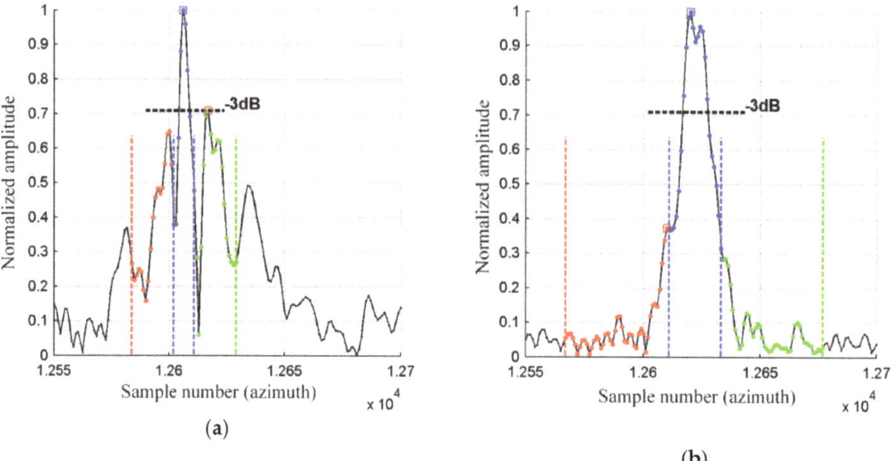

Figure 17. Normalized amplitude of the selected corner reflector image: (**a**) with the INS/GPS-based MOCO, (**b**) with the MINS-based MOCO.

Compared to the image obtained using the INS/GPS system, in the case of the MINS-based image the amplitude of sidelobes is lower (which is an advantage), while the main lobe is apparently wider. As a result, the SR of the MINS-based image (SR = 0.278 m) is slightly better than in the INS-based image (SR = 0.304) and theoretically worse than the result obtained with INS/GPS-based MOCO (SR = 0.119). In the case of the INS/GPS method, however, it should be noted that high-level sidelobes deteriorate the practical resolution, which can be seen in Figure 16. Visualization of the normalized amplitude in 2D is shown in Figures 18 and 19.

MINS-based MOCO allowed the best (lowest) values of PSLR and ISLR to be obtained among three analyzed navigation systems (PSLR = −8.57 dB, ISLR = −10.68 dB). The improvement with respect to the INS/GPS-based image (PSLR = 2.99 dB, ISLR = −1.52 dB) results from the smoothness of the navigation data calculated by the MINS system. The improvement with respect to the INS-based image (PSLR = −8.45 dB, ISLR = −10.50 dB) results from lower navigation errors, which led to a better fitting of the assumed reference function in the azimuth compression of the SAR algorithm. It should be noted that after evaluation of MOCO efficiency, the speckle noise of the SAR image can be reduced using techniques presented in [42,43].

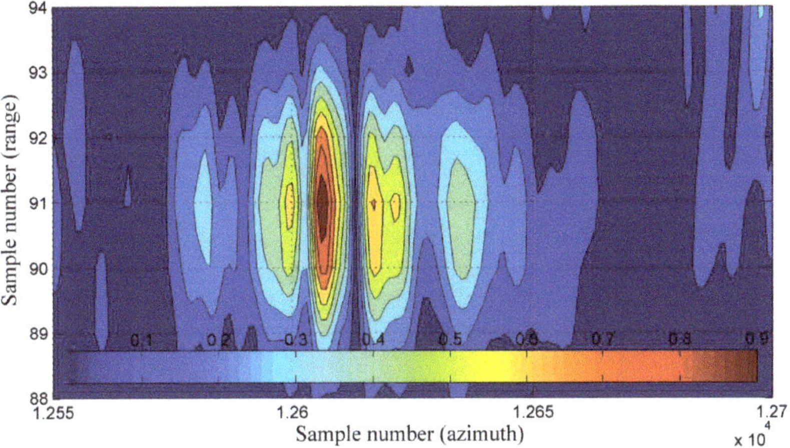

Figure 18. Normalized amplitude of corner reflector with INS/GPS-based MOCO.

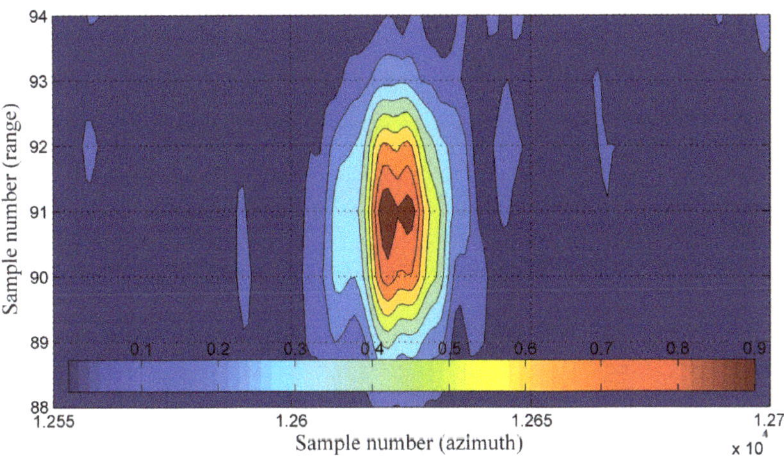

Figure 19. Normalized amplitude of corner reflector with MINS-based MOCO.

Similar tests were carried out for other UAV flights. Figure 20 shows the trajectory obtained using the MINS system for another analyzed flight (called here flight no 2). The time duration of this trajectory was approximately 24 s. Table 3 contains the values of quality indicators calculated for images obtained during this flight and using three MOCO methods.

The results show that among all three verified MOCO methods, the worst values of the considered quality indicators were obtained using the INS/GPS method. Better results are ensured using the INS-based MOCO, however, it should be noted that this image has geometric distortions. The best results were obtained using the MINS system. The corresponding image has the lowest values of SR, PSLR, and ISLR. The proposed method also allowed the best contrast and entropy to be obtained. The results obtained for two different flights led to the same conclusions. The comparison of the image of the arrow and corresponding signal amplitudes obtained for the classic INS and for the proposed MINS system are presented in Figures 21 and 22.

Figure 20. MINS trajectories for flight no 2 (multicolor r line) and INS trajectory (blue line).

Table 3. Parameters of the quality of the selected SAR image.

Source of Corrections	Parameter of Image Quality				
	IC (↑)	E (↓)	SR [m] (↓)	PSLR [dB] (↓)	ISLR [dB] (↓)
none	4.69	13.83	–	–	–
INS	6.00	13.01	0.195	−4.12	−2.58
INS/GPS	5.78	12.96	0.366	−1.92	−0.40
MINS	6.50	12.91	0.187	−6.80	−5.46

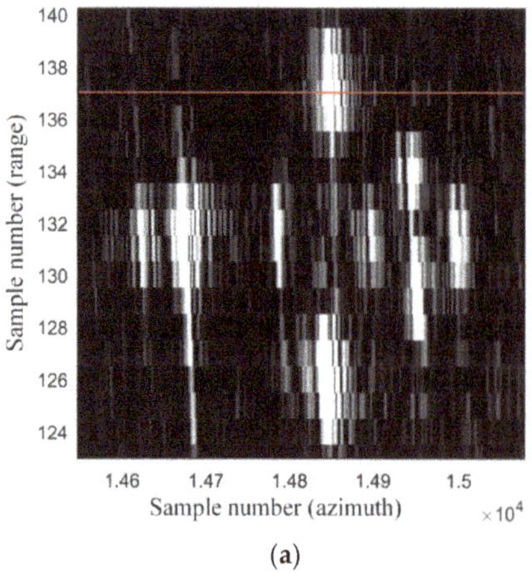

(**a**)

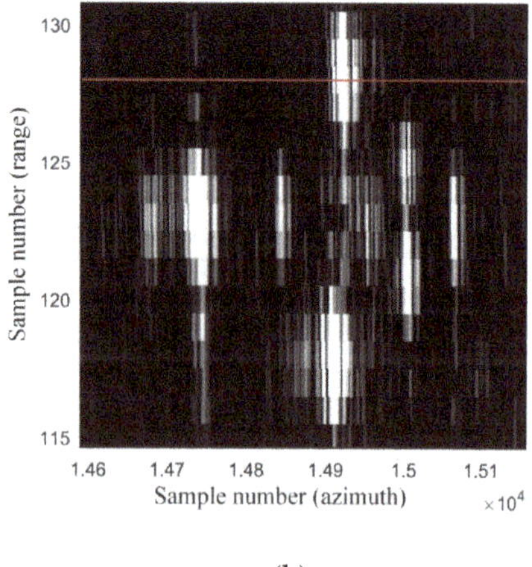

(**b**)

Figure 21. Image of the arrow made of corner reflectors for flight no 2: (**a**) with the INS-based MOCO, (**b**) with the MINS-based MOCO; red line—an analyzed row.

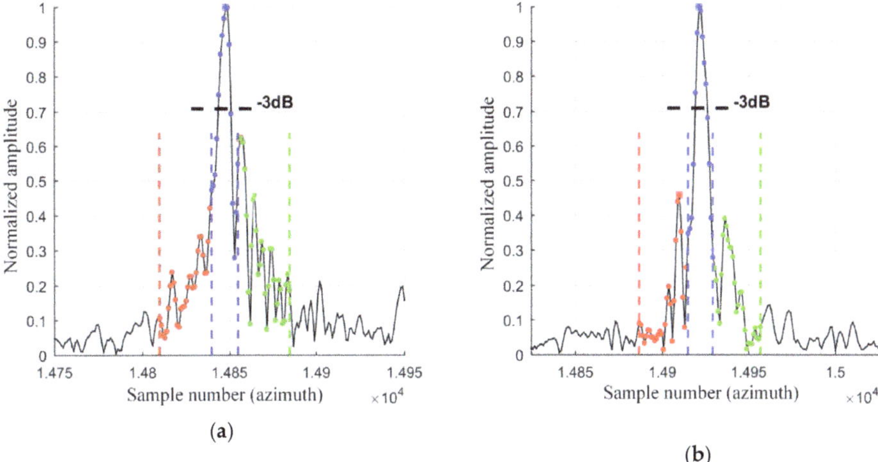

Figure 22. Normalized amplitude of the selected corner reflector image: (**a**) with the INS-based MOCO, (**b**) with the MINS-based MOCO.

5. Conclusions

The article discusses a method of calculating UAV navigation elements (position, velocity, and attitude) using the proposed multi-instance INS (MINS). The motivation for its development was the analysis of the pros and cons of the INS- and INS/GPS-based MOCO methods used in SAR algorithms and the search for a new method which profits from the advantages of existing MOCO procedures but avoids their drawbacks.

The results obtained from the proposed MINS system are smooth, similar to the classical INS output. Moreover, thanks to the initialization of new INS instances, errors in MINS are kept at a low level, similar to the INS/GPS systems.

The presented MINS system was tested using real measurement navigation and radar data. Based on the obtained results, it can be concluded that the radar images calculated using MINS data combine the positive features of INS- and INS/GPS-based images. Thanks to the MINS error control, a significant reduction of geometric distortions was obtained, analogous to the results achieved with the use of the INS/GPS system. On the other hand, thanks to the INS instance switching procedure, it is possible to avoid abrupt changes of the position and velocity data, which is a drawback of the INS/GPS system. Moreover, the proposed mechanism allows for high contrast and entropy to be maintained and for the improvement of the PSRL and ISLR in relation to the INS-based images.

The proposed MINS system is based on chosen ideas presented in [15,19,21]. The authors added an INS instance switching algorithm that uses overlaps. Thanks to this procedure, each synthetic aperture uses corrections based on only one INS instance; therefore, there are no abrupt changes in the navigation elements which are characteristic of the INS/GPS integrated navigation system. As a result, the duration of the measurement session is not limited (contrary to the system presented in [21]) and, at the same time, the errors of measured flight trajectory are periodically corrected.

Further works related to the MINS algorithm are possible, and they should concern the use of a more complex filter than presented in [34]. By enriching the dynamics model with additional state variables, such as accelerometer and gyroscope biases and scale factor errors, as well as changing the structure of the loosely integrated INS/GPS system into a tightly integrated one, where GPS satellite pseudoranges and range rates are used instead of the position and velocity, it would be possible to improve the accuracy of the INS/GPS system. This would also positively influence the accuracy of the MINS system and consequently the quality of the radar terrain images.

The MINS-based MOCO can be an alternative to autofocus methods, especially in real-time systems, which aim to quickly obtain high-quality images. The advantage of the proposed method is the fact that the image synthesis is performed without iterations, and navigation corrections are calculated in a parallelly-working subsystem, independently of the SAR system calculations.

Author Contributions: Conceptualization, M.L., P.K.; methodology, M.L., P.K.; software and validation, M.L., P.K.; investigation, M.L., P.K.; writing-original draft preparation, M.L.; writing-review and editing, P.K. Both authors have read and agreed to the published version of the manuscript.

Funding: This project was supported by the National Centre for Research and Development, Poland, in the frame of the Applied Research Program under Research Project PBS/B3/15/2012.

Acknowledgments: We would like to thank our colleagues from MUT and WB Electronics S.A. who participated in data collection.

Conflicts of Interest: The authors declare no conflict of interest. The funders had no role in the design of the study; in the collection, analyses, or interpretation of data; in the writing of the manuscript, or in the decision to publish the results.

References

1. Moreira, A.; Prats-Iraola, P.; Younis, M.; Kreiger, G.; Hajnsek, I.; Papathanassiou, K. A tutorial on synthetic aperture radar. *IEEE Geosci. Remote Sens. Mag.* **2013**, *1*, 6–43. [CrossRef]
2. Blacknell, D.; Freeman, A.; Quegan, S.; Ward, I.; Finley, I.; Oliver, C.; Wood, J. Geometric accuracy in airborne SAR images. *IEEE Trans. Aerosp. Electron. Syst.* **1989**, *25*, 241–258. [CrossRef]
3. Fornaro, G. Trajectory deviations in airborne SAR: Analysis and compensation. *IEEE Trans. Aerosp. Electron. Syst.* **1999**, *35*, 997–1009. [CrossRef]
4. Kirk, J.C. Motion compensation for synthetic aperture radar. *IEEE Trans. Aerosp. Electron. Syst.* **1975**, *11*, 338–348. [CrossRef]
5. Li, J.; Wang, P.; Li, C.; Chen, J.; Yang, W. Precise estimation of flight path for airborne SAR motion compensation. *IEEE Geosci. Remote Sens. Symp.* **2014**, 1121–1123. [CrossRef]
6. Li, Y.; Liu, C.; Wang, Y.; Wang, Q. A robust motion error estimation method based on raw data. *IEEE Trans. Geosci. Remote Sens.* **2012**, *50*, 2780–2790. [CrossRef]
7. Purchla, M.; Malanowski, M. Simple motion compensation algorithm for unfocused synthetic aperture radar. In Proceedings of the SPIE: Photonics Applications in Astronomy, Communications, Research and High Energy Physics Experiments II, Wilga, Poland, 22 July 2004. [CrossRef]
8. Samczyński, P.; Malanowski, M.; Gromek, D.; Gromek, A.; Kulpa, K.; Krzonkalla, J.; Mordzonek, M.; Nowakowski, M. Effective SAR image creation using low cost INS/GPS. In Proceedings of the 15th International Radar Symposium (IRS), Gdańsk, Poland, 16–18 June 2014. [CrossRef]
9. Zhang, C.B.; Yeo, Y.S.; Kooi, P.S.; Deng, L.P.; Tsao, C. SAR real time motion compensation: Average cancellation method for aircraft motion error extraction. In Proceedings of the ICCS, Singapore, 14–18 November 1994. [CrossRef]
10. Fornaro, G.; Franceschetti, G.; Perna, S. Motion compensation errors: Effects on the accuracy of airborne SAR images. *IEEE Trans. Aerosp. Electron. Syst.* **2005**, *41*, 1338–1352. [CrossRef]
11. Guo, H.; Li, Y.; Qu, Q.; Lie, P. Studying atmospheric turbulence effects on aircraft motion for airborne SAR motion compensation requirements. In Proceedings of the IEEE International Conference on Imaging Systems and Techniques, Manchester, UK, 16–17 July 2012. [CrossRef]
12. Moreira, J.R. A new method of aircraft motion error extraction from radar raw data for real time motion compensation. *IEEE Trans. Geosci. Remote Sens.* **1990**, *28*. [CrossRef]
13. Sun, L.; Yao, D.; Tian, W.; Zeng, T. Research on antenna stabilization technology of micro SAR system. In Proceedings of the IET International Radar Conference, Xi'an China, 14–16 April 2013. [CrossRef]
14. Oliver, C.; Quegan, S. *Understanding Synthetic Aperture Radar Images*; SciTech Publishing Inc.: Raleigh, NC, USA, 2004.
15. Kennedy, T. Strapdown inertial measurement units for motion compensation for synthetic aperture radars. *IEEE Aerosp. Electron. Syst. Mag.* **1988**, *3*, 32–35. [CrossRef]

16. Labowski, M.; Kaniewski, P.; Serafin, P. Inertial navigation system for radar terrain imaging. In Proceedings of the IEEE/ION Position Location and Navigation Symposium (PLANS), Savannah, GA, USA, 11–14 April 2016; pp. 942–948. [CrossRef]
17. Titterton, D.H.; Weston, J.L. *Strapdown Inertial Navigation Technology*; The Institution of Electrical Engineers: London, UK, 2004.
18. Groves, P.D. *Principles of GNSS, Inertial and Multisensor Integrated Navigation Systems*, 2nd ed.; Artech House: Boston, MA, USA, 2012.
19. Fuxiang, C.; Zheng, B. Analysis and simulation of GPS/SINU integrated system for airborne SAR motion compensation. In Proceedings of the International Conference on Radar, Beijing, China, 15–18 October 2001. [CrossRef]
20. Gong, X.; Fang, J. Analyses and comparisons of some nonlinear Kalman filters in POS for airborne SAR motion compensation. In Proceedings of the International conference on Mechatronics and Automation, Harbin, China, 5–8 August 2007. [CrossRef]
21. Fang, J.; Gong, X. Predictive iterated Kalman filter for INS/GPS integration and its application to SAR motion compensation. *IEEE Trans. Instrum. Meas.* **2010**, *59*, 909–915. [CrossRef]
22. Fuxiang, C.; Zheng, B.; Jianping, Y. Motion compensation for airborne SAR. In Proceedings of the 16th World Computer International Conference on Signal Processing Proceeding, Beijing, China, 21–25 August 2000; pp. 1864–1867. [CrossRef]
23. Chen, Y.; Li, G.; Zhang, Q.; Zhang, Q.; Xia, X. Motion compensation for airborne SAR via parametric sparse representation. *IEEE Trans. Geosci. Remote Sens.* **2017**, *55*, 551–562. [CrossRef]
24. Cumming, I.; Wong, F. *Digital Processing of Synthetic Aperture Radar Data—Algorithms and Implementation*; Artech House: London, UK, 2005.
25. Martorella, M.; Berizzi, F.; Haywood, B. Contrast maximization based technique for 2-D ISAR autofocusing. *IEEE Proc. Radar Sonar Navig.* **2005**, *152*, 253–262. [CrossRef]
26. Samczynski, P.; Kulpa, K. Concept of the coherent autofocus map-drift technique. In Proceedings of the 2006 International Radar Symposium, Krakow, Poland, 24–26 May 2006; pp. 1–4. [CrossRef]
27. Samczynski, P.; Kulpa, K. Non iterative map-drift technique. In Proceedings of the 2008 International Radar Symposium, Krakow, Poland, 2–5 September 2008; pp. 76–81. [CrossRef]
28. Wahl, D.; Eichel, P.; Ghighlia, D.; Jakowatz, C. Phase gradient autofocus—A robust tool for high resolution SAR phase correction. *IEEE Trans. Aerosp. Electron. Syst.* **1994**, *30*, 827–835. [CrossRef]
29. Wang, J.; Liu, X. SAR Minimum-entropy autofocus using an adaptive-order polynomial model. *IEEE Geosci. Remote Sens. Lett.* **2006**, *3*, 512–516. [CrossRef]
30. Xie, P.; Zhang, M.; Zhang, L.; Wang, G. Residual motion error correction with backprojection multisquint algorithm for airborne synthetic aperture radar interferometry. *Sensors* **2019**, *19*, 2342. [CrossRef] [PubMed]
31. Li, N.; Niu, S.; Guo, Z.; Liu, Y.; Chen, J. Raw data-based motion compensation for high-resolution sliding spotlight synthetic aperture radar. *Sensors* **2018**, *18*, 842. [CrossRef]
32. Bezvesilniy, O.; Gorovy, I.; Vavriv, D. Estimation of phase errors in SAR data by local-quadratic map-drift autofocus. In Proceedings of the 19th International Radar Symposium, Warsaw, Poland, 23–25 May 2012; pp. 376–381. [CrossRef]
33. Huang, Y.; Liu, F.; Chen, Z.; Li, J.; Hong, W. An improved map-drift algorithm for unmanned aerial vehicle SAR imaging. *IEEE Geosci. Remote Sens. Lett.* **2020**, 1–5. [CrossRef]
34. Labowski, M.; Kaniewski, P. Motion compensation for radar terrain imaging based on INS/GPS system. *Sensors* **2019**, *19*, 3895. [CrossRef]
35. Kulakova, V.I.; Nozdrin, S.A.; Sokharev, A.Y. Micronavigation system to support a radar with synthetic aperture aboard a small UAV. *Gyroscopy Navig.* **2019**, *10*, 245–255. [CrossRef]
36. Labowski, M.; Kaniewski, P.; Konatowski, S. Estimation of flight path deviations for SAR radar installed on UAV. *Metrol. Meas. Syst.* **2016**, *23*, 383–391. [CrossRef]
37. Kaniewski, P.; Komorniczak, W.; Lesnik, C.; Cyrek, J.; Susek, W.; Serafin, P.; Labowski, M. S-band and Ku-band SAR system development for UAV-based applications. *Metrol. Meas. Syst.* **2019**, *26*, 53–64. [CrossRef]
38. Kaniewski, P.; Lesnik, C.; Serafin, P.; Labowski, M. Chosen results of flight tests of WATSAR system. In Proceedings of the 17th International Radar Symposium (IRS), Krakow, Poland, 10–12 May 2016; pp. 1–5. [CrossRef]

39. Wang, B. Range-Doppler processing on SAR images. In *Digital Signal Processing Techniques and Applications in Radar Image Processing*; John Willey & Sons: Hoboken, NJ, USA, 2008; Volume 1, pp. 226–284.
40. Lu, X.; Sun, H. Parameter assessment for SAR image quality evaluation system. In Proceedings of the 1st Asian and Pacific Conference on Synthetic Aperture Radar, Huangshan, China, 5–9 November 2007. [CrossRef]
41. Zhang, H.; Li, Y.; Su, Y. SAR image quality assessment using coherent correlation function. In Proceedings of the International Congress on Image and Signal Processing, Chongqing, China, 16–18 October 2012. [CrossRef]
42. Trouve, E.; Chambenoit, Y.; Classeau, N.; Bolon, P. Statistical and operational performance assessment of multitemporal SAR image filtering. *IEEE Trans. Geosci. Remote Sens.* **2003**, *41*, 2519–2530. [CrossRef]
43. Ouahabi, A. A review of wavelet denoising in medical imaging. In Proceedings of the 8th International Workshop on Systems, Signal Processing and Their Applications (IEEE/WoSSPA), Algiers, Algeria, 12–15 May 2013; pp. 19–26. [CrossRef]

Article

A Robust Algorithm Based on Phase Congruency for Optical and SAR Image Registration in Suburban Areas

Lina Wang [1,2], Mingchao Sun [1], Jinghong Liu [1,2,*], Lihua Cao [1,2] and Guoqing Ma [3]

[1] Changchun Institute of Optics, Fine Mechanics and Physics, Chinese Academy of Sciences, Changchun 130033, China; wanglina18@mails.ucas.ac.cn (L.W.); sunmingchao@ciomp.ac.cn (M.S.); caolh@ciomp.ac.cn (L.C.)
[2] University of Chinese Academy of Sciences, Beijing 100049, China
[3] College of Earth Exploration Science and Technology, Jilin University, Changchun 130012, China; maguoqing@jlu.edu.cn
* Correspondence: liujinghong@ciomp.ac.cn

Received: 3 September 2020; Accepted: 10 October 2020; Published: 13 October 2020

Abstract: Automatic registration of optical and synthetic aperture radar (SAR) images is a challenging task due to the influence of SAR speckle noise and nonlinear radiometric differences. This study proposes a robust algorithm based on phase congruency to register optical and SAR images (ROS-PC). It consists of a uniform Harris feature detection method based on multi-moment of the phase congruency map (UMPC-Harris) and a local feature descriptor based on the histogram of phase congruency orientation on multi-scale max amplitude index maps (HOSMI). The UMPC-Harris detects corners and edge points based on a voting strategy, the multi-moment of phase congruency maps, and an overlapping block strategy, which is used to detect stable and uniformly distributed keypoints. Subsequently, HOSMI is derived for a keypoint by utilizing the histogram of phase congruency orientation on multi-scale max amplitude index maps, which effectively increases the discriminability and robustness of the final descriptor. Finally, experimental results obtained using simulated images show that the UMPC-Harris detector has a superior repeatability rate. The image registration results obtained on test images show that the ROS-PC is robust against SAR speckle noise and nonlinear radiometric differences. The ROS-PC can tolerate some rotational and scale changes.

Keywords: optical and synthetic aperture radar (SAR); image registration; phase congruency (PC); radiometric difference

---

1. Introduction

The rapid development of sensor technology provided multiple remote sensing images for the observation of the Earth. Optical images ensure facilitated interpretation and are similar to human vision; however, they are affected easily by the weather. The synthetic aperture radar (SAR) is an active microwave imaging system that effectively compensates for the shortcomings of optical imaging systems and operates irrespective of the time of day and weather conditions. Optical and SAR images can be used together to form complementary information, which has important application value, such as image fusion [1,2], pattern recognition [3], and change detection [4,5]. Image registration is a preliminary work of these applications. It refers to aligning two or more images of the same scene acquired by different times, viewpoints, or sensors. Registration accuracy seriously affects these applications. Optical and SAR registration is still a challenging task owing to the speckle noise of SAR and the large radiation differences between optical and SAR images [6,7].

Generally, image registration methods can be roughly divided into two categories—namely, area-based methods and feature-based methods [8]. In area-based methods, which are also known as intensity-based methods, first, a template is defined, and subsequently, the geometric transformation model is estimated by optimizing a similarity measurement between the SAR and optical images, such as mutual information [9,10], normalized cross-correlation [11], and cross-cumulative residual entropy [12]. Area-based methods deliver high accuracy, as the entire intensity information is utilized. However, due to its high computational loads and sensitivity to the geometry and radiation differences, they are limited in their applications of optical and SAR image registration.

Feature-based methods usually first extract features such as points [13], edges [14,15], and contours [16] from input images. Then, a distinctive feature descriptor is designed. Finally, the transformation model is estimated by establishing the corresponding relationship between the features. Feature-based methods are recommended for optical and SAR image registration because they process images with their significant features rather than all intensity information, thereby achieving high precision and robustness to geometry and radiation differences. Feature-based methods are mainly composed of three steps: feature detection, feature description, and feature matching.

The most representative feature-based method is the scale-invariant feature transform (SIFT), owing to its efficient performance and invariance to scale and rotation [17]. Subsequently, a variety of improved methods have been reported. To improve matching efficiency, principal component analysis (PCA) is applied to reduce the dimension of the descriptor [18]. To reduce time, a speeded-up robust feature uses the determinant value of the Hessian matrix to detect feature points and an integral graph to accelerate the operation [19]. Affine SIFT simulates the parameters of affine transformation to achieve full affine invariance and considerably expands the scope of application of image registration [20]. A uniform robust SIFT is proposed to extract uniformly distributed and robust feature points [21]. Adaptive binning SIFT is proposed to increase the particularity and robustness of descriptors [22].

However, speckle noise in SAR images and the intensity difference between optical and SAR images make it difficult to obtain good results when these methods are applied directly to image registration. Numerous scholars have proposed improved methods for optical and SAR image registration. An improved SIFT is realized using optical and SAR satellite image registration by exploring their spatial relationship [23]. An automatic SAR and optical image registration method, from rough to accurate, is proposed with the use of SIFT features [24]. A novel gradient definition, yielding an orientation and a magnitude that is robust to speckle noise, is specifically dedicated to SAR images [25]. Further, to overcome the difference in image intensity between remote image pairs and increase the number of correct correspondences, a new gradient definition and an enhanced feature matching method by combining the position, scale, and orientation of each keypoint are proposed [26]. The gradients in the descriptor are computed by a multiscale Gabor odd filter (GOF)-based ratio operator, and the proposed GOF-based descriptor is formed for the SIFT features [27]. Xiang et al. proposed a robust SIFT-like algorithm (OS-SIFT) to register high-resolution optical and SAR images, in which the consistent gradient magnitudes in the SAR and optical images are computed using a multi-scale ratio of exponentially weighted averages (ROEWA) operator and a multi-scale Sobel operator, respectively [28].

Although numerous methods have achieved improvements in gradient redefinition and descriptor construction when encountering optical and SAR images with large nonlinear radiation differences, the matching performance of feature descriptors based on gradient information is not ideal, and there are still many mismatches. Recently, various registration methods based on phase congruency (PC) information have been widely used in multi-sensor images, because PC has been confirmed as an illumination and contrast invariant measure of the features [29–31].

An image descriptor, namely, the histogram of oriented phase (HOP) based on the PC concept and PCA is present, and it is more robust to image scale variations and contrast and illumination changes [32]. Ye et al. proposed a novel feature descriptor named the histogram of oriented phase congruency (HOPC) for multimodal image registration [33]. Further, they proposed a local phase-based invariant feature for remote sensing image matching, which consists of a feature detector called minimum moment of PC (MMPC)-Lap and a feature descriptor called the local HOPC (LHOPC) [34]. Similar to gradients,

PC also reflects the significance of the features of local image regions. Chen et al. proposed an optical and SAR image registration method by combining a new Gaussian-Gamma-shaped bi-windows-based gradient operator and the histogram of oriented gradient pattern [35]. To address large geometric differences and speckle noise in SAR images, a novel optical-to-SAR image registration algorithm was proposed using a new structural descriptor [36]. A dense descriptor named the histograms of oriented magnitude and phase congruency was proposed to register multi-sensor images. It is based on the combination of the magnitude and PC information of local regions, and successfully captures the common features of images with nonlinear radiation changes [37]. A novel image registration method, which combines nonlinear diffusion and PC structural descriptors, has been proposed for the registration of SAR and optical images [38]. To overcome nonlinear radiation distortions, Li et al. [39] proposed a radiation invariant feature transform (RIFT) algorithm to register multi-sensor images, including optical and SAR images. The RIFT uses PC instead of image intensity for feature point detection and it proposes a maximum index map (MIM) for feature description. Further, the RIFT not only largely improves the stability of feature detection but also overcomes the limitation of gradient information for feature description.

Although a number of PC-based image registration methods have been proposed in the past few years, there are limitations that cannot be ignored when these methods are applied to optical and SAR image registration with large radiation differences. These limitations are listed below.

- Several methods detect keypoints directly from a PC map (PCM) or the moment of the PCM for feature matching. However, because of SAR speckle noise, some unreasonable points are detected in the SAR image; further, because of significant nonlinear intensity differences, the feature detection result of one image generally has no corresponding feature points in the other image. Several classical methods are tested using a pair of optical and SAR images, as shown in Figure 1. These limitations lead to the low repeatability of the feature point, which is not conducive to feature matching.
- The extracted points are not uniformly distributed. When calculating the PC of a whole image, the noise threshold T is estimated using the Rayleigh distribution mode, which is a fixed value. This leads to a noise threshold larger than the actual noise in the dark region. The feature information is drowned by the noise. As shown in Figure 1, the features are always concentrated in the bright region, especially in the SAR image. The nonuniform distribution of feature points leads to limited registration accuracy on large images or high-contrast images.
- Because of the different imaging mechanisms for optical and SAR sensors, the acquired images have different expressions for the same objects, thereby resulting in large radiation differences between image pairs. Such nonlinear radiation differences reduce the correlation between corresponding points, which often leads to difficulties in feature description.

In this paper, we address the above limitations by developing a robust optical and SAR image registration method based on PC (ROS-PC). The proposed method mainly contains the following two works.

First, a uniform Harris feature detection method based on multi-moment of the PCM (UMPC-Harris) is proposed. In the UMPC-Harris, we take the corners and edge points as keypoints. The edge structure feature has a high similarity and better resistance to radiation difference between optical and SAR images [30,36,39], thus, extracting feature points on the edge can ensure enough number of features and robustness to radiation difference. Besides, corner features can increase the number of homologous points. Therefore, the multi-moment of the PCM is constructed by using maximum and minimum moment maps. Harris operator is used on the multi-moment to detect corners and edge points. Finally, the overlapping block and voting strategy are introduced to detect uniformly distributed and reliable keypoints.

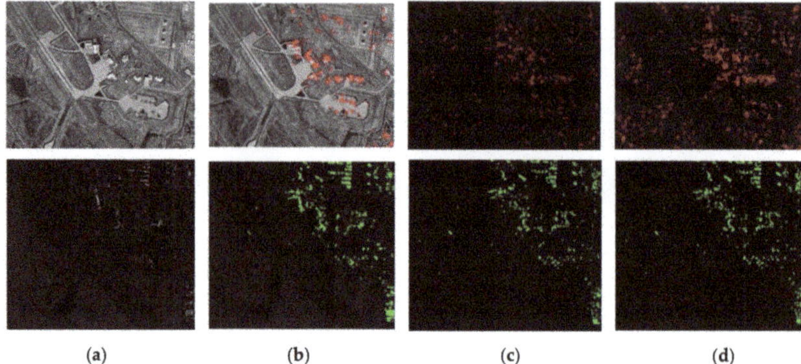

Figure 1. Comparison results of keypoints detection in optical and synthetic aperture radar (SAR) images (top row depicts the optical image, and bottom row depicts SAR images). (**a**) Original images; (**b**) Harris on the original images; (**c**) Harris on the minimum moment of the phase congruency map (PCM); (**d**) Harris on the maximum moment of the PCM.

Second, since PC is not suitable for constructing descriptors directly, the feature descriptor is derived for a keypoint by utilizing the histogram of phase congruency orientation on multi-scale max amplitude index maps (HOSMI). The proposed HOSMI descriptor is utilizing the MIM instead of the PCM because it is more robust to intensity radiation distortions than the PCM [39]. Furthermore, in remote sensing images, many salient features usually appear in different scales [38]. Therefore, we construct the phase congruency orientation maps and max amplitude index maps, respectively. In the local region of each keypoint, the histograms of phase congruency orientation on multi-scale max amplitude index maps are calculated. Finally, the descriptor is constructed by combining the feature vectors of all patched in order. Compared with state-of-the-art, the main contribution of this study can be summarized as follows:

- The UMPC-Harris feature detection method is proposed based on the multi-moment of the PCM, a voting strategy, and an overlapping block strategy. The detector can obtain enough reliable and uniformly distributed feature points.
- The HOSMI feature description method is proposed based on the histograms of phase congruency orientation on multi-scale max amplitude index maps. The descriptor is more robust against nonlinear radiation variation and speckle noise.

The rest of this paper is organized as follows: Section 2 starts with a review of PC theory, and followingly introduces the ROS-PC in detail, including the UMPC-Harris feature detector and HOSMI feature descriptor. In Section 3, through several experiments, the repeatability rate of keypoints by UMPC-Harris, the robustness of ROS-PC, and the sensitivity of ROS-PC to scale and rotation changes are evaluated and discussed. Finally, the conclusions are provided in Section 4.

2. Methodology

The PC has been confirmed to be robust to nonlinear radiometric differences, which can capture the common features between multi-sensor images [37,39,40]. The ROS-PC method is based on PC. This section first reviews the PC theory briefly and then presents the design processing of the UMPC-Harris detector and HOSMI descriptor.

2.1. Review of PC Theory

According to Kovesi's approach, PC can be computed by convolving an image with a log-Gabor filter (LGF) to extract local phase information. The LGF is efficient for detecting features over multiple scales and orientations. In the frequency domain, LGF is defined as:

$$LGF(\omega) = \exp\left(\frac{-(\log(\omega/\omega_0))^2}{2(\log(\kappa/\omega_0))}\right), \tag{1}$$

where ω_0 is the central frequency of the filter, κ is the related-width parameter of the filter that varies with ω_0, which ensures that κ/ω_0 is a constant.

The filter is transformed from the frequency to the spatial domain using an inverse Fourier transform. In the spatial domain, the 2-D LGF is represented as:

$$LGF(x,y) = LGF^{even}_{s,o}(x,y) + i \times LGF^{odd}_{s,o}(x,y). \tag{2}$$

Considering the coordinates of an input image $I(x,y)$, the convolution responses $e_{s,o}(x,y)$ and $o_{s,o}(x,y)$ at scale s and orientation o are obtained, and then, the convolution results of even and odd symmetric wavelets form the response arrays as follows:

$$[e_{s,o}(x,y), o_{s,o}(x,y)] = \left[I(x,y) * LGF^{even}_{s,o}, I(x,y) * LGF^{odd}_{s,o}\right], \tag{3}$$

where $LGF^{even}_{s,o}$ and $LGF^{odd}_{s,o}$ refer to the even-symmetric (cosine) and odd-symmetric (sine) wavelets of the LGF at scale s and orientation o, respectively. Further, $e_{s,o}(x,y)$ and $o_{s,o}(x,y)$ are the convolution responses of $LGF^{even}_{s,o}$ and $LGF^{odd}_{s,o}$ at scale s and orientation o, respectively.

The corresponding amplitude $A_{s,o}(x,y)$ and phase $\varphi_{s,o}(x,y)$ at scale s and orientation o are given by:

$$A_{s,o}(x,y) = \sqrt{e_{s,o}(x,y)^2 + o_{s,o}(x,y)^2}, \tag{4}$$

$$\varphi_{s,o}(x,y) = \arctan(o_{s,o}(x,y),\ e_{s,o}(x,y)). \tag{5}$$

Considering the negative effect of image noise, the improved PC (called PC_2) and the phase deviation function are, respectively, defined as [41]:

$$PC_2 = \frac{\sum_o \sum_s W_o(x,y) \lfloor A_{s,o}(x,y) \Delta\varphi_{s,o}(x,y) - T \rfloor}{\sum_o \sum_s A_{s,o}(x,y) + \varepsilon}, \tag{6}$$

$$\Delta\varphi_{s,o}(x,y) = \cos(\varphi_{s,o}(x,y) - \overline{\varphi}_{s,o}(x,y)) - \left|\sin(\varphi_{s,o}(x,y) - \overline{\varphi}_{s,o}(x,y))\right|, \tag{7}$$

where $W_o(x,y)$ is the weighting function, T is the estimated noise threshold, ε is a small constant to prevent division by zero, and $\overline{\varphi}_{s,o}(x,y)$ is the mean phase angle. The function $\lfloor \cdot \rfloor$ denotes that the enclosed quantity is equal to itself when its value is positive, and zero otherwise. PC_2 denotes the PC magnitude map of the input image.

Further, to obtain the information of PC varying with orientation o in the image, phase congruency is calculated independently in each orientation. Thus, serval PCMs according to the orientation angle are obtained [42].

$$PC_2 = \sum_o PC(\theta_o), \tag{8}$$

where θ_o denotes the angle corresponding to orientation o, and $PC(\theta_o)$ represents a PCM at orientation angle θ_o. The moment of PC is calculated using these intermediate quantities as:

$$a = \sum_o (PC(\theta_o)\cos(\theta_o))^2, \tag{9}$$

$$b = 2\sum_{o} (PC(\theta_o)\cos(\theta_o)) \cdot (PC(\theta_o)\sin(\theta_o)), \tag{10}$$

$$c = \sum_{o} (PC(\theta_o)\sin(\theta_o))^2. \tag{11}$$

The maximum moment max_ψ and the minimum moment min_ψ of PC are defined as:

$$max_\psi = \frac{1}{2}(a + c + \sqrt{b^2 + (a-c)^2}), \tag{12}$$

$$min_\psi = \frac{1}{2}(a + c - \sqrt{b^2 + (a-c)^2}). \tag{13}$$

The maximum and minimum moments of the PCM represent the edge and corner strength map, respectively.

2.2. The Proposed UMPC-Harris Feature Detector

Keypoints with high repeatability and uniform distribution can obtain sufficient matches, thus improving the image registration accuracy [38]. The subsection presents a novel feature detector UMPC-Harris, which is based on voting strategy, Harris on the multi-moment of PCMs, and overlapping block strategy for the detection of corners and edge points. The purpose of this UMPC-Harris detector is to detect sufficient, reliable, and well-distributed keypoints in optical and SAR images. Figure 2 presents the main process of the UMPC-Harris detector, which contains three steps.

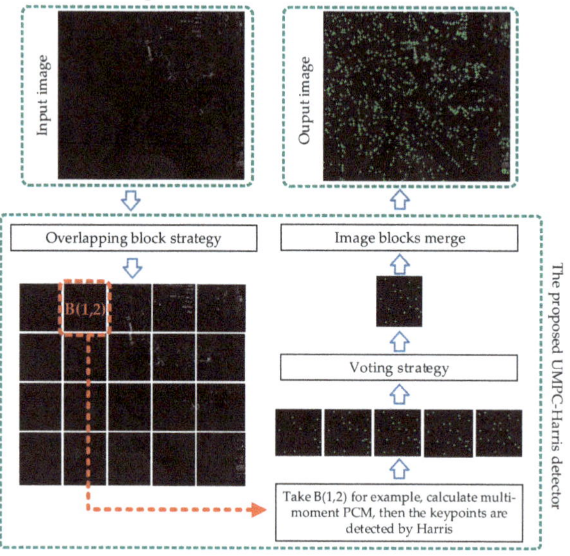

Figure 2. Main process of UMPC-Harris detector.

First, the input image is divided into $S_n \times S_m$ blocks. Further, to avoid missing feature information on the block boundary, an overlap region with n_{op} pixels is added between adjacent blocks. The choice of parameters S_n and S_m is a tradeoff between the amount of computation and the uniform distribution of the keypoints. When more blocks are divided, the keypoints will become more uniform, while increasing the number of calculations. The size and local complexity of the image should be considered for the selection of S_n and S_m.

Second, according to the description in Figure 2, we take block (1,2) as an example to illustrate the construction of multi-moment of the PCM. According to the definition of the maximum and minimum moments, the moment of the PCM M_k is defined as:

$$M_k = \frac{1}{2}(a+c) + \frac{k_t}{2}\sqrt{b^2 + (a-c)^2}, \tag{14}$$

where k_t is a variable between −1 and 1. The moment map contains the maximum and minimum moment map, and we can use max_ψ and min_ψ to describe the above equation as:

$$M_k = \frac{1}{2}\left(max_\psi + min_\psi\right) + \frac{k_t}{2}\left(max_\psi - min_\psi\right), \tag{15}$$

where M_k represents the moment of the PCM with parameter k_t, and it is obvious that if k_t is set to −1, M_k is the minimum moment map $M_k = min_\psi$, and if k_t is set to 1, M_k is the maximum moment map $M_k = max_\psi$. The number of moments is n, and the step h is $\frac{2}{n-1}$.

Third, the points detected by Harris on the maximum and minimum moments of the PCM represent edge points and corners, respectively. Because the edge feature has a high similarity and better resistance to radiation difference between optical and SAR images, thus, extracting feature points on the edge can ensure enough number of features and robustness to radiation difference. Besides, corner features can increase the number of homologous points. Thus, we combine the corners and edge points as keypoints. However, corner features are sensitive to SAR speckle noise and the repeatability rate of the edge points is poor, and therefore, if all of them are considered as keypoints, there could be some unreasonable keypoints. Therefore, we extract Harris corners on the multi-moment of the PCMs, respectively, and we consider the points appearing many times as the final keypoints. Stable and reliable keypoints are found based on the voting strategy.

2.3. Feature Description

After a set of keypoints are detected by the UMPC-Harris method, feature descriptors need to be designed for each keypoint to achieve image registration. The orientation and maximum amplitude index of PC have been proved suitable for describing the similar local features of multi-sensor images [34,37]. In this subsection, we first introduce the construction process of the multi-scale max index maps and the orientation map of phase congruency. Finally, the HOSMI descriptor is established to increase the distinction of the features.

2.3.1. Multi-Scale Max Index Maps

Max index map is more suitable for multimodal image registration and is more robust to intensity radiation distortions compared to the gradient amplitude map. Furthermore, in remote sensing images, many salient features appear in different scales [38]. Inspired by this, multi-scale MIMs are formed by calculating the index of the maximum amplitude at each scale to improve the significance of descriptors. Figure 3 shows the construction process of the four-scale MIMs.

First, the input image $I(x, y)$ is convoluted with LGF to obtain $s \times o$ PC amplitude maps in four scales and six orientations. Second, in the six amplitude maps of the same scale, we can find the maximum amplitude $max\{A_{s,o}(i,j)\}_1^o$ and the corresponding orientation o, where the superscripts and subscripts indicate orientations ranging from 1 to o for a pixel p with coordinates $A_{s,o}(i,j)$. Third, the coordinate of the pixel p in MIM is represented by the corresponding orientation o. Thus, the MIM is an image with all elements from 1 to o. Finally, multi-scale MIMs are constructed by calculating the index of the maximum amplitude on four scales. Therefore, the information in the fine scale and coarse scale of the input image can be obtained, which can effectively enhance the saliency of features in the image.

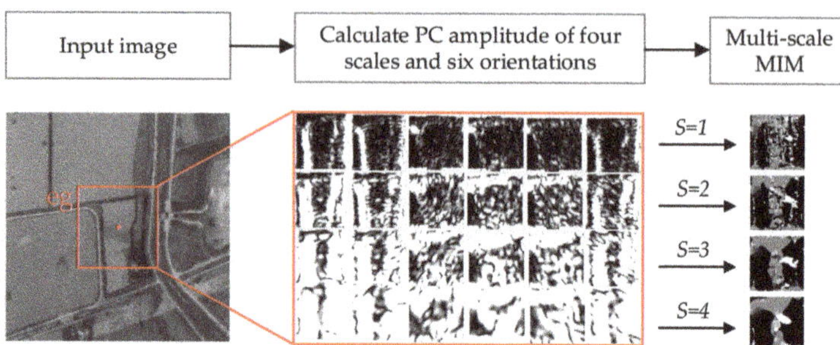

Figure 3. Construction process of four-scale maximum index maps (MIMs).

2.3.2. Orientation of Phase Congruency

The orientation of PC represents the important directions of feature variation, and it has been proved robust to nonlinear radiation distortions [32]. Therefore, similar to the gradient and gradient orientations in the SIFT algorithm, we need to find orientation information in addition to multiscale MIMs.

The PC is calculated by the odd and even symmetric wavelets of LGF, wherein the odd-symmetric wavelet is a smooth derivative filter, which can compute the image derivative in a single direction [34]. For the calculation of PC, the convolution results of the odd-symmetric are obtained according to six orientations. The six convolution results are projected into x and y axes, respectively, and the projections of the x and y axes are obtained. The orientation of the PC can be calculated by the arctangent functions defined as:

$$O_x = \sum_{o} (o_{s,o}(\theta_o)\cos(\theta_o)), \tag{16}$$

$$O_y = \sum_{o} (o_{s,o}(\theta_o)\sin(\theta_o)), \tag{17}$$

$$O_{pc} = \arctan(O_y, O_x), \tag{18}$$

where θ_o represents the angle corresponding to the orientation o, and $e_{s,o}(\theta_o)$ is the convolution result of the odd-symmetric in angle θ_o. Further, O_x and O_y are the sum of the projection of the convolution result in the x and y directions, respectively. The PC orientation O_{pc} can be obtained by the arctangent function. Notably, the PC orientation is limited to $[0^\circ, 180^\circ)$, which can handle gradient inversion in optical and SAR images. Figure 4 shows the calculation process of the PC orientation.

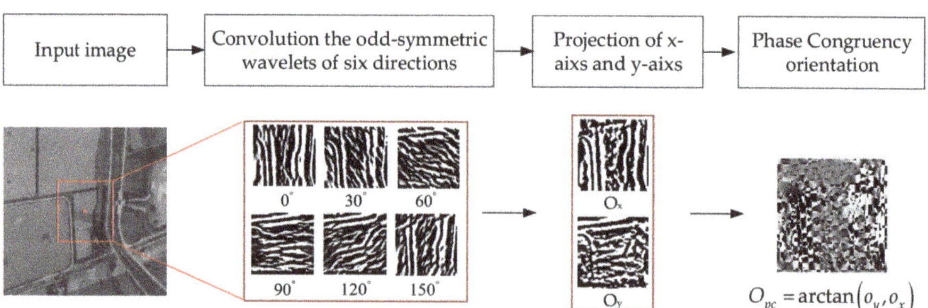

Figure 4. Calculation process of PC orientation.

2.3.3. The Proposed HOSMI Feature Descriptor

The proposed HOSMI descriptor is constructed using the histograms of the PC orientation on the multi-scale MIM. Figure 5 presents the main processing chain of the proposed HOSMI descriptor.

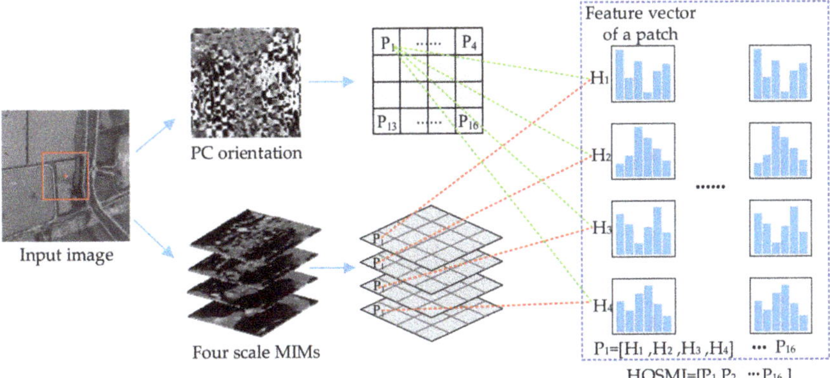

Figure 5. Main processing chain of proposed HOSMI descriptor.

As shown in Figure 5, HOSMI is calculated based on a grid of patches, where local histograms of PC orientation are formed on each scale MIM. The main steps of the feature descriptor are listed below:

1. Apply the LGF to the local region Lx of each keypoint, and then, calculate the odd and even convolution results of four scales and six orientations.
2. Calculate the amplitude map over four scales and six orientations. In each scale, the corresponding orientation to the maximum amplitude forms the multi-scale MIMs; the detailed calculation process is shown in Figure 3.
3. Obtain the PC orientation using the odd convolution results; the detailed calculation process is shown in Figure 4. The PC orientation is restricted to an interval $\left[0^{\circ}, 180^{\circ}\right)$, which can handle gradient inversion in optical and SAR images, and large intensity differences between the optical and SAR images can be reduced.
4. Divide the PC orientation map and the multi-scale MIMs of each keypoint into $n_p \times n_p$ patches. If the local region Lx is selected with a size of $m \times m$ pixels, the size of the patch is $(m/n_p) \times (m/n_p)$ pixels. The feature vector of each patch is calculated in order, and then a descriptor is constructed by combining the feature vectors of all patches.

    - To calculate the feature vector of a patch, PC orientation is formed using n_o bins covering the 180 degrees range of orientations. The sample added to the histogram is the element of the corresponding location on the MIM. To interpolate the peak position for better accuracy, a parabola is fitted to the three histogram values closest to each peak. The feature vector of patch P is calculated on four scales; therefore, the dimension of the feature vector of a patch is $s \times n_o$. In Figure 5, we take the first patch P_1 as an example. The scale used in the PC method is set to 4, and the feature vector of the patch is constructed as $P_1 = [H_1, H_2, H_3, H_4]$, where $H_1 \sim H_4$ are the histograms of the four scales.
    - To obtain the feature descriptor of a keypoint, the feature vectors of all patches are combined into one feature vector. The feature descriptor is normalized by the $L2$ norm to achieve better invariance to illumination and shadowing. The dimension of the feature descriptor of a keypoint is $s \times n_o \times n_p \times n_p$. As shown in Figure 5, if the local region of a keypoint is divided into 4×4 patches, the feature descriptor is constructed by the 16 patches, as in $HOSMI = [P_1, P_2, \cdots, P_{16}]$.

5. Construct a local feature descriptor HOSMI for optical and SAR image registration.

A pair of corresponding points in the optical and SAR images are selected to construct descriptors. To verify the similarity of descriptors, we draw descriptors into stem images in Figure 6.

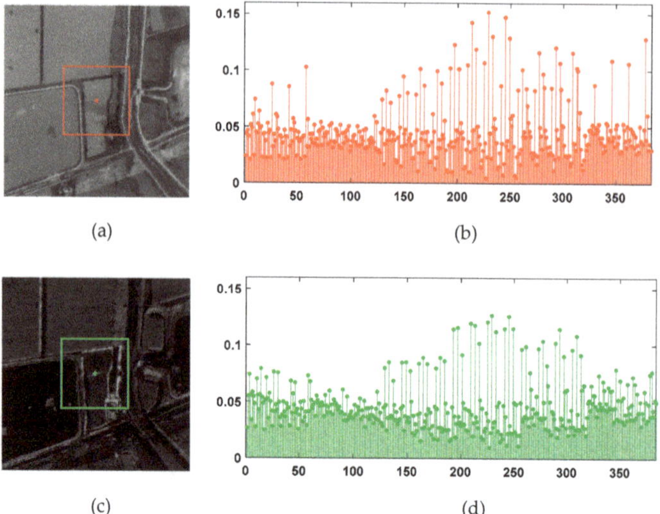

Figure 6. Comparison of similarity of feature vector between a pair of corresponding points in optical and SAR images. (**a**) Optical image; (**b**) Feature vector of the optical keypoint; (**c**) SAR image; (**d**) Feature vector of the SAR keypoint.

Figure 6 shows the HOSMI descriptors of a pair of keypoints between the optical and SAR images. This pair of optical and SAR images has a strong difference in intensity and in gradient inversion, and there is obvious scattering in the SAR image. These differences introduce great challenges to the robustness of the descriptors. The square region represents the local region (96 × 96 pixels) around the keypoint, which is used for computing the feature descriptor. As shown in Figure 6, the similarity of the feature vector is high and radiation changes have a low effect on the proposed descriptor.

3. Experimental Results and Discussion

In this section, we evaluate the performance of the feature detector on simulated images with different SAR noise levels and radiometric (non-uniform intensity) changes. Then, eight pairs of optical and SAR images are used to test the ROS-PC and analyze the experimental results. The registration performances are evaluated via objective and subjective approaches. One approach is to use the evaluation criteria, and the other is to use a chessboard mosaic image and enlarged submaps. Finally, experiments are conducted to evaluate the tolerance of rotation and scale changes from the ROS-PC method. All experiments were conducted with the MATLAB R2017b software on a computer with an Intel Core i5-7200U CPU and 16.0 GB memory.

3.1. Performance Experiments of Proposed UMPC-Harris Detector

We test the performance of the proposed UMPC-Harris detector on simulated images with different noise levels and radiometric (non-uniform intensity) changes. The UMPC-Harris is compared to three other state-of-the-art detectors, Harris, SAR-Harris, and m+M-Harris.

3.1.1. Evaluation Criteria of Feature Detector

Repeatability rate: Given a pair of simulated optical and SAR images to be registered, the keypoints are detected on the two images. Further, two points are regarded as a pair of corresponding keypoints, only if their coordinates are satisfied:

$$\left\|p_{so}(x,y)-p_{ss}(x,y)\right\|_2 \leq T, \tag{19}$$

where $p_{so}(x,y)$ and $p_{ss}(x,y)$ denote the coordinates of the corresponding keypoints in the simulated optical and SAR images, respectively. The function $\|\cdot\|_2$ denotes the Euclidean distance between points $p_{so}(x,y)$ to $p_{ss}(x,y)$. T is the threshold of the Euclidean distance, which is set to 2 pixels in this experiment. The repeatability rate is defined as:

$$R_{rep} = \frac{2N_{cor}}{n_{so}+n_{ss}}, \tag{20}$$

where N_{cor} is the number of pairs of the corresponding keypoints, and n_{so} and n_{ss} are the number of keypoints in the simulated optical and SAR images, respectively. The repeatability rate is a number between 0 and 1. The larger the repeatability rate, the better is the robustness of the detector.

3.1.2. Experimental Data and Parameter Settings of Feature Detector

a. Experimental Data

We used the high-resolution (HR) optical images from the official website of Changguang Satellite Technology Company as the experimental data. The resolution of these images is better than 1 m/pixel. We selected three images with 1000 × 1000 pixels, captured at the Kabul International Airport, Afghanistan, in June 2018; these images are named as Group 1 to 3, as shown in Figure 7.

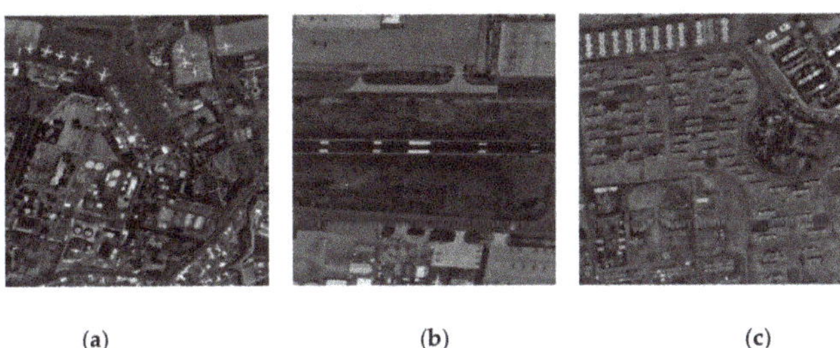

(a) (b) (c)

Figure 7. High-resolution (HR) optical images. (**a**) Group 1; (**b**) Group 2; (**c**) Group 3.

b. Parameter Settings

For the Harris detector, the threshold of the Harris operator is normalized between 0 and 1, and it is set to 0.1 in the following experiments. For the SAR-Harris detector, the first scale is set as $\sigma = 2$, the constant between two adjacent scales is set as $k = 2^{1/3}$, the number of the scale layer is set to 8, and the arbitrary parameter is set as $d = 0.04$. Based on our previous experience, the threshold in the keypoint detection of the simulated optical image is set between 1 and 5, and that of the SAR image is set between 5 and 10. Other parameters used in the experiments follow the parameter settings suggested in Reference [25]. The m+M-Harris and UMPC-Harris are both based on PC. For fairness, the parameters of the PC method are tuned to the same value at each noise level. The PC is calculated in four scales and six orientations. The wavelength of the smallest filters is set from 3 to 5 pixels, according to different images. The scaling factor between successive filters is set to 1.6. In the experiment, the parameters are selected as $S_n = 4$, $S_m = 5$, $n_{op} = 20$, and the number of moments of the PCM is selected as $n = 5$ and $h = 0.5$. The array of parameters k_t is $[-1, -0.5, 0, 0.5, 1]$.

3.1.3. Influence of Noise Level on Proposed UMPC-Harris Detector

To assess the robustness of the feature detector to noise, the HR optical images are utilized to generate simulated optical and SAR images by adding Gaussian noise and speckle noise, respectively. The simulated optical image is obtained by adding Gaussian white noise with 0 mean and 0.01 variance. In the simulated SAR images, the noise level is defined to describe the degree of multiplicative noise. The multiplicative noise with a different number of looks is simulated from one-look to nine-look in the simulated SAR image. It decreases with an increase in the SAR number of looks. A high number of looks refers to a small noise level in SAR images. The simulated images are shown in Figure 8.

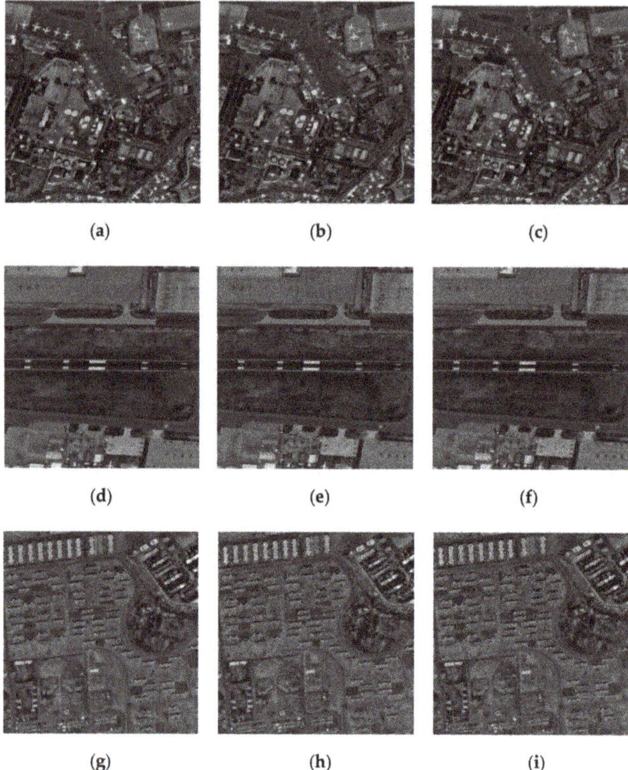

Figure 8. Simulated images. (**a**) Group 1 HR optical image; (**b**) Group 1 simulated optical image; (**c**) Group 1 simulated SAR image (five-look); (**d**) Group 2 HR optical image; (**e**) Group 2 simulated optical image; (**f**) Group 2 simulated SAR image (five-look); (**g**) Group 3 HR optical image; (**h**) Group 3 simulated optical image; (**i**) Group 3 simulated SAR image (five-look).

With the SAR noise level ranging from 1 to 9, the repeatability rate of the UMPC_Harris detector is compared to three other detectors, Harris, Sar-Harris, and m+M-Harris. For fairness, each detector extracts approximately 600 pairs of points by adjusting the threshold, and the average value of ten calculations is taken as the experimental result. The curves of the repeatability rate with the SAR noise level in the three groups are shown in Figure 9.

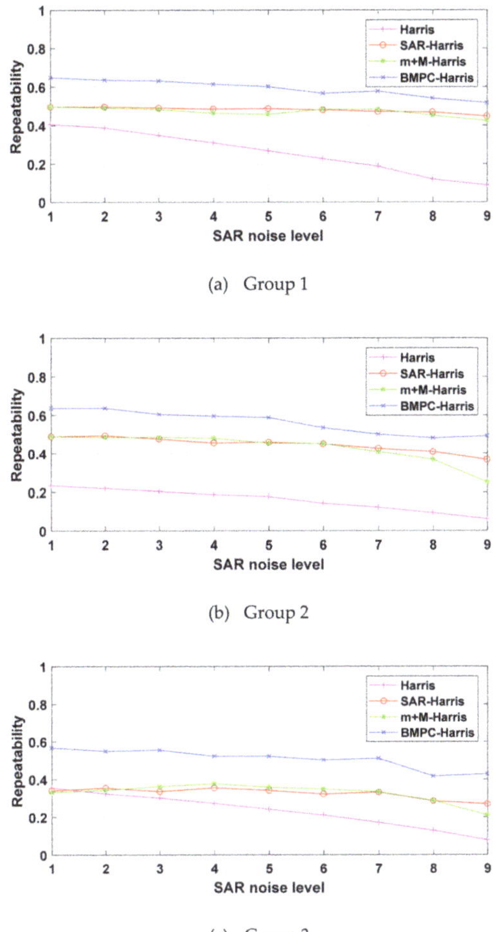

(a) Group 1

(b) Group 2

(c) Group 3

Figure 9. Repeatability rate with different SAR noise level. (**a**) Group 1; (**b**) Group 2; (**c**) Group 3.

The repeatability rate of the UMPC-Harris is the highest among the four detectors in the three groups, and it is more robust to noise than the other detectors. SAR-Harris and m+M-Harris have similar repeatability rates as that of the SAR noise level. Their repeatability rates are between 0.5 and 0.3. When the SAR noise level is high, the repeatability rate of the UMPC-Harris is still higher than 0.4. When the SAR noise level is small, the differences between the two simulated images caused by noise are small, and hence, the three methods show good performance, except for the Harris detector. The repeatability rate of the Harris detector is lower than 0.4, and it decreases rapidly with an increase in the SAR noise level. It is difficult for the Harris detector to deal with multiplicative noise in SAR images directly.

3.1.4. Influence of Radiometric Changes on Proposed UMPC-Harris Detector

To assess the robustness of the feature detector to nonlinear radiometric differences, HR optical images are utilized to generate an image with non-uniform radiometric differences. This is achieved by multiplying the HR optical image by a variable coefficient according to the change of the image column. The results are shown in Figure 10.

Figure 10. Results of original image and non-uniform radiometric differences.

The four detectors are tested on these images. Each detector extracts approximately 600 pairs of keypoints by adjusting the threshold properly, and the experiment results are shown in Figure 11. Furthermore, the repeatability rate of the four detectors are presented in Table 1.

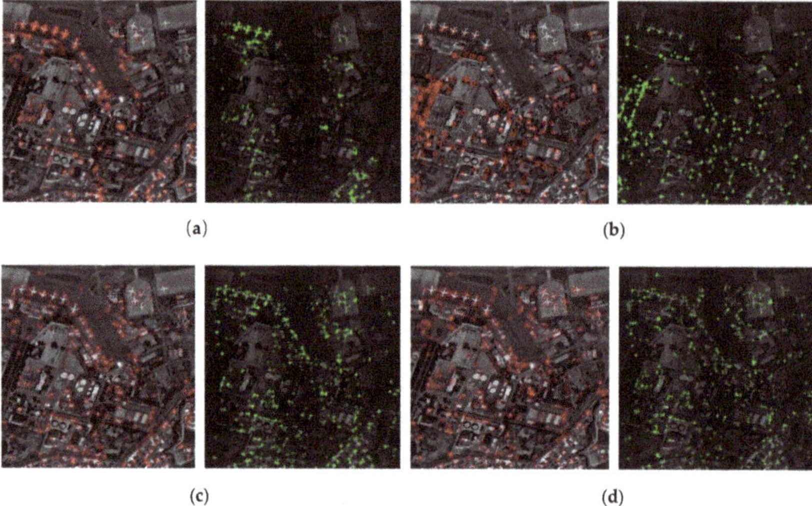

Figure 11. Comparison of keypoint detection. (**a**) Harris; (**b**) SAR-Harris; (**c**) m+M-Harris; (**d**) UMPC-Harris.

Table 1. Repeatability rate of the detectors on the images with non-uniform radiometric differences.

Method	Harris	SAR-Harris	m+M-Harris	UMPC-Harris
Repeatability (%)	59.98	71.83	68.80	90.96

The repeatability rate of the UMPC-Harris is the highest among the four detectors. One can see that the keypoints detected by UMPC-Harris are distributed more uniformly over the image than other detectors, which illustrates that the UMPC-Harris is more robust to radiometric variation and it further indicates that the UMPC-Harris can be applied to keypoint detection for images with radiometric differences.

3.1.5. Results and Discussion of the Proposed UMPC-Harris Detector

The results of corner detection by UMPC-Harris on a pair of optical and SAR images are shown in Figure 12. They include images of a suburban in Weinan, Shaanxi, China. The size of the optical image and SAR image are 863 × 761 and 858 × 761, respectively. The optical image is obtained by Google Earth, and the SAR image is obtained by Airborne SAR. The resolutions are both 3.2 m/pixel. The comparison results indicate that the UMPC-Harris detector is able to extract uniformly distributed in the entire image and avoid missing keypoints at the border of the block.

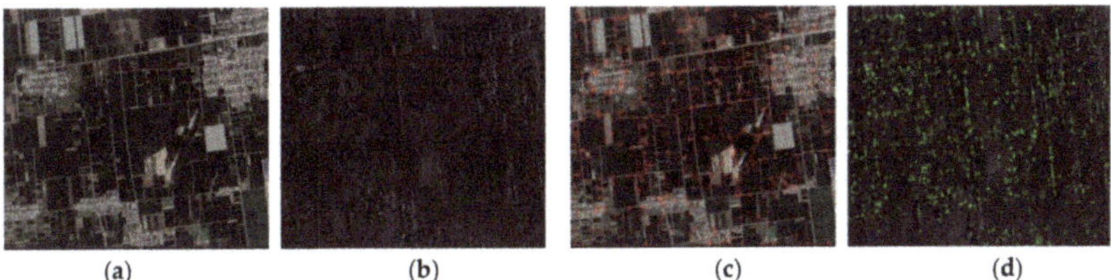

Figure 12. Corner detection results of UMPC-Harris. (**a**) Optical image; (**b**) SAR image; (**c**) UMPC-Harris on optical image; (**d**) UMPC-Harris on SAR image.

The comparison of the repeatability rates of the four detectors indicates that the proposed UMPC-Harris detector has the highest repeatability rate. It is more robust to noise and it has better resistance to radiation differences. The reasons are listed below.

- The UMPC-Harris aims to extract feature points on the multi-moment of the PCM. Stable and valuable keypoints are selected by voting on the Harris corner, which appears repeatedly on the PCM. The combination of the effective corners and edge points not only ensures the high repeatability of the features, but also a large number of features, which lays a foundation for subsequent feature matching.
- Keypoints are well-distributed in the entire image, further points with obvious local features in the dark regions can be detected. This ensures that the keypoints are not limited to the bright region, thereby improving the accuracy of image registration.

3.2. *Performance Experiments of Proposed ROS-PC Registration Algorithm*

To evaluate the performance of the ROS-PC, it is compared with two other state-of-the-art algorithms, namely OS-SIFT [28] and RIFT [39]. The OS-SIFT is a feature-based method, and ROEWA and Sobel operators are used to calculate consistent gradients for optical and SAR images. The RIFT is a PC-based method that detects corners and edge points on the PC map, and it proposes a MIM for feature description. Both the OS-SIFT and RIFT exhibit good performances on multi-sensor image registration. The comparative programs are obtained through their respective academic home pages. Subjective and objective criteria are used to evaluate the performance of the registration algorithm.

3.2.1. Evaluation Criteria of the Registration Algorithm

The checkboard mosaic image and enlarged sub-images are displayed to observe the effect and details of image registration. For each test image pair, each algorithm is executed ten times, and the average of the ten results is computed as the final result. The following evaluation criteria are used to analyze the performance of the algorithm objectively and quantitatively.

Root mean square error (RMSE): This criterion is used to measure the accuracy of the image registration algorithm and is computed by the following method.

First, approximately 20 pairs of corresponding points are manually selected from the optical and SAR images to estimate the affine transformation matrix H. The coordinates of *ith* correctly matched keypoints are $\{(x_i^o, y_i^o), (x_i^s, y_i^s)\}$. The RMSE is computed as [26]:

$$RMSE = \sqrt{\frac{1}{N_{cor}} \sum_{i=1}^{N_{cor}} (x_i^o - (x_i^s)')^2 + (y_i^o - (y_i^s)')^2} \tag{21}$$

where N_{cor} is the number of correctly matched keypoints after the fast sample consensus (FSC) [43], $((x_i^s)', (y_i^s)')$ and it denotes the transformed coordinates of (x_i^s, y_i^s) by the estimated transformation matrix H.

Number of correct matches (NCM): The NCM is the number of correctly matched keypoints after the FSC. If the NCM of an image pair is less than four, the matching is considered to have failed.

A small RMSE indicates that the accuracy of optical and SAR image registration is high. A large NCM indicates that there exist more correctly matched keypoints, thereby resulting in a more accurate transformation matrix H.

3.2.2. Datasets and Parameter Settings of the Registration Algorithm

a. Datasets

In our experiment, eight pairs of optical and SAR images are used to test the ROS-PC; these pairs are referred to as Pairs A-H. Table 2 lists the information for the test images.

Table 2. Information for the test images.

Pair	Sensor	Resolution	Date	Size (Pixel)
A	Google Earth	1 m	9 October 2012	923 × 704
	TerraSAR-X	1 m	23 December 2010	900 × 795
B	Google Earth	1 m	27 March 2020	932 × 684
	Airborne SAR	1 m	June 2020	867 × 740
C	Google Earth	3 m	24 June 2020	1019 × 699
	Airborne SAR	3 m	April 2018	1016 × 697
D	Google Earth	3 m	1 July 2017	1741 × 1075
	Airborne SAR	3 m	April 2018	1744 × 1078
E	Google Earth	3.2 m	25 April 2020	874 × 768
	Airborne SAR	4 m	April 2018	692 × 612
F	Google Earth	2.5 m	19 February 2020	1019 × 701
	Airborne SAR	2.5 m	April 2018	1020 × 711
G	Google Earth	2.2 m	19 February 2020	968 × 662
	Airborne SAR	2.2 m	April 2018	1010 × 676
H	Google Earth	2.5 m	19 February 2020	858 × 758
	Airborne SAR	2.5 m	April 2018	863 × 761

Numerous factors are considered in the selection of test images, including different SAR sensors, date, resolution, and size. The optical images of the eight pairs are obtained from Google Earth, and the SAR image contains a satellite SAR image and seven airborne SAR images. To verify the robustness of the ROS-PC, the image contains different features, as shown in Figure 13.

Pair A includes images of an airport in Tucson, AZ, USA; in this pair, there exists a slight rotation and translation difference. Pair B is obtained in Zhengzhou, Henan, China, and it includes images of a small village. There is a slight rotation and translation in this pair, and because of the arc-shaped roof, there is a large radiation difference over the houses. Some houses even are difficult to recognize on SAR images. Pair C is also obtained in Zhengzhou, Henan, China, and it includes images of a field, and the features of the field vary significantly with the date. Some obvious features exist in the SAR image but not in the optical image. The remaining five pairs of images are obtained in Weinan, Shaanxi, China. Pair D includes images of a large scene, that includes a river, small buildings, multiple fields, and several roads. There are no scale and rotation changes in this pair, and the features exhibit a little temporal difference. This pair of images is used for the rotation and scale variation experiments. Pair E mainly includes images of a lake. There are certain scale variations and time differences. Pair F includes images of a terrace. The fields and terraces in the image are divided into two parts by a road.

The feature intensity of the terraces is stronger than that of the fields. Pair G includes images of a scene including a river and some fields, there exists a large time span between the two images. Pair H includes images of a complex scene with some different structure buildings.

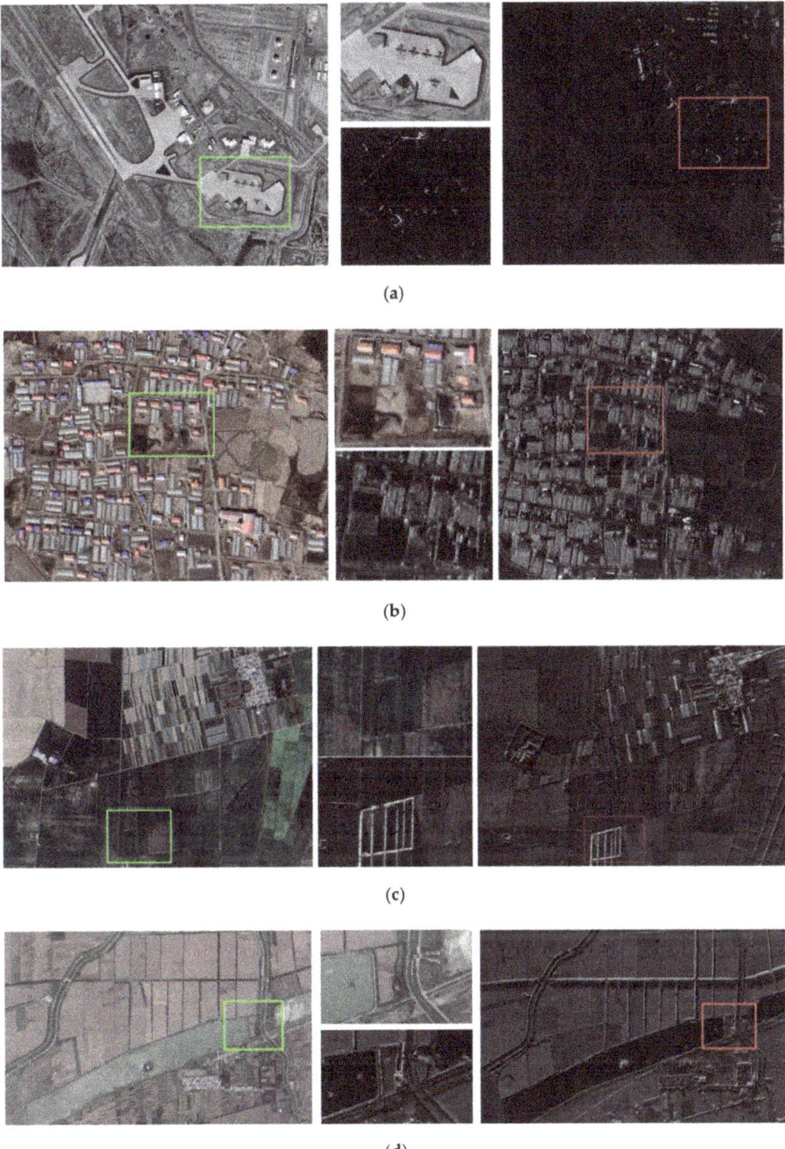

(a)

(b)

(c)

(d)

Figure 13. *Cont.*

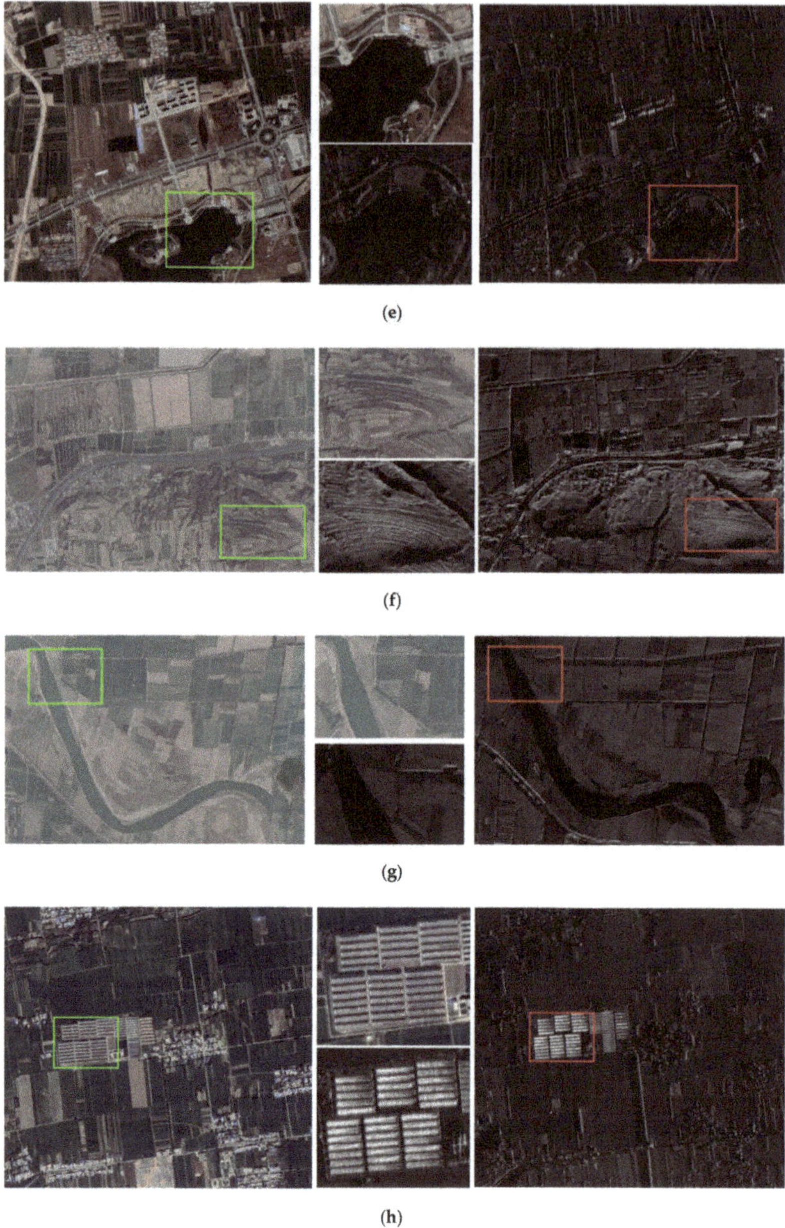

Figure 13. Eight pairs of test images and enlarged view of the main features. (**a**) Pair A; (**b**) Pair B; (**c**) Pair C; (**d**) Pair D; (**e**) Pair E; (**f**) Pair F; (**g**) Pair G; (**h**) Pair H.

b. Parameter Settings

The parameter settings of the UMPC-Harris detector are described in Section 3.1.2. The proposed ROS-PC method contains three parameters n_o, n_p and m, respectively. Parameter n_o and n_p are related to the dimensions of the descriptor, so they should not be too large. Parameter m is the size of the local region used for feature description. If the local region is too small, it contains less information, which does not reflect the difference of features. On the contrary, if the local region is too large, not only the amount of calculation will increase but also the effect of geometric distortion will be received. In the feature descriptor, the parameters are set as $n_o = 6$, $n_p = 4$, and $m = 96$. Therefore, the dimension of the feature descriptor of a keypoint is 384. The parameter settings of comparative algorithms OS-SIFT and RIFT follow the References [28,39]. For a fair, the thresholds of keypoints detection are properly adjusted to obtain similar numbers of keypoints (approximately 1000~1200).

For the feature matching, the sum of squared differences (SSD) is selected for the feature matching metric. If the distance between the two feature vectors is less than the threshold, a pair of keypoints is considered as a potential match, and the threshold is set to 3 pixels. Generally, the matching pairs contain many false matches. The FSC algorithm is used to remove false matches.

3.2.3. Comparison of Experimental Results and Discussion

To evaluate the optical and SAR image registration performance of the ROS-PC, the algorithm is compared with OS-SIFT and RIFT. The OS-SIFT utilizes two different operators to calculate the gradients for SAR and optical images. Multiple image patches are aggregated to construct a gradient location orientation histogram-like descriptor. It is an advanced gradient-based method. The RIFT is a radiation-insensitive feature matching method based on PC and MIM, which is considerably more robust to nonlinear radiation distortions than traditional gradient maps. The registration results of eight pairs of optical and SAR images are shown in Figures 14–21.

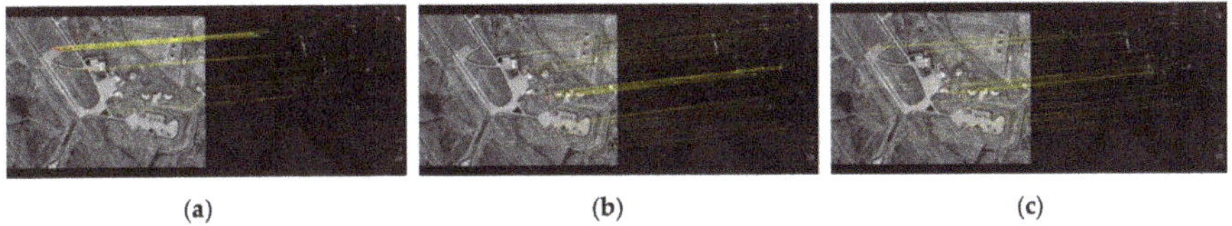

(a) (b) (c)

Figure 14. Registration results of Pair A. (**a**) OS-SIFT; (**b**) (RIFT); (**c**) ROS-PC.

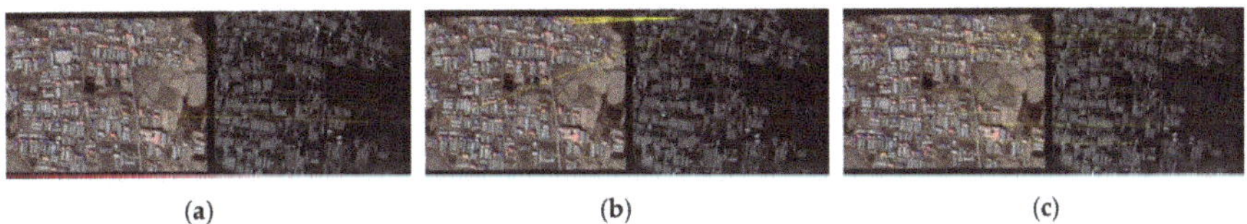

(a) (b) (c)

Figure 15. Registration results of Pair B. (**a**) OS-SIFT; (**b**) RIFT; (**c**) ROS-PC.

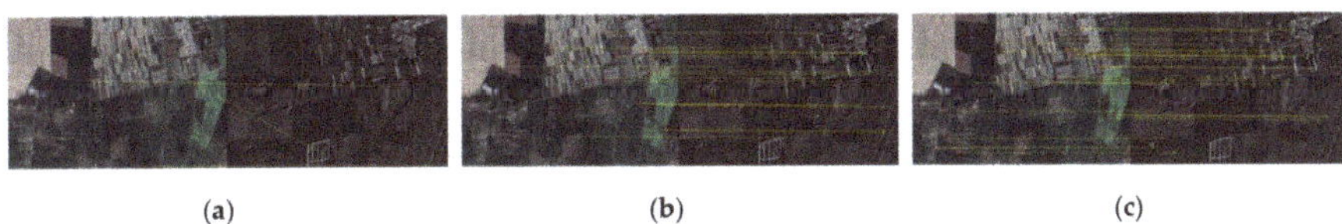

(a) (b) (c)

Figure 16. Registration results of Pair C. (**a**) OS-SIFT; (**b**) RIFT; (**c**) ROS-PC.

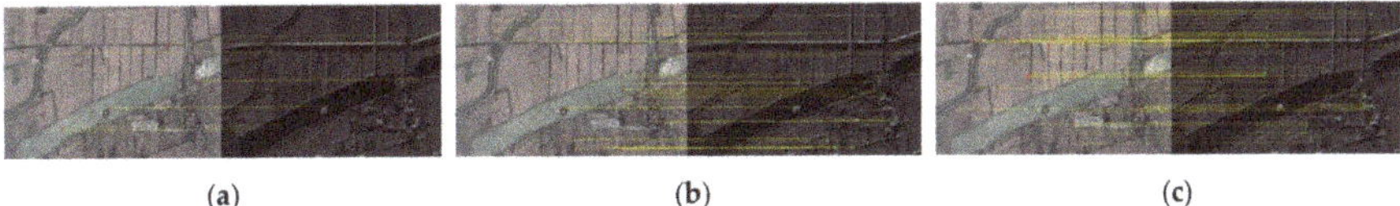

(a) (b) (c)

Figure 17. Registration results of Pair D. (**a**) OS-SIFT; (**b**) RIFT; (**c**) ROS-PC.

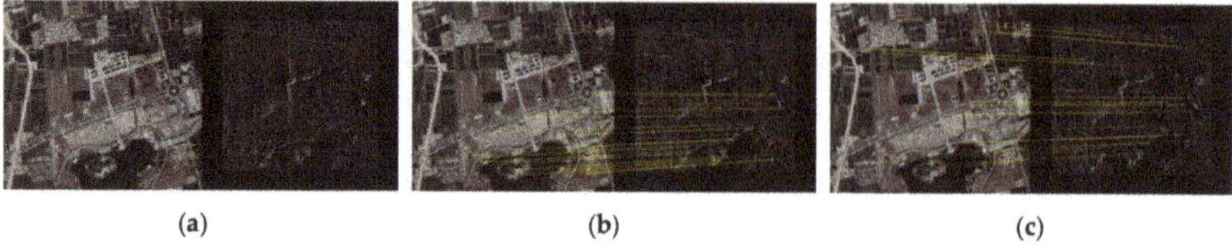

(a) (b) (c)

Figure 18. Registration results of Pair E. (**a**) OS-SIFT; (**b**) RIFT; (**c**) ROS-PC.

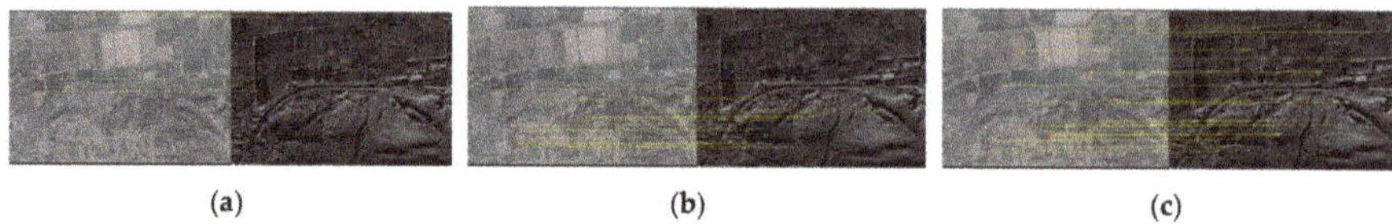

(a) (b) (c)

Figure 19. Registration results of Pair F. (**a**) OS-SIFT; (**b**) RIFT; (**c**) ROS-PC.

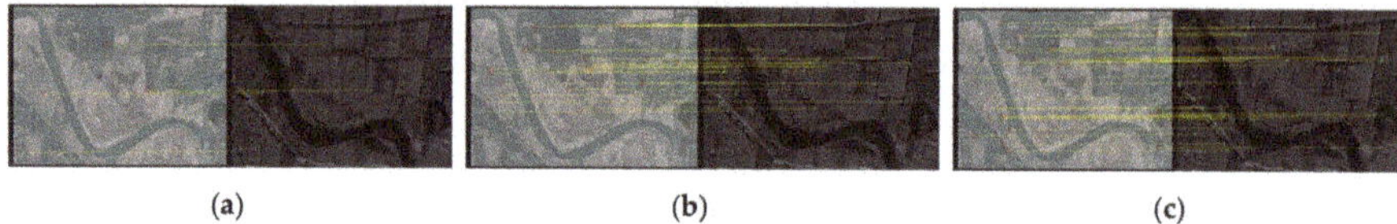

(a) (b) (c)

Figure 20. Registration results of Pair G. (**a**) OS-SIFT; (**b**) RIFT; (**c**) ROS-PC.

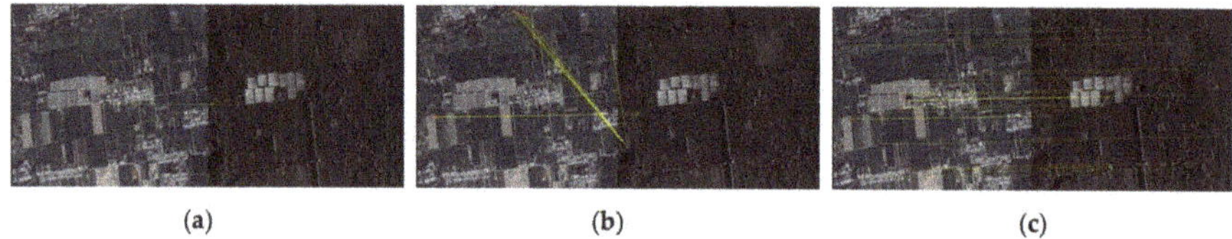

(a) (b) (c)

Figure 21. Registration results of Pair H. (**a**) OS-SIFT; (**b**) RIFT; (**c**) ROS-PC.

Eight groups of images with different features are selected to verify the robustness of ROS-PC algorithm. It can be found that the ROS-PC has the best performance among the three methods, owing to the advantages of the proposed UMPC-Harris detector and HOSMI descriptor. For images with date and season differences, shown in Figures 14 and 16, the ROS-PC shows better robustness and obtains some matched keypoints with the time difference. For images with large radiation differences, shown in Figures 15 and 21, ROS-PC can still obtain some correctly matched keypoints, which are well-distributed in the image. For images with multiple objects, shown in Figures 17–20, ROS-PC can extract matching keypoints from each object, and they are uniformly distributed, which ensures the accuracy of registration. To sum up, the ROS-PC is a robust algorithm, which is suitable for optical and SAR image registration.

To further observe the registration accuracy of the ROS-PC, the checkboard mosaic images and enlarged sub-images of each pair are displayed in Figure 22.

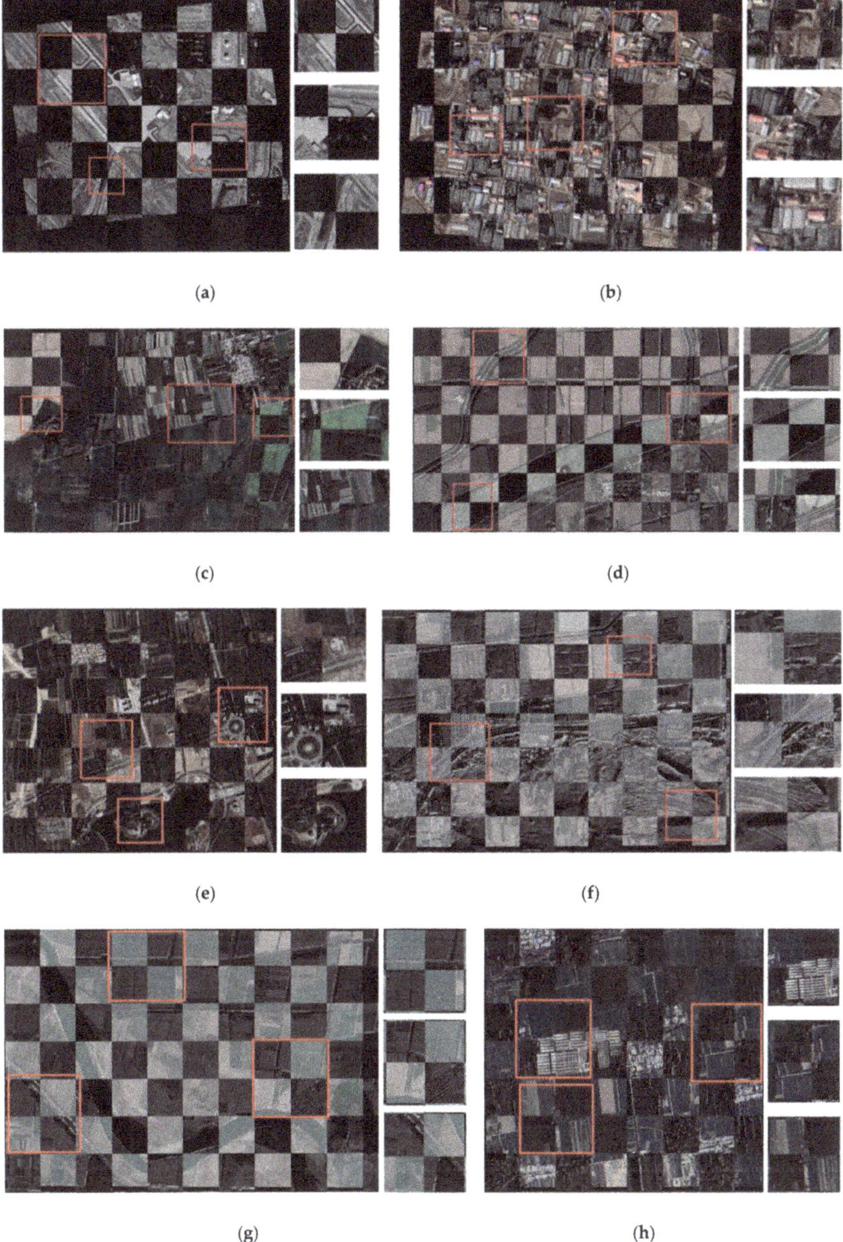

Figure 22. Checkboard mosaic images and enlarged sub-images of ROS-PC. (**a**) Pair A; (**b**) Pair B; (**c**) Pair C; (**d**) Pair D; (**e**) Pair E; (**f**) Pair F; (**g**) Pair G; (**h**) Pair H.

The sub-image is an enlarged view of the intersection of the checkboard mosaic images, where the common features in the optical and SAR images are displayed clearly. In each pair, three sub-images with different features are selected.

Comparisons of RMSE, NCM, and the running time of the eight pairs are presented in Table 3.

Table 3. Comparison of root mean square error (RMSE), number of correct matches (NCM), and time for different methods on eight pairs of test images.

Method / Pair	OS-SIFT			RIFT			ROS-PC		
	RMSE	**NCM**	**Time(s)**	**RMSE**	**NCM**	**Time(s)**	**RMSE**	**NCM**	**Time(s)**
A	1.8507	30	52.84	2.2953	39	10.4	1.9326	41	73.46
B	16.1443	6	65.74	—	—	—	2.8296	17	77.36
C	—	—	—	5.6334	17	11.94	2.6961	31	74.45
D	4.8541	9	81.12	2.6507	33	24.33	1.6741	58	97.49
E	20.7450	4	52.90	2.7754	38	9.93	1.9116	38	75.56
F	22.2591	5	67.62	4.6117	19	13.58	1.7773	30	74.12
G	5.6215	5	57.6	2.9563	23	12.25	1.5264	35	71.26
H	—	—	—	—	—	—	1.7612	33	70.59

In the eight pairs of test images, most features of suburban areas, such as airports, houses, fields, terraces, roads, rivers, and lakes are included. The optical and SAR images exhibit nonlinear radiation distortion, which leads to intensity differences or gradient inversion. Next, we analyze and discuss the registration results of each pair of images.

For Pair A, all three methods can successfully achieve optical and SAR image registration because of the HR and less noise. The ROS-PC performs best on the RMSE and NCM, benefiting from the high repeatability rate of keypoints and the robustness of the feature description method. For Pair B, the RIFT fails to register the two images correctly because of the serious nonlinear radiometric difference. This is because the houses in the image have arc-shaped roofs, which induce the optical and SAR images with intensity differences. Although several correctly matched keypoints are detected by the OS-SIFT, the number and accuracy are significantly lower than the ROS-PC. For Pair C, the OS-SIFT fails to register the two images because of the gradient difference and obvious scattering in the SAR image. However, the ROS-PC is suitable for describing similar local information based on the PC orientation and the multi-scale MIMs of the keypoints. For Pair D, the image contains many features, and there are little scale, rotation, and date difference. The ROS-PC remains superior to the other two methods, as it is robust to nonlinear radiation differences and noise. For pair E and F, the image contains two types of features. ROS-PC can obtain correctly matching keypoints from each object, and they are well-distributed. For pair G, owing to the time difference, there are many unmatched keypoints of the river in the image, which causes extra difficulties in feature matching. However, the ROS-PC still successfully completes more NCM and achieves higher accuracy. For pair H, the radiation difference between optical and SAR images is large, because there are many buildings in the image, which leads to the failure of the other two algorithms. To sum up, the ROS-PC has the most uniform distribution, the largest NCM, and the best RMSE among the three algorithms. Therefore, the ROS-PC is more robust to noise and scattering in SAR images and the radiation difference between optical and SAR images.

Gradient-based descriptors such as OS-SIFT are more sensitive to nonlinear radiation differences because the gradient-based descriptors rely on a linear relationship between images, and therefore, they are not appropriate for significant nonlinear intensity differences caused by radiation distortion. The speckle noise and scattering in SAR images pose significant challenges in image registration. The RIFT performs better than the gradient-based descriptors because it uses PC to capture MIM. The RIFT uses the MIM to express the shape and structure information of objects and, therefore, it is robust to nonlinear radiation distortion. However, the repeatability rate of the corner detector in RIFT is not as good as that of UMPC-Harris, and the descriptor in RIFT has limited significance and robustness to noise and scattering in SAR images. Therefore, our ROS-PC yields the smallest RMSE and the largest NCM among all eight pairs because of two reasons, which are listed below.

- The UMPC-Harris can obtain a higher repeatability rate of keypoints than SAR-Harris and m + M-Harris between SAR and optical images.

- The HOSMI descriptor uses four-scale and six-orientation LGFs to capture the multi-scale max index and orientation feature information of PC, which is robust to nonlinear radiation variations of optical and SAR images. Further, it can effectively overcome the noise and scattering of SAR images.

After comparisons of the running time in Table 3, it can be found that the ROS-PC is the most time-consuming. The reason is that the algorithm is based on the principle of PC, which is slow by nature. Second, in the process of feature detection, the overlapping block and voting strategies need to additional calculation than the other methods. Third, the descriptor is constructed over the four scales and the dimension is larger. This paper only focuses on a robust registration method for optical and SAR images, and the running time is not the focus. Therefore, reducing the computation time and improving the efficiency of the algorithm is a problem we need to study in the future. Moreover, computational efficiency can be further improved by optimizing the algorithm and implementing the ROS-PC in C/C++.

3.3. *Influence of Rotation and Scale Variations on the Proposed ROS-PC*

The previous experimental results show that the algorithm is robust to the radiation distortion between optical and SAR images; however, the ROS-PC is not designed for scale and rotation deformations. The large-angle rotation between remote sensing images can be corrected using sensor geographic information. Further, by employing remote sensing image ground resolution information, remote sensing images can be assigned to the same scale by resampling. Then, the ROS-PC could be used for fine matching, which can handle slight rotation and scale differences between optical and SAR images. In this subsection, the influence of rotation and scale variation on our algorithm is evaluated based on the NCM for Pair D.

3.3.1. Rotation Experiments of the Proposed ROS-PC

We tested the effect of rotation changes on the ROS-PC. Assuming the optical image remains unchanged, the SAR image is rotated from $-12°$ to $16°$. The optical and SAR image registration results of the rotation variation are shown in Figure 23. The relationship between the NCM and the rotation angle is listed in Table 4.

Table 4. NCM with different rotation angles.

Rotation Angle	$-12°$	$-9°$	$-6°$	$-3°$	$0°$	$4°$	$8°$	$12°$	$16°$
NCM	8	19	28	35	58	32	31	20	15

Figure 23 and Table 4 indicate that the ROS-PC can tolerate rotations between optical and SAR images below $9°$, which is sufficient for images that have been corrected by sensor geographic information.

3.3.2. Scale Experiments of the Proposed ROS-PC

We test the robustness of the ROS-PC to scale changes. The optical image in Pair-D remains unchanged, and the SAR image is resized from 0.6 to 1.4 with an interval of 0.1. The optical and SAR image registration results of the scale variation are shown in Figure 24. The relationship between the NCM and the scale factor is listed in Table 5.

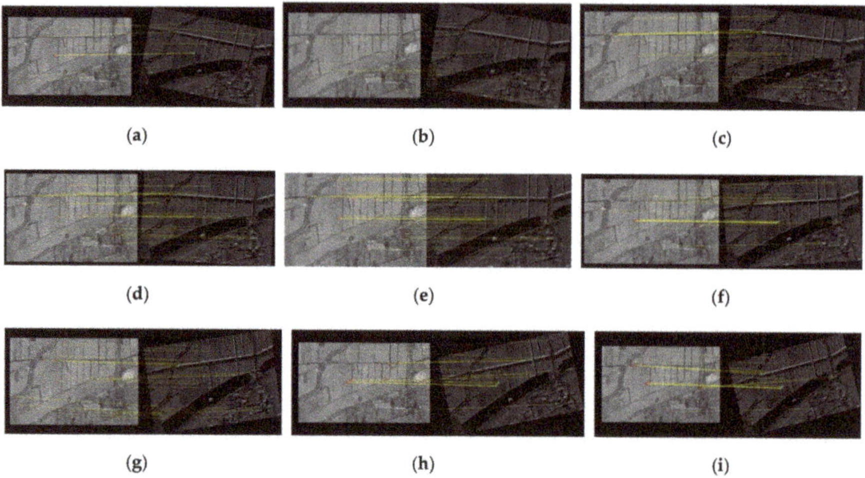

Figure 23. Registration results of optical and SAR images with different rotation angles (degree). (**a**) −12°, (**b**) −9°, (**c**) −6°, (**d**) −3°, (**e**) 0°, (**f**) 4°, (**g**) 8°, (**h**) 12°, and (**i**) 16°.

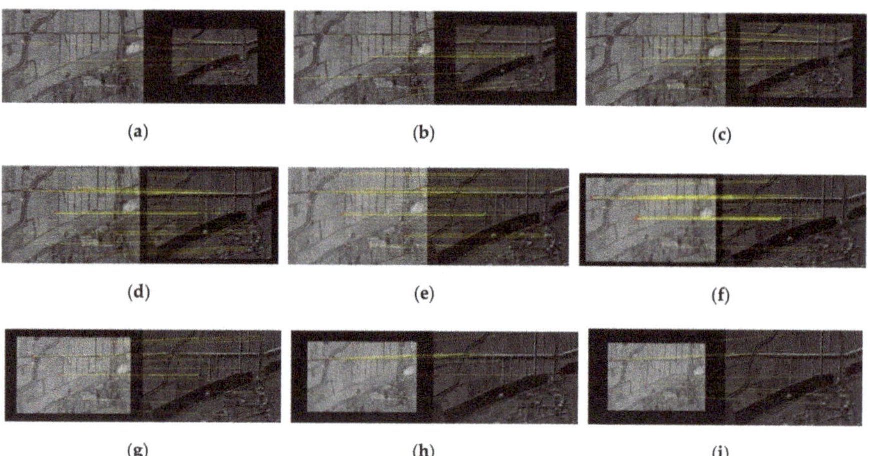

Figure 24. Registration results of optical and SAR images with different scales. (**a**) 0.6, (**b**) 0.7, (**c**) 0.8, (**d**) 0.9, (**e**) 1.0, (**f**) 1.1, (**g**) 1.2, (**h**) 1.3, and (**i**) 1.4.

Table 5. NCM with different scale factors.

Scale	0.6	0.7	0.8	0.9	1	1.1	1.2	1.3	1.4
NCM	16	27	39	55	58	39	23	13	13

Figure 24 and Table 5 indicates that the ROS-PC can tolerate the scale difference between optical and SAR images in the range of 0.7–1.2, which is sufficient for images that have been assigned a similar scale by resampling.

4. Conclusions

In this study, a PC-based optical and SAR image registration algorithm ROS-PC is proposed to address the matching difficulties caused by complex nonlinear radiation differences and speckle noise.

We designed a novel feature detector named UMPC-Harris, comprising an overlapping block strategy, Harris on multi-moment of the PCM, and the vote strategy to obtain uniformly distributed keypoints and increase the repeatability rate of the keypoints. The experimental results on simulated images demonstrated that the proposed UMPC-Harris method achieved a good performance in keypoints detection. The proposed HOSMI descriptor is constructed using the histograms of the PC orientation on the multi-scale MIM. The image registration experiments prove that the ROS-PC is robust to nonlinear radiation variations of optical and SAR images and it can tolerate a small amount of rotation and scale changes.

Author Contributions: Conceptualization, L.W. and M.S.; methodology, L.W.; software, L.W.; validation, L.W., M.S., and J.L.; formal analysis, L.W. and M.S.; writing—original draft preparation, L.W.; writing—review and editing, L.C. and G.M.; All authors have read and agreed to the published version of the manuscript.

Funding: This work was supported in part by the National Key R&D Program of China under Grant NO. 2017YFC0822402.

Acknowledgments: The authors would like to thank Dr. Xiang for providing the OS-SIFT and Pair A images and the Beijing Institute of Radio Measurement for providing the SAR images in Pair B-H.

Conflicts of Interest: The authors declare no conflict of interest.

References

1. Zhang, Q.; Liu, Y.; Blum, R.S.; Han, J.G.; Tao, D.C. Sparse representation based multi-sensor image fusion for multi-focus and multi-modality images: A review. *Inf. Fusion* **2018**, *40*, 57–75. [CrossRef]
2. Kulkarni, S.C.; Rege, P.P. Pixel level fusion techniques for SAR and optical images: A review. *Inf. Fusion* **2020**, *59*, 13–29. [CrossRef]
3. Tapete, D.; Cigna, F. Detection of archaeological looting from space: Methods, achievements, and challenges. *Remote Sens.* **2019**, *11*, 2389. [CrossRef]
4. Song, S.L.; Jin, K.; Zuo, B.; Yang, J. A novel change detection method combined with registration for SAR images. *Remote Sens. Lett.* **2019**, *10*, 669–678. [CrossRef]
5. Zhang, S.L.; Chen, J.Q.; Liu, X.; Li, J. Change Detection of Huangqi Lake Based on Modified Active Contour Using Sentinel-1 SAR Image. In Proceedings of the 2018 Progress in Electromagnetics Research Symposium (PIERS), Toyama, Japan, 1–4 August 2018; pp. 2291–2295.
6. Li, K.; Zhang, X.Q. Review of Research on Registration of SAR and Optical Remote Sensing Image Based on Feature. In Proceedings of the 2018 IEEE 3rd International Conference on Signal and Image Processing (ICSIP 2018), Shenzhen, China, 13–15 July 2018; pp. 111–115.
7. Wang, F.; You, H.J. Robust registration method of SAR and optical remote sensing images based on cascade. *J. Infrared Millim. Waves* **2015**, *34*, 486–492.
8. Zitova, B.; Flusser, J. Image registration methods: A survey. *Image Vis. Comput.* **2003**, *21*, 977–1000. [CrossRef]
9. Suri, S.; Reinartz, P. Mutual-information-based registration of TerraSAR-X and Ikonos imagery in urban areas. *IEEE Trans. Geosci. Remote Sens.* **2010**, *48*, 939–949. [CrossRef]
10. Shu, L.X.; Tan, T.N. SAR and SPOT Image Registration Based on Mutual Information with Contrast Measure. In Proceedings of the 2007 IEEE International Conference on Image Processing, San Antonio, TX, USA, 16 September–19 October 2007; pp. 2681–2684.
11. Shi, W.; Su, F.Z.; Wang, R.R.; Fan, J.F. A Visual Circle Based Image Registration Algorithm for Optical and SAR Imagery. In Proceedings of the 2012 IEEE International Geoscience and Remote Sensing Symposium (IGARSS 2012), Munich, Germany, 22–27 July 2012; pp. 2109–2112.
12. Wang, F.; Vemuri, B.C. Non-rigid multi-modal image registration using cross-cumulative residual entropy. *Int. J. Comput. Vis.* **2007**, *74*, 201–215. [CrossRef]
13. Yu, L.; Zhang, D.R.; Holden, E.J. A fast and fully automatic registration approach based on point features for multi-source remote-sensing images. *Comput. Geosci.* **2008**, *34*, 838–848. [CrossRef]
14. Liu, S.Y.; Jiang, J. Registration algorithm based on line-intersection-line for satellite remote sensing images of urban areas. *Remote Sens.* **2019**, *11*, 26. [CrossRef]

15. Sui, H.G.; Xu, C.; Liu, J.Y.; Hua, F. Automatic optical-to-SAR image registration by iterative line extraction and Voronoi integrated spectral point matching. *IEEE Trans. Geosci. Remote Sens.* **2015**, *53*, 6058–6072. [CrossRef]
16. Li, H.; Manjunath, B.S.; Mitra, S.K. Contour-Based Multisensor Image Registration. In Proceedings of the Conference Record of the Twenty-Sixth Asilomar Conference on Signals, Systems & Computers, Pacific Grove, CA, USA, 26–28 October 1992; pp. 182–186.
17. Lowe, D. Distinctive image features from scale-invariant keypoints. *Int. J. Comput. Vis.* **2004**, *20*, 91–110. [CrossRef]
18. Ke, Y.; Sukthankar, R.; Society, I.C. PCA-SIFT: A More Distinctive Representation for Local Image Descriptors. In Proceedings of the 2004 IEEE Computer Society Conference on Computer Vision and Pattern Recognition (CVPR 2004), Washington, DC, USA, 27 June–2 July 2004; pp. 506–513.
19. Bay, H.; Tuytelaars, T.; Gool, L.V. SURF: Speeded Up Robust Features. In Proceedings of the 9th European Conference on Computer Vision (ECCV 2006), Graz, Austria, 7–13 May 2006; pp. 404–417.
20. Morel, J.M.; Yu, G. ASIFT: A new framework for fully affine invariant image comparison. *SIAM J. Imaging Sci.* **2009**, *2*, 438–469. [CrossRef]
21. Sedaghat, A.; Mokhtarzade, M.; Ebadi, H. Uniform robust scale-invariant feature matching for optical remote sensing images. *IEEE Trans. Geosci. Remote Sens.* **2011**, *49*, 4516–4527. [CrossRef]
22. Sedaghat, A.; Ebadi, H. Remote sensing image matching based on adaptive binning SIFT descriptor. *IEEE Trans. Geosci. Remote Sens.* **2015**, *53*, 5283–5293. [CrossRef]
23. Fan, B.; Huo, C.L.; Pan, C.H.; Kong, Q.Q. Registration of optical and SAR satellite images by exploring the spatial relationship of the improved SIFT. *IEEE Trans. Geosci. Remote Sens. Lett.* **2013**, *10*, 657–661. [CrossRef]
24. Gong, M.; Zhao, S.; Jiao, L.; Tian, D.; Wang, S. A novel coarse-to-fine scheme for automatic image registration based on SIFT and mutual information. *IEEE Trans. Geosci. Remote Sens.* **2014**, *52*, 4328–4338. [CrossRef]
25. Dellinger, F.; Delon, J.; Gousseau, Y.; Michel, J.; Tupin, F. SAR-SIFT: A SIFT-like algorithm for SAR images. *IEEE Trans. Geosci. Remote Sens.* **2015**, *53*, 453–466. [CrossRef]
26. Ma, W.; Wen, Z.; Wu, Y.; Jiao, L.; Gong, M.; Zheng, Y.; Liu, L. Remote sensing image registration with modified SIFT and enhanced feature matching. *IEEE Geosci. Remote Sens. Lett.* **2017**, *14*, 3–7. [CrossRef]
27. Paul, S.; Pati, U.C. A Gabor odd filter-based ratio operator for SAR image matching. *IEEE Trans. Geosci. Remote Sens. Lett.* **2019**, *16*, 397–401. [CrossRef]
28. Xiang, Y.; Wang, F.; You, H. OS-SIFT: A robust SIFT-like algorithm for high-resolution optical-to-SAR image registration in suburban areas. *IEEE Trans. Geosci. Remote Sens.* **2018**, *56*, 3078–3090. [CrossRef]
29. Govindaraj, P.; Sudhakar, M.S. A new 2D shape retrieval scheme based on phase congruency and histogram of oriented gradients. *Sig. Image Video Process.* **2019**, *13*, 771–778. [CrossRef]
30. Jiang, S.; Wang, B.N.; Zhu, X.Y.; Xiang, M.S.; Fu, X.K.; Sun, X.F. Registration of SAR and Optical Images by Weighted SIFT Based on Phase Congruency. In Proceedings of the IEEE International Geoscience and Remote Sensing Symposium (IGARSS 2018), Valencia, Spain, 22–27 July 2018; pp. 8885–8888.
31. Cui, S.; Zhong, Y.F. Multi-Modal Remote Sensing Image Registration Based on Multi-Scale Phase Congruency. In Proceedings of the 10th IAPR Workshop on Pattern Recognition in Remote Sensing (PRRS 2018), Beijing, China, 19–20 August 2018; pp. 1–5.
32. Ragb, H.K.; Asari, V.K. Histogram of oriented phase (HOP): A new descriptor based on phase congruency. In *Mobile Multimedia/Image Processing, Security, and Applications 2016*; SPIE: Bellingham, WA, USA, 19 May 2016; p. 98690V1-10.
33. Ye, Y.; Shan, J.; Bruzzone, L.; Shen, L. Robust registration of multimodal remote sensing images based on structural similarity. *IEEE Trans. Geosci. Remote Sens.* **2017**, *55*, 2941–2958. [CrossRef]
34. Ye, Y.; Shan, J.; Hao, S.; Bruzzone, L.; Qin, Y. A local phase based invariant feature for remote sensing image matching. *ISPRS J. Photogramm. Remote Sens.* **2018**, *142*, 205–221. [CrossRef]
35. Chen, M.; Habib, A.; He, H.Q.; Zhu, Q.; Zhang, W. Robust feature matching method for SAR and optical images by using Gaussian-gamma-shaped Bi-windows-based descriptor and geometric constraint. *Remote Sens.* **2017**, *9*, 25. [CrossRef]
36. Paul, S.; Pati, U.C. Automatic optical-to-SAR image registration using a structural descriptor. *IET Image Process.* **2020**, *14*, 62–73. [CrossRef]

37. Fu, Z.; Qin, Q.; Luo, B.; Sun, H.; Wu, C. HOMPC: A local feature descriptor based on the combination of magnitude and phase congruency information for multi-sensor remote sensing images. *Remote Sens.* **2018**, *10*, 1234. [CrossRef]
38. Fan, J.W.; Wu, Y.; Li, M.; Liang, W.K.; Cao, Y.C. SAR and optical image registration using nonlinear diffusion and phase congruency structural descriptor. *IEEE Trans. Geosci. Remote Sens.* **2018**, *56*, 5368–5379. [CrossRef]
39. Li, J.; Hu, Q.; Ai, M. RIFT: Multi-modal image matching based on radiation-variation insensitive feature transform. *IEEE Trans. Image Process.* **2020**, *29*, 3296–3310. [CrossRef]
40. Ye, Y.; Li, S. HOPC: A Novel Similarity Metric Based on Geometric Structural Properties for Multi-Modal Remote Sensing Image Matching. In Proceedings of the ISPRS Annals of the Photogrammetry, Remote Sensing and Spatial Information Sciences, Prague, Czech Republic, 12–19 July 2016; pp. 1–18.
41. Kovesi, P. Image features from phase congruency. *J. Comput. Vis. Res.* **1999**, *1*, 1–26.
42. Kovesi, P. Phase Congruency Detects Corners and Edges. In Proceedings of the International Conference on Digital Image Computing: Techniques and Applications (DICTA 2003), Macquarie University, Sydney, Australia, 10–12 December 2003; pp. 309–318.
43. Wu, Y.; Ma, W.; Gong, M.; Su, L.; Jiao, L. A novel point-matching algorithm based on fast sample consensus for image registration. *IEEE Geosci. Remote Sens.* **2015**, *12*, 43–47. [CrossRef]

Article

Modality-Free Feature Detector and Descriptor for Multimodal Remote Sensing Image Registration

Song Cui, Miaozhong Xu, Ailong Ma * and Yanfei Zhong

The State Key Laboratory of Information Engineering in Surveying, Mapping and Remote Sensing, Wuhan University, Wuhan 430079, China; cuisong@whu.edu.cn (S.C.); mzxu6319@whu.edu.cn (M.X.); zhongyanfei@whu.edu.cn (Y.Z.)

* Correspondence: maailong007@whu.edu.cn; Tel.: +86-27-6877-9969

Received: 29 July 2020; Accepted: 7 September 2020; Published: 10 September 2020

Abstract: The nonlinear radiation distortions (NRD) among multimodal remote sensing images bring enormous challenges to image registration. The traditional feature-based registration methods commonly use the image intensity or gradient information to detect and describe the features that are sensitive to NRD. However, the nonlinear mapping of the corresponding features of the multimodal images often results in failure of the feature matching, as well as the image registration. In this paper, a modality-free multimodal remote sensing image registration method (SRIFT) is proposed for the registration of multimodal remote sensing images, which is invariant to scale, radiation, and rotation. In SRIFT, the nonlinear diffusion scale (NDS) space is first established to construct a multi-scale space. A local orientation and scale phase congruency (LOSPC) algorithm are then used so that the features of the images with NRD are mapped to establish a one-to-one correspondence, to obtain sufficiently stable key points. In the feature description stage, a rotation-invariant coordinate (RIC) system is adopted to build a descriptor, without requiring estimation of the main direction. The experiments undertaken in this study included one set of simulated data experiments and nine groups of experiments with different types of real multimodal remote sensing images with rotation and scale differences (including synthetic aperture radar (SAR)/optical, digital surface model (DSM)/optical, light detection and ranging (LiDAR) intensity/optical, near-infrared (NIR)/optical, short-wave infrared (SWIR)/optical, classification/optical, and map/optical image pairs), to test the proposed algorithm from both quantitative and qualitative aspects. The experimental results showed that the proposed method has strong robustness to NRD, being invariant to scale, radiation, and rotation, and the achieved registration precision was better than that of the state-of-the-art methods.

Keywords: image registration; nonlinear radiation distortions; phase congruency; multimodal remote sensing image

---

1. Introduction

Image registration is an essential and fundamental task for remote sensing interpretation. It is aimed at registering images obtained from different sensors, different perspectives, different times or different imaging conditions [1], and is an essential preliminary task for image fusion [2], 3D modeling [3], and change detection [4]. With the rapid advance of remote sensing systems, more and more data sources can now be acquired. As a result, the complementary information between multimodal remote sensing images can significantly improve the capacity and effectiveness of the interpretation. However, the efficiency and accuracy of the image registration result greatly affects the performance of the subsequent processing [1]. Nevertheless, for multimodal remote sensing images, a large number of nonlinear radiation distortions (NRD) will be present, as a result of the different physical imaging mechanisms, which brings significant challenges to the image registration.

Generally speaking, image registration methods can be divided into two categories according to the factors (areas and features) on which they are based. The area-based methods adopt the intensity value of the image itself, while the transformation model for the registration is calculated by optimizing a similarity measure between the image to be registered and the reference image. Correlation [5–7], mutual information [8–13], or frequency-domain information [14,15] can be used as metrics to measure whether the images are registered. However, these area-based methods often reach a locally optimal solution for the optimization of the model transformation, especially when there is NRD among the images. At the same time, the optimization process for the image conversion parameters is of high computational complexity. The feature-based methods have high robustness to geometric distortion and NRD, so that they are commonly adopted in the registration of multimodal remote sensing images. The feature-based methods achieve the matching goal by identifying the reliable characteristic correspondence between the images, and they are not directly based on the image intensity [16]. The features considered by the feature-based methods include point features [17–19], line characteristics [20], and structural features [21,22]. Scale-invariant feature transform (SIFT) [23] is a classic feature point matching method. A number of improved versions of SIFT have since been developed, including affine SIFT (ASIFT) [24], speeded-up robust features (SURF) [25], the SIFT-like algorithm for synthetic aperture radar (SAR) imaging (SAR-SIFT), and principal component analysis SIFT (PCA-SIFT). When SIFT or one of its improved algorithms is used to describe the features, the estimated reference direction must be specified, to make the descriptions more unique and robust to the rotation. Image registration based on features is confronted with two problems: (1) in the feature description stage, the estimation of the principal orientation is often prone to error when it is based on the local features of the image, and a lot of corresponding points will be removed due to incorrect principal orientation estimation; and (2) in the feature matching stage, due to the significant NRD, a feature detection result for an image can often not be found, so that there are many abnormalities in the matching result.

In recent years, research into deep learning has exploded in the computer vision field [26,27]. In the study of medical image registration, deep learning has been used to model the relationship between different modal images [28,29]. However, studies of remote sensing image registration based on deep learning are relatively rare, especially for multimodal remote sensing images [30–33]. The main reasons for this are as follows: (1) Compared with natural or medical images, remote sensing images have an extensive range, complex distortion, and weak uniqueness of targets. It is also common in remote sensing images that the same ground object type presents different forms, or the same form corresponds to different ground object types. (2) Compared with the large number of natural image datasets that can be used for training, manually annotated remote sensing image datasets are very rare. The labeling of remote sensing images needs considerable expert knowledge and manpower. Furthermore, the application of a model trained on natural images to remote sensing registration tasks is impractical. (3) Deep learning is essentially a method of supervised learning, but it is almost impossible to use a model trained on one modal image to register another modal image. For example, the performance of the LiDAR and optical image registration task can be unsatisfactory when using the SAR and optical image training model. Therefore, there is a need to develop a universal handcrafted descriptor for the multimodal remote sensing image registration task, from the perspective of the physical radiation mechanism and the imaging geometric model, in the case of limited training data.

In this paper, we propose a scale-radiation-rotation-invariant feature transform (SRIFT) algorithm for the registration of multimodal remote sensing images. The contributions of this paper can be summarized as follows:

(1) A modality-free multimodal remote sensing image registration method is proposed, which can handle the scale, radiation, and rotation distortions at the same time. The structural characteristics are captured, in which the same kind of ground object will present a similar structural distribution. Thus, the corresponding features of the images are mapped into a unified space to establish a one-to-one relationship.

(2) The nonlinear diffusion scale (NDS) space is constructed using a nonlinear diffusion filter, instead of a Gaussian filter, to preserve more structure and detail information, which is of great importance for feature extraction for the registration. The structural characteristics are captured by computing the local orientation and scale phase congruency (LOSPC) value in the NDS space of the image. The minimum and maximum moment maps of LOSPC are then used to detect the remaining points. Rotation invariance depends on a rotation-invariant coordinate (RIC) system in the feature description stage. The points in the neighborhood are statistically calculated through a continually changing local coordinate system, which itself realizes rotation invariance, without the need for the estimated orientation to be assigned.

The rest of this paper is organized as follows. The related work is introduced in Section 2. Section 3 details the process of image registration using the SRIFT algorithm. In Section 4, the experimental results obtained using both simulated and real multimodal remote sensing images are provided for the experimental verification and analysis. A summary and our conclusion are presented in Section 5.

2. Related Work

For the registration of homologous images, the pixel basis of the above methods is the intensity or the gradient of the image; however, for multimodal images, the structural information presented in the image by the same point on the ground will be entirely different, due to the NRD caused by the different imaging mechanisms. As a result, the above methods will match many pseudo-corresponding features.

With regard to multimodal remote sensing image registration, scholars have put forward various descriptors on the basis of structural information [34,35]. Compared with gradient information, structural information is less sensitive to nonlinear intensity changes [36]. For example, the edge-oriented histogram (EOH) descriptor designates the shape and contour information of the local image centered on each keypoint, instead of the gradient [37]. The partial intensity invariant feature descriptor (PIIFD) was proposed to solve the problem of the relative gradient direction of the corresponding points [38]. A descriptor for the distribution of internal geometric structures was introduced for images captured on a log-polar grid, which is known as the local self-similarity (LSS) descriptor [39]. A dense LSS (DLSS) method was used to process the registration of optical and SAR images in [40]. The histogram of oriented phase congruency (HOPC) [41] algorithm was extended on the basis of the phase congruency algorithm, and its description process adds phase direction statistics to increase the robustness. Radiation-invariant feature transform (RIFT) [42] also considers the phase congruency, and presents a maximum index map, instead of the gradient, in the feature description. The phase congruency-based structural descriptor (PCSD) [43] has also been successful in optical and SAR image registration without rotational distortion.

In this section, the HOPC and RIFT algorithms are introduced as typical registration methods for multimodal remote sensing images.

The HOPC method successfully introduces an oriented phase congruency algorithm into the automatic registration of remote sensing images, while innovatively mapping the two images acquired under different physical mechanisms into a unified space. Therefore, the corresponding features that cannot be mapped one by one can be mapped in this space. However, there are three significant deficiencies in HOPC: (1) it needs relatively accurate geographic information (or rough geometric correction) for the images in the execution process. However, many multimodal remote sensing images do not contain accurate geographic information. (2) As a template matching algorithm, it is sensitive to geometric distortion, such as rotation and scale. (3) It uses the Harris corner detector, which is very sensitive to NRD when extracting feature keypoints.

The RIFT algorithm was developed on the basis of the limitations of the HOPC algorithm. It also uses a phase congruency algorithm for reference in the stage of mapping the corresponding features, but it adopts a novel descriptor for the feature description. The RIFT algorithm is a method of describing the features per-pixel, which also takes rotation invariance into account. However, its rotation invariance and robustness are not outstanding, because of the possibility of losing the spatial information, and the time-consuming nature of its calculation process. Furthermore, it is not scale-invariant. The main disadvantages of the RIFT algorithm are as follows: (1) the RIFT algorithm adopts a "convolution sequence ring" to deal with the rotation distortion of the images, which may lose some spatial information, resulting in insufficient unique feature vectors being generated and unfavorable feature matching; and (2) the RIFT algorithm does not consider the scale invariance, so it is susceptible to scale changes.

Although image registration has been the subject of numerous studies over the last few decades, there is still no unified registration framework that can automatically register multiple multimodal images while considering the scale and rotation distortions.

3. Image Registration Based on SRIFT

In this section, we describe the proposed SRIFT method in detail. Figure 1 presents a registration flowchart based on SRIFT. The highlighted feature extraction and matching parts in the second column of the figure represent the main innovations of this paper. The first two steps of the algorithm involve solving the problem of feature extraction, and the last step involves the feature description. Initially, the NDS space is constructed using a nonlinear diffusion filter, instead of a Gaussian filter, to preserve more structural and detail information, in order to achieve scale invariance. The structural characteristics are then captured by computing the LOSPC values, in which the same kind of ground object will present similar structural distributions. Finally, the RIC system is applied, which itself realizes rotation invariance, without the need for the estimated reference orientation.

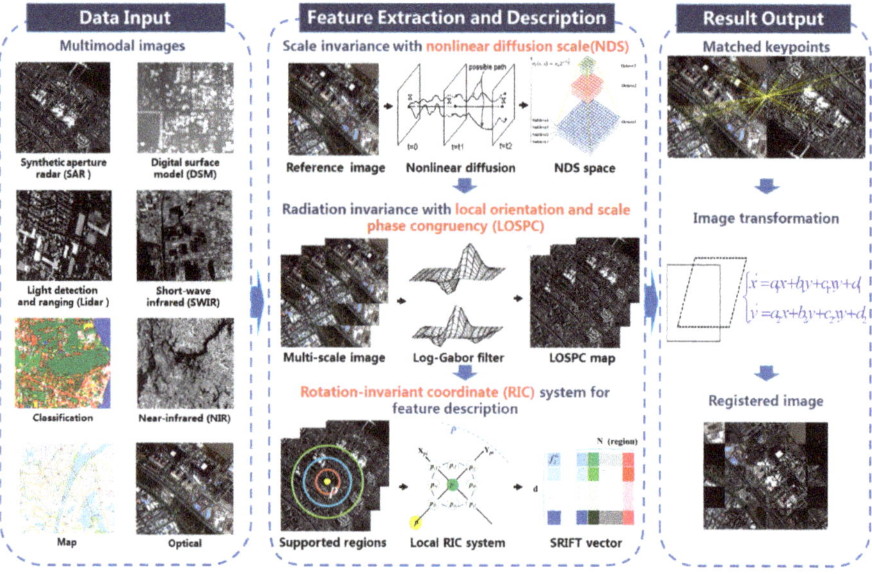

Figure 1. Multimodal image registration flowchart based on scale-radiation-rotation-invariant feature transform (SRIFT).

The fundamental reason that restricts the accurate registration of multiple source images is the nonlinear mapping of the corresponding features caused by the existence of NRD. Therefore, as long as the corresponding features can correspond one-to-one, the registration problem for the different source images can be transformed into a registration problem for the same source images. Hence, in the third column of Figure 1, we can use a large number of methods that have been well studied by predecessors in the field of homologous image registration for the image transformation and resampling.

3.1. Scale Invariance Through the Nonlinear Diffusion Scale (NDS) Space

In the construction process of the scale space, the SIFT algorithm uses a Gaussian space, which is generated by convolving the original image with a Gaussian filter of different scales. However, some structure and contour information in the image will be lost in the filtering process because the Gaussian filtering is a kind of smoothing operator, and the loss of information will adversely affect the feature extraction of multimodal remote sensing images with NRD.

The detailed structural information needs to be included as much as possible when the scale space is established. Inspired by anisotropic diffusion [44], a nonlinear diffusion function, instead of a Gaussian function, is adopted to generate the scale space of the image in the proposed algorithm:

$$\frac{\partial f(x,y)}{\partial t} = f_t = div(c(x,y,t)\nabla f) = c(x,y,t)\Delta f + \nabla c \bullet \nabla f \tag{1}$$

where t is a scale parameter, div is a bifurcation operator, ∇ is the gradient operator,Δ is the Laplace operator, and $c(x,y,t)$ is the diffusion coefficient. In the particular case where the nonlinear diffusion is assumed to be isotropic, $c(x,y,t)$ is a constant, and the above formula is the same as a Gaussian function. Since there is no analytical solution to the nonlinear diffusion equation, a numerical method is needed to approximate the solution. Equation (1) generates the following relationship after being applied as an additive operator splitting strategy:

$$f^{k+1} = \left(I - \tau \sum_{l=1}^{m} A_l(f^k)\right)^{-1} f^k \tag{2}$$

where I is an identity matrix, τ signifies the time step, and l represents the direction. Along the l-th coordinate axis, matrix A_l is established accordingly.

The same tactic is applied as is used in SIFT, where the scale space is discretized into a series of O octaves and S sublevels. By using the original image as an initial condition, the multi-scale space for the multimodal image is generated as a series of smoothed images. The scale values are equal to:

$$s = \sigma^2/2 \tag{3}$$

where σ values are calculated from the following expression:

$$\sigma_i(o,s) = \sigma_0 2^{o+\frac{s}{S}}, o \in [0,\ldots,O-1], s \in [0,\ldots S+2], i \in [0,\ldots W-1] \tag{4}$$

where σ_0 is the base scale level; o and s are the indices of octave O and sublevel S, respectively; and W is the total number of smoothed images. It is notable that the image is downsampled when the last sublevel is reached in each octave, and the downsampled image is used as an initial image for the next octave, as shown in Figure 2.

Through the construction of the NDS space, the attained multi-scale images can preserve the structural and detail information. Therefore, it is anticipated that the proposed SRIFT method will be able to detect many more keypoints.

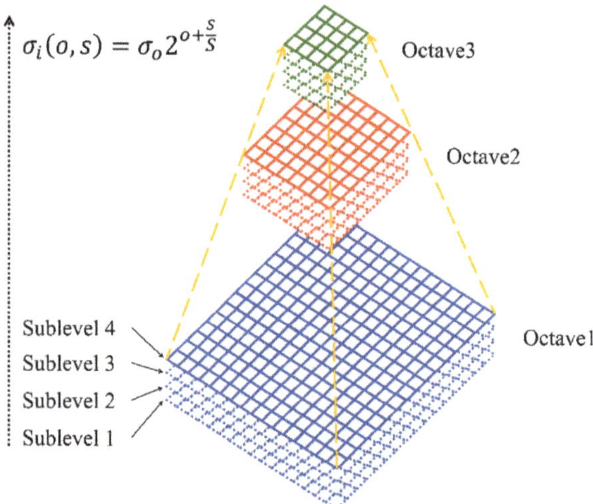

Figure 2. Creation of the nonlinear diffusion scale (NDS) space.

3.2. *Radiation Invariance Through Local Orientation and Scale Phase Congruency (LOSPC)*

It should be noted that the orientation and scale referred to in this section are different from the orientation referred to in Section 3.1 and the scale referred to in Section 3.3. The orientation and scale referred to in this section are aimed at giving the spatial mapping more structural information, and the orientation referred to in Section 3.1 and the scale referred to in Section 3.3 are taken into consideration to make the feature description more stable.

3.2.1. Frequency Domain Spatial Mapping via Phase Congruency

The feature extraction can be carried out after the establishment of the multi-scale space. To solve the NRD problem, the LOSPC algorithm is proposed to construct a unified and describable space, which is a prerequisite for the subsequent extraction of the corresponding features.

Feature extraction in the spatial domain of multimodal images often fails due to the distortion of the grayscale and gradient information. Instead, in the frequency domain, an image is decomposed into amplitude and phase components, where the same kind of ground object will present similar structural distribution features in the multimodal images. Phase congruency can be used to measure the degree of local phase information consistency at various angles [45]. Instead of considering the locations with the maximum intensity gradient as being the edges, the phase congruency model regards the edge points as being where the Fourier components are maximally in phase [46]. The phase information is the measurement describing the structural distribution features of the image in the frequency domain.

A 2-D phase congruency operator is developed for the calculation of the phase congruency of any point in the plane, which is a theory that can also be applied to image processing [47]. We define an image as $I(x,y)$, and then the odd-symmetric part $O_{so}(x,y)$ and even-symmetric part $E_{so}(x,y)$ can be obtained by convolving the image $I(x,y)$ with the log-Gabor wavelet transform:

$$\begin{gathered}[E_{so}(x,y), O_{so}(x,y)] = \\ \left[I(x,y) \otimes L^{even}(x,y,s,o), I(x,y) \otimes L^{odd}(x,y,s,o)\right]\end{gathered} \tag{5}$$

where, in scale o and orientation s, $L^{even}(x,y,s,o)$ and $L^{odd}(x,y,s,o)$ stand for the even-symmetric and the odd-symmetric log-Gabor wavelets, respectively. The amplitude and the phase parts of image $I(x,y)$ can be expressed as:

$$A_{so}(x,y) = \sqrt{E_{so}(x,y)^2 + O_{so}(x,y)^2} \tag{6}$$

$$\phi_{so}(x,y) = \arctan(O_{so}(x,y)/E_{so}(x,y)) \tag{7}$$

When all the scales o and orientations s are considered, the results of the two-dimensional phase congruency are calculated as follows:

$$PC(x,y) = \frac{\sum\limits_{s}\sum\limits_{o} w_o(x,y)\lfloor A_{so}(x,y)\Delta\Phi_{so}(x,y) - T\rfloor}{\sum\limits_{s}\sum\limits_{o} A_{so}(x,y) + \xi} \tag{8}$$

where $w_o(x,y)$ is a weight function; ξ is a constant with a minimal number; the $\lfloor\ \rfloor$ action is to prevent negative values, which means that when the value is negative, its result is 0; and $A_{so}(x,y)\Delta\Phi_{so}(x,y)$ is a phase deviation function, which is defined as:

$$\begin{gathered} A_{so}(x,y)\Delta\Phi_{so}(x,y) = (E_{so}(x,y)\overline{\phi}_E(x,y) \\ +O_{so}(x,y)\overline{\phi}_O(x,y)) - \left|E_{so}(x,y)\overline{\phi}_O(x,y) + O_{so}(x,y)\overline{\phi}_E(x,y)\right| \end{gathered} \tag{9}$$

where

$$\overline{\phi}_E(x,y) = \sum_{s}\sum_{o} E_{so}(x,y)/C(x,y) \tag{10}$$

$$\overline{\phi}_O(x,y) = \sum_{s}\sum_{o} O_{so}(x,y)/C(x,y) \tag{11}$$

$$C(x,y) = \sqrt{\left(\sum_{s}\sum_{o} E_{so}(x,y)\right)^2, \left(\sum_{s}\sum_{o} O_{so}(x,y)\right)^2} \tag{12}$$

Through the above formulas, each pixel in the image can acquire a statistical value for the phase congruency, which contains orientation and scale information and is based on the structural distribution. In different multimodal remote sensing images, the same ground object will have the same structural distribution. Although it has undergone different physical radiation mechanisms, the value of the phase congruency will be the same.

3.2.2. Feature Point Extraction

In the feature extraction stage, it is necessary to design a feature extraction method based on the statistical index of the phase congruency. Equation (8) is used to calculate the phase congruency of the image pixel by pixel, so an edge graph can be obtained, which is robust to the various multimodal remote sensing images. However, when the image is calculated with log-Gabor filters in different directions, the phase congruency should change with the direction of the filter. Unfortunately, this information is not recorded. Therefore, in order to prevent the phase congruency information changing with the direction, it is necessary to calculate the moment of the phase congruency in each direction, and to record the values with the change of direction. In moment statistics, the principal axis indicates that the moment at this axis is the smallest [46]. The moment perpendicular to the principal axis is the maximum moment. If the maximum moment is large, it is likely to be an edge point in the image, while if the minimum moment is large, it is likely to be a corner point in the image.

According to the moment analysis algorithm, the following three statistics can be obtained from the typical moment calculation formula:

$$a = \sum_{o}(PC(\theta_o)\cos(\theta_o))^2 \tag{13}$$

$$b = 2\sum_{o}(PC(\theta_o)\cos(\theta_o))(PC(\theta_o)\sin(\theta_o)) \tag{14}$$

$$c = \sum_{o} (PC(\theta_o)\sin(\theta_o))^2 \tag{15}$$

The angle of the principal axis ψ can then be calculated by the following formula:

$$\psi = \frac{1}{2}\arctan\left(\frac{b}{a-c}\right) \tag{16}$$

After obtaining the principal axis, the minimum moment m_ψ and maximum moment M_ψ can are calculated as follows:

$$m_\psi = \frac{1}{2}\left(c + a - \sqrt{b^2(a-c)^2}\right) \tag{17}$$

$$M_\psi = \frac{1}{2}\left(c + a + \sqrt{b^2 + (a-c)^2}\right) \tag{18}$$

By detecting the extreme values of the maximum moment and the minimum moment on the image, a group of feature points can be obtained, which are called keypoints. The feature extraction step has now been completed. The detection of feature points is undertaken in the frequency domain, instead of the traditional approach of the feature points being detected directly from the image intensity or gradient value, so that the proposed method can better deal with the NRD in multimodal remote sensing images. These keypoints are then used in the subsequent feature matching.

3.3. *Rotation Invariance Through a Rotation-Invariant Coordinate (RIC) System*

After extracting a large number of stable feature keypoints, these points then need to be described. The process of description should consider the change of features with the transformation of the various influencing factors, and should highlight the uniqueness of the features, to ensure the uniqueness of the subsequent feature matching. In order to achieve rotation invariance in the feature description stage, inspired by the work of "aggregating gradient distributions into intensity orders" [48], a RIC system is proposed. In the feature description, the sample points neighboring a keypoint are statistically calculated through a constantly changing local coordinate system, which itself realizes rotation invariance, without the need for the dominant direction. Firstly, we select several candidate support regions, according to a certain proportion. The difference between local orientation and scale phase congruency (DLOSPC) histogram is then calculated in the local RIC system at each sub-region. Finally, by connecting each DLOSPC vector in the image neighborhood of the multiple support regions, the descriptor is constructed. A flowchart of the construction of the descriptor is shown in Figure 3.

3.3.1. The Local Rotation-Invariant Coordinate (RIC) System

In order to realize the rotation invariance of the proposed algorithm, the descriptor of each support region is calculated by the local RIC system. Specifically, a RIC system is built around each keypoint. p_i is a point in one support region of the keypoint p, where the line connecting p and p_i is set as the y-axis, and the direction of the vector $\overrightarrow{pp_i}$ is the y-axis direction. We then construct a local Cartesian (x-y) coordinate system. For the sample points p_i, the pixels in the field participate in the calculation in a rotation-invariant manner; that is to say, the local structure in the field of the sample points is retained. Therefore, the feature description in the locally invariant coordinate system is rotation-invariant. A local RIC system for keypoint p is then constructed. p_i is set as the origin, i.e., the first pixel along the direction of the y-axis is set to p_{i1}, and then the pixels in the eight fields of p_i are marked as $p_{i2}, p_{i3}, \ldots p_{i8}$, as shown in Figure 3b.

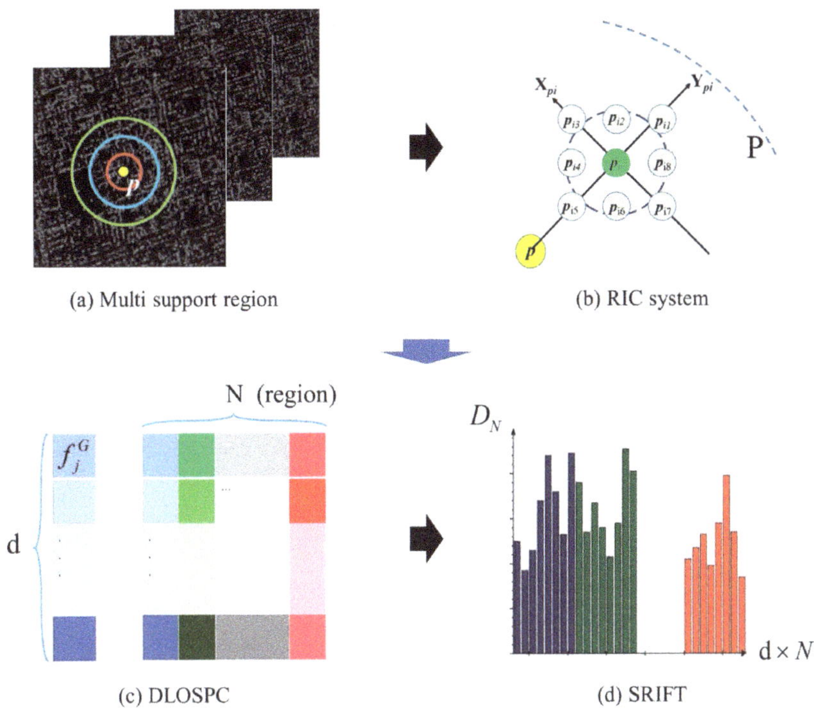

Figure 3. A flowchart of the construction of the descriptor. (**a**) Multiple supported regions. (**b**) Local RIC system. (**c**) difference between local orientation and scale phase congruency (DLOSPC). (**d**) Vectors of SRIFT.

3.3.2. The Difference between Local Orientation and Scale Phase Congruency (DLOSPC) in the RIC System

For each sample point p_i, its difference of local orientation scale phase congruency can be computed in the local RIC system. The calculation formula is as follows:

$$Dx(p_i) = I(p_{i3}) - I(p_{i7}), \tag{19}$$

$$Dy(p_i) = I(p_{i1}) - I(p_{i5}), \tag{20}$$

where p_{ij} are point p_i's neighboring points along the x-axis and y-axis in the local x-y coordinate system, and $I(p_{ij})$ stands for the intensity at p_{ij} on the LOSPC map. The difference $D(p_i)$ and orientation $\theta(p_i)$ can then be computed as:

$$D(p_i) = \sqrt{Dx(p_i)^2 + Dy(p_i)^2}, \tag{21}$$

$$\theta(p_i) = \tan^{-1}(Dy(p_i)/Dx(p_i)). \tag{22}$$

Note that $\theta(p_i)$ is converted into the range of $[0, 2\pi)$, along with the values of $Dx(p_i)$ and $Dy(p_i)$. The gradient of p_i is then constructed as a d-dimensional vector represented as $F_G(p_i) =$

$(f_{G1}, f_{G2}, \ldots, f_{Gd})$. To do this, $[0, 2\pi)$ is divided into d equivalent boxes as $dir_i = (2\pi/d) \times (i-1), i = 1, 2, \ldots, d$, and then $\theta(p_i)$ is assigned to the different boxes by linear distance weighting $D(p_i)$:

$$f_j^G = \begin{cases} D(p_i)\dfrac{\left(2\pi/d - \alpha\left(\theta(p_i), dir_j\right)\right)}{\frac{2\pi}{d}}, \alpha\left(\theta(p_i), dir_j\right) < \frac{2\pi}{d} \\ 0, \text{otherwise} \end{cases} \tag{23}$$

where $\alpha\left(\theta(p_i), dir_j\right)$ is the angle between $\theta(p_i)$ and dir_j.

3.3.3. Construction of the Keypoint Descriptors Through Multiple Supported Regions

As shown in Figure 3, it is not sufficient to distinguish correct matches from a large number of wrong matches with a single support region. Furthermore, two non-corresponding keypoints may exhibit similarity in some support regions. However, the two corresponding keypoints should have a similar appearance in all the support regions of different sizes, although there may be some small differences due to positioning errors of the keypoints and area detection. That is, when multiple support regions are used, mismatches can be better handled than when only a single support region is used.

N nested support regions are selected with the different radius r_i, centered on a keypoint, as shown in Figure 3a. The minimum support region is defined as $A \in I\Re^{2\times 2}$, and then the other support regions can be expressed as $A_i = (1/r_i)A$, where r_i represents the size of the i-th support region. r_i is defined as $r_i = 1 + 0.5 \times (i-1)$ in this paper, so that the radius increments of the support regions are equal.

The cumulative vectors are combined to form a vector in each support region, and then the cumulative vectors of the different support regions are connected together to describe the features of the keypoint. $F(R_i)$ is used to represent the cumulative vector for a support region:

$$F(R_i) = \sum_{p \in R_i} F_G(p), \tag{24}$$

Finally, all the vectors calculated in the N support regions are connected together to form the final descriptor $\{F_1, F_2 \ldots F_N\}$.

4. Experiments and Analyses

The performance of the proposed SRIFT method was tested in both simulated and real-data experiments. In the simulated experiments, the robustness of the algorithm to geometric distortion was tested by artificially adding various scale and rotation distortions. In the real-data experiments, the registration performance obtained by the SRIFT method was compared to that of eight state-of-the-art image registration methods (SIFT [23], ASIFT [24], SAR-SIFT [49], PSO-SIFT [50], DLSS [40], HOPC [41], PCSD [43], and RIFT [51]), with different modal image pairs.

4.1. Experimental Description

The aim of the simulated experiments was to analyze the ability of SRIFT to resist different image geometric distortions, including scale, rotation, and combined distortions.

4.1.1. Simulated Dataset Construction and Evaluation

The simulated dataset experiments involved two sets of images, in which the ground truth had been geometrically corrected and manually checked, so that the positioning accuracy was better than one pixel, as shown in Figures 4 and 5. Figure 4a,b show the first case of a SAR and optical image pair from Shanghai, China, which were acquired by the GF-3 and GF-2 remote sensing satellites, respectively.

Figure 5a,b show the second case of a SAR and optical image pair from Leshan, Sichuan province, China, which were acquired by the Sentinel-1 and Sentinel-2 remote sensing satellites, respectively.

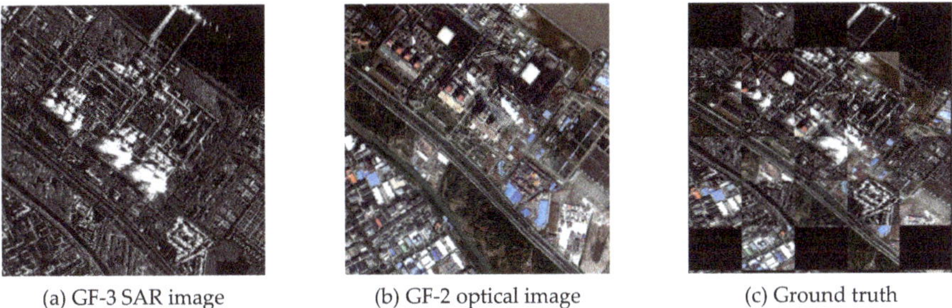

(a) GF-3 SAR image (b) GF-2 optical image (c) Ground truth

Figure 4. The original simulated data and the ground truth for the GF-3 and GF-2 images.

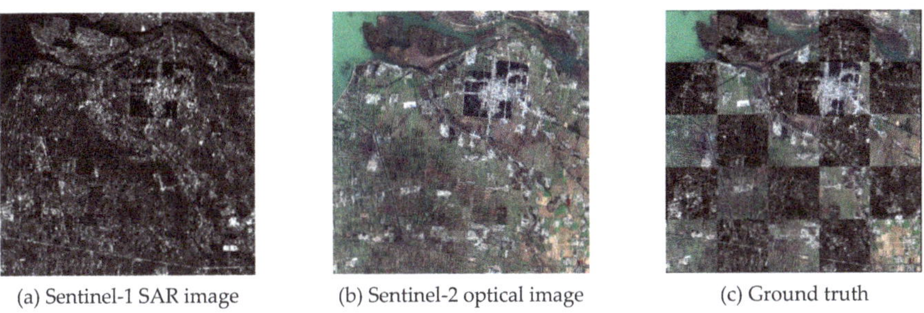

(a) Sentinel-1 SAR image (b) Sentinel-2 optical image (c) Ground truth

Figure 5. The original simulated data and the ground truth for the Sentinel-1 and Sentinel-2 images.

The experimental parameters were set as follows. The simulated datasets were generated by scale and rotation transforms with regard to the ground-truth images, where the images were resampled by 0.5, 0.3, and 0.25 times for the scale transforms and rotated by 10, 30, and 90 degrees, respectively. Therefore, the geometric distortion parameters of the images were known and could be used to calculate the accuracy of the registration results.

If the sensed image I can be regarded as an initial condition, and transform matrices of the simulated dataset can be denoted as T, then image I can be transformed into an image I^T. When the image I has m keypoints $\{p_1, p_2, \ldots, p_m\}$, the corresponding keypoints in the image I^T are $\{p_1^T, p_2^T, \ldots, p_m^T\}$, and are regarded as the ground truth.

The transform matrices calculated after registration are appropriately denoted as $\widetilde{T}$, and the root-mean-square error (RMSE) is used to evaluate the registration accuracy of the simulated data images. The higher the RMSE, the worse the accuracy. The RMSE is defined as:

$$RMSE = \sqrt{\frac{1}{\mathrm{m}}\sum_{i=1}^{\mathrm{m}} (p_i^{\widetilde{T}} - p_i^T)^2} \tag{25}$$

where $(p_i^{\widetilde{T}} - p_i^T)$ is the residual error, which is calculated between the transformation parameters $\widetilde{T}$ and the transformation parameters T. $\widetilde{T}$ is solved by the corresponding points after registration, while T is given by the ground truth. The RMSE is measured in pixels. The keypoints in which the residual error is less than two pixels are regarded as correctly matched. The number of correctly matched (NCM) corresponding keypoints is an important evaluation metric for image matching.

4.1.2. The Overall Performance Comparison

Qualitative evaluation: the overall performances for the corresponding points obtained by the proposed SRIFT method on the simulated datasets are shown in Figures 6 and 7, in which the best results of the state-of-the-art comparison methods for each distortion case are selected for display. As can be seen, the corresponding points obtained by SRIFT are abundant, evenly distributed, and accurately located, and can thus be used to calculate the image transformation model, and then register the image with the calculated model to obtain the registration result.

From the experimental results, it can be seen that, in the process of keypoint extraction, the SRIFT algorithm can extract points characterized by a uniform distribution and sufficient quantity, but after the feature matching and error elimination, the regions with rich shape and texture, and high structural uniqueness, retain more keypoints. In contrast, flat regions with less texture or regions with less structural uniqueness and more repeated textures have fewer corresponding keypoints. Compared with the registration methods based on intensity or gradient, the SRIFT algorithm is essentially a kind of structural descriptor. The description vector of each keypoint is the geometric statistics in an image patch of a specific size centered on this keypoint. In other words, a SRIFT vector describes all the structural information of an image block centered on the keypoint, with a specific sized region (the region sizes are described in Section 3.3.3) as the radius. When the structural information of two image blocks is highly similar, their center points are the registered keypoints.

Quantitative evaluation: Tables 1 and 2 list the RMSEs of the simulated data image registration results of the SRIFT method, as well as those of the other state-of-the-art image registration methods.

As can be seen from Tables 1 and 2, the registration accuracy of the SRIFT algorithm is consistently the highest. Compared with SIFT and ASIFT, the SRIFT algorithm has a better feature extraction and description ability for multimodal images. SAR-SIFT, PSO-SIFT, DLSS, HOPC, PCSD, and RIFT are specially designed for multimodal image registration. Compared with the template matching algorithms such as DLSS and HOPC, SRIFT can resist the various scale and rotation distortions. Compared with feature matching algorithms such as SAR-SIFT, PSO-SIFT, PCSD, and RIFT, SRIFT can overcome more complex image geometric distortions. For a more detailed analysis of the results of the state-of-the-art algorithms, see Section 4.2.3.

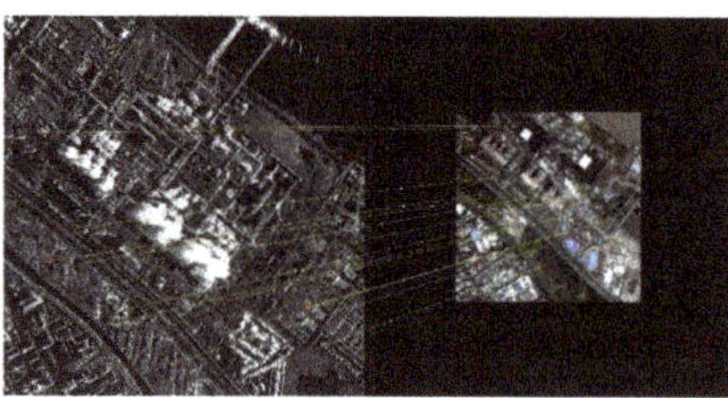

(a) 0.5 times scale by phase congruency-based descriptor (PCSD)

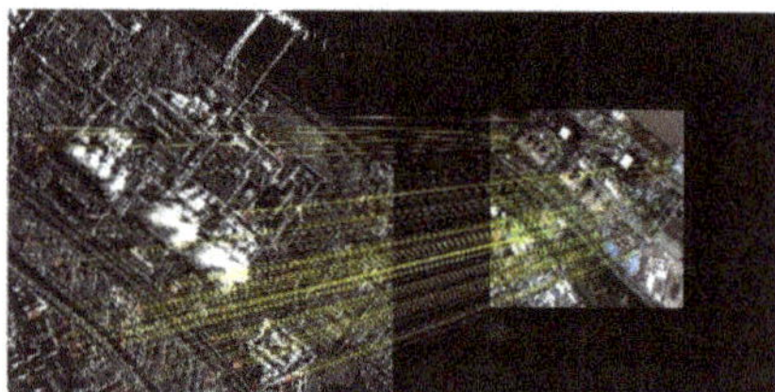

(b) 0.5 times scale by SRIFT

Figure 6. *Cont.*

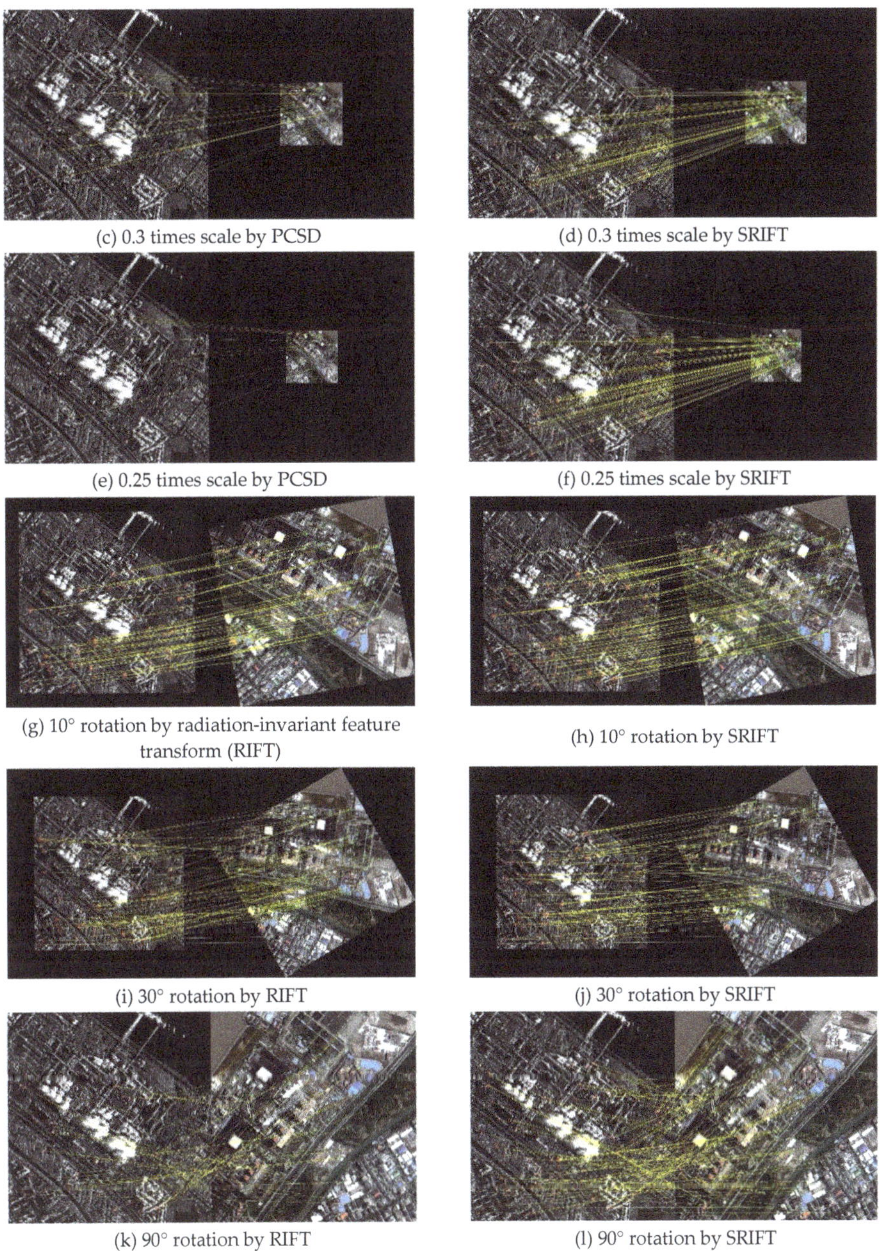

(c) 0.3 times scale by PCSD
(d) 0.3 times scale by SRIFT
(e) 0.25 times scale by PCSD
(f) 0.25 times scale by SRIFT
(g) 10° rotation by radiation-invariant feature transform (RIFT)
(h) 10° rotation by SRIFT
(i) 30° rotation by RIFT
(j) 30° rotation by SRIFT
(k) 90° rotation by RIFT
(l) 90° rotation by SRIFT

Figure 6. The simulated experiment results for the GF-3 and GF-2 images.

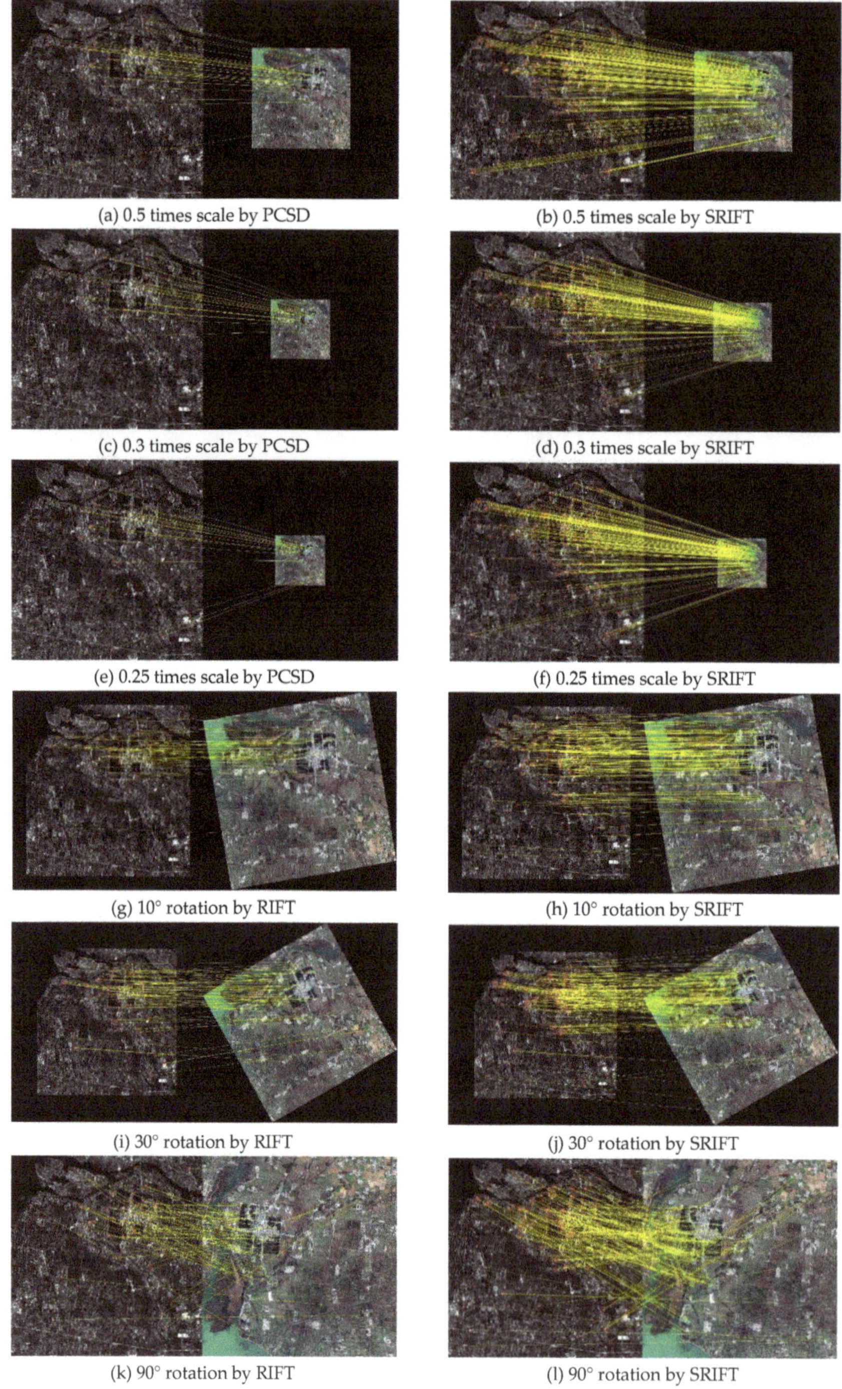
(a) 0.5 times scale by PCSD
(b) 0.5 times scale by SRIFT
(c) 0.3 times scale by PCSD
(d) 0.3 times scale by SRIFT
(e) 0.25 times scale by PCSD
(f) 0.25 times scale by SRIFT
(g) 10° rotation by RIFT
(h) 10° rotation by SRIFT
(i) 30° rotation by RIFT
(j) 30° rotation by SRIFT
(k) 90° rotation by RIFT
(l) 90° rotation by SRIFT

Figure 7. The simulated experiment results for the Sentinel-1 and Sentinel-2 images.

Table 1. Root-mean-square error (RMSE) comparison for simulated dataset 1 (GF-3 SAR image and GF-2 optical image).

Transform	RMSE (Pixels)								
	SIFT	ASIFT	SAR-SIFT	PSO-SIFT	DLSS	HOPC	PCSD	RIFT	SRIFT
Scale									
0.5	-	-	3.89	2.16	-	-	1.93	-	1.36
0.3	-	-	-	-	-	-	2.85	-	1.77
0.25	-	-	-	-	-	-	2.22	-	1.69
Rotation									
10	-	-	2.98	2.88	1.91	1.87	2.06	1.82	1.62
30	-	-	3.33	3.10	-	-	2.96	1.97	1.82
90	-	-	3.12	2.79	-	-	-	1.92	1.49

Table 2. RMSE comparison for simulated dataset 2 (Sentinel-1 SAR image and Sentinel-2 optical image).

Transform	RMSE (Pixels)								
	SIFT	ASIFT	SAR-SIFT	PSO-SIFT	DLSS	HOPC	PCSD	RIFT	SRIFT
Scale									
0.5	-	-	3.65	1.99	-	-	1.89	-	1.21
0.3	-	-	-	-	-	-	2.23	-	1.43
0.25	-	-	-	-	-	-	1.98	-	1.36
Rotation									
10	-	-	2.22	2.14	1.98	1.99	1.96	1.95	1.47
30	-	-	3.84	3.27	-	-	2.66	1.97	1.72
90	-	-	3.01	2.52	-	-	-	1.99	1.36

4.1.3. The Ability of the Algorithm to Resist the Scale and Rotation Distortion

To test the robustness of SRIFT to image scale and rotation distortion, simulated images were generated by resizing the image using a scale change factor from 1 to 10 with different intervals and rotating the image using a rotation change factor from 0 to 360 with an interval of 30. The matching performance of the proposed method with scale and rotation distortion is shown in Figure 8.

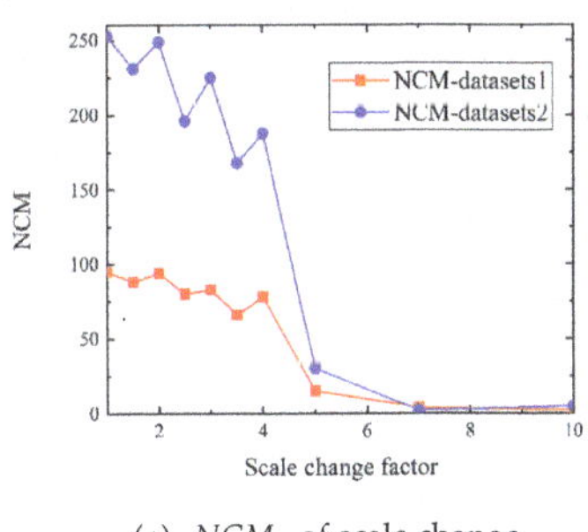

(a) *NCM* of scale change

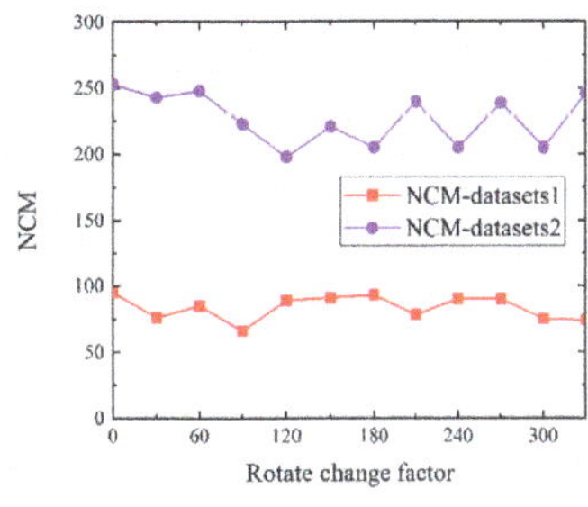

(b) *NCM* of rotation change

Figure 8. SRIFT matching performance with scale and rotation distortion.

As shown in Figure 8a, when the image scale difference is between one time and four times, the precision of the proposed method does not decrease significantly when the scale factor increases. The proposed method maintains a good performance when the scale factor is less than four times. We can, therefore, conclude that the proposed method is robust for scale difference. However, when the scale difference increases to more than five times, the correct matching point number plummets. From Figure 8b, for rotation distortion at any angle, the proposed SRIFT algorithm extracts relatively abundant corresponding points.

4.2. Experiments with Real Images

In the real-data experiments, the registration results obtained by the proposed method were compared to those obtained by eight state-of-the-art methods: SIFT, ASIFT, SAR-SIFT, PSO-SIFT, DLSS, HOPC, PCSD, and RIFT.

4.2.1. Real Datasets

In the real-data experiments, nine sets of multimodal images were selected: SAR/optical, DSM/optical, LiDAR/optical, NIR/optical, SWIR/optical, classification/optical, and map/optical images, which are shown in Figure 9 and described in Table 3.

Figure 9a shows the first SAR/optical image pair covering an urban area. These images were acquired by the GF-3 and GF-2 remote sensing satellites, respectively. The resolution of the SAR image was set to 1 m, referring to the 4-m resolution of the optical image through panchromatic and multispectral fusion. The image sizes are 4865 × 3504 and 3979 × 3619 pixels, respectively.

Figure 9b shows the second SAR/optical image pair covering a mountainous and water area. These images were acquired by the GF-3 and GF-1 remote sensing satellites, respectively. The resolution of the SAR image was set to 8 m, referring to the original 8-m resolution of the multispectral optical image. The image sizes are both 6000 × 6000 pixels.

(a) No. 1 synthetic aperture radar (SAR)/optical (b) No. 2 SAR/optical (c) No. 3 SAR/optical (d) No. 4 digital surface model (DSM)/optical (e) No. 5 light detection and ranging (LiDAR)/optical (f) No. 6 near-infrared (NIR)/optical

Figure 9. *Cont.*

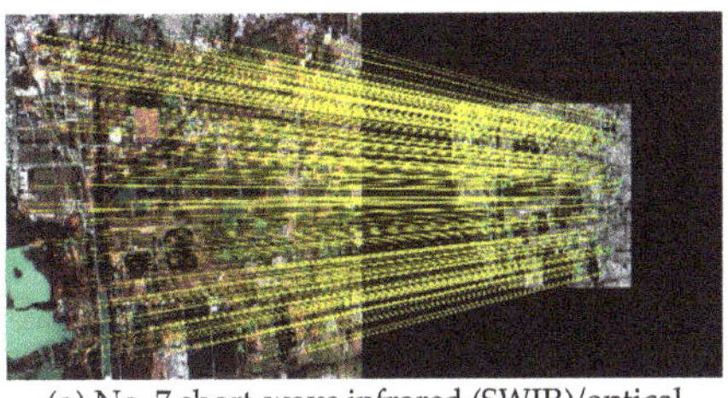

(g) No. 7 short-wave infrared (SWIR)/optical

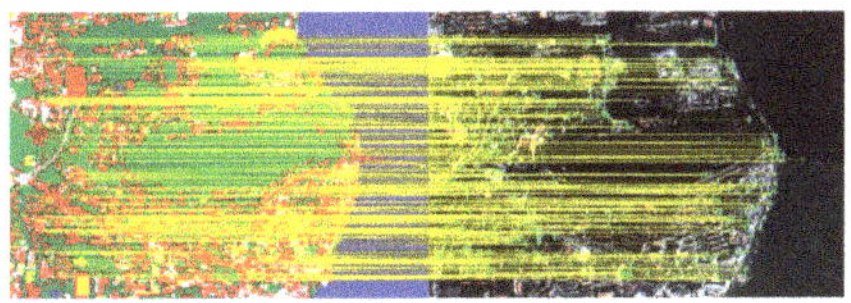

(h) No. 8 Classification/optical

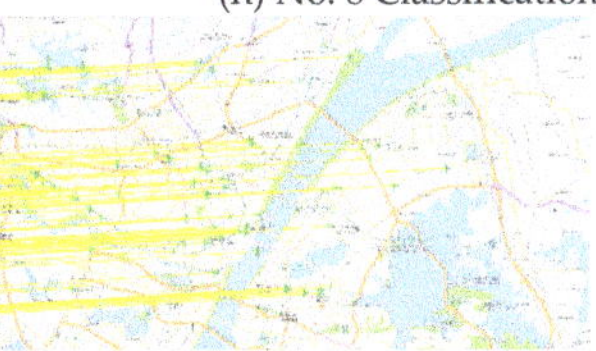

(i) No. 9 Map/optical

Figure 9. Feature point matching for the real multimodal remote sensing image pairs.

Table 3. The real datasets.

No.	Data Source	Image Type	Size	GSD	Date	Image Content	Description
1	GF-3 GF-2	SAR Visible	4865 × 3504 3979 × 3619	4 m 4 m		Urban	SAR/optical
2	GF-3 GF-1	SAR MSS	6000 × 6000 6000 × 6000	8 m 8 m		Mountains and water	SAR/optical
3	Sentinel-1 Sentinel-2	SAR Visible	2000 × 2000 2000 × 2000	10 m 10 m	04/11/2017 06/11/2017	Mountains and water	SAR/optical
4	Manual UAV	DSM Visible	1200 × 1200 1200 × 1200	1 m 1 m	15/05/2017 15/05/2017	Urban	DSM/optical
5	LiDAR Airborne	Intensity Hyperspectral	349 × 349 349 × 349	2.5 m 2.5 m	22/06/2012 23/06/2012	Urban	LiDAR/optical
6	GF-2 Google Earth	NIR Visible	1202 × 1011 1014 × 950	3.2 m 4 m	11/04/2016	Farmland and water	NIR/optical
7	Sentinel-2 Sentinel-2	SWIR Visible	1000 × 1000 1000 × 1000	20 m 10 m		Farmland and water	SWIR/optical
8	Manual GF-1	Classification Visible	640 × 400 640 × 400	4 m 4 m	05/03/2013 05/03/2013	Urban	Classification/optical
9	Google Earth Google Earth	Map Visible	1867 × 1018 1867 × 1018	4 m 4 m		Urban	Map/optical

Figure 9c shows the third SAR/optical image pair, again covering a mountainous and water area. These images were acquired by the Sentinel-1 and Sentinel-2 remote sensing satellites, respectively, in November 2017. The resolution of the SAR image was set to 10 m, referring to the original 10-m resolution of the multispectral optical image. The image sizes are both 2000 × 2000 pixels.

Figure 9d shows the DSM/optical image pair covering an urban area. These images were acquired by manual production and an unmanned aerial vehicle (UAV), respectively, in May 2017. The resolution of the DSM was set to 1 m, referring to the original 1-m resolution of the optical image. The image sizes are both 1200 × 1200 pixels.

Figure 9e shows the LiDAR/optical image pair covering an urban area. These images were acquired by a LiDAR system and aerial photography, respectively, in June 2017. The image resolutions are 2.5 m, and the image sizes are 349 × 349 pixels.

Figure 9f shows the NIR/optical image pair covering a farmland and water area. The NIR image was acquired by the GF-2 remote sensing satellite in April 2016, and the optical image was downloaded from Google Earth. The image resolutions are 3.2 m and 4 m, and the sizes are 1202×1011 and 1014 × 950 pixels, respectively.

Figure 9g shows the SWIR/optical image pair covering a farmland and water area. These images were both acquired by the Sentinel-2 remote sensing satellite. The image resolutions are 20 m and 1 m, respectively, and the image sizes are 1000 × 1000 pixels.

Figure 9h shows the classification/optical image pair covering an urban area. These images were respectively acquired by manual production and the GF-1 remote sensing satellite in March 2013. The resolution of the classification image was set to 4 m, referring to the original 4-m resolution of the multispectral optical image. The image sizes are both 640 × 400 pixels.

Figure 9i shows the map/optical image pair covering an urban area. These images were downloaded from Google Earth. The image resolutions are 4 m, and the image sizes are 1867 × 1018 pixels.

There is severe distortion between all these images, especially radiation distortion. They therefore pose a significant challenge for the image registration algorithms.

4.2.2. Ground-Truth Setting and Evaluation Metrics

Ground truth is essential for the quantitative evaluation of registration. However, due to the different sensors and/or different perspectives in multimodal images, it is often impossible to achieve a one-to-one correspondence between pixels. Therefore, precise geometric correction is required for every set of image pairs in real datasets, and then the geometric correction transformation parameters are approximated as the ground truth. In detail, a certain number of precise corresponding points are artificially selected in each pair of images, and the image transformation parameters are then solved using these points. The images are registered using these parameters, and the registration result is manually checked. If the result is not accurate, the artificial selection is repeated until the two image pairs are accurately registered. When evaluating the registration accuracy, the image transformation parameters estimated by the registration algorithm are used to calculate the residuals of the artificially selected corresponding points.

For the real-data experiments, three evaluation metrics were selected to evaluate the registration performance. The precision is expressed as $Precision = NCM/NM$, where NM is the total number of corresponding keypoints. The definitions of NCM and RMSE have been given in Section 4.1

4.2.3. Registration Performance Comparison

Qualitative comparison: Figure 9 intuitively shows the corresponding point-line diagrams obtained by the SRIFT algorithm in the real-image experiments, where the number and distribution of the corresponding points reflect the robustness and applicability of the algorithm. Figure 10 uses checkerboard mosaicked images of the nine groups for the qualitative evaluation, where the continuousness of the sub-region edges of the images directly reflect the accuracy of the registration.

Through the analysis of the above experimental results, the following conclusions can be drawn. The proposed SRIFT registration algorithm was able to obtain satisfactory results on all nine datasets of various multimodal remote sensing images. When dealing with the multimodal image registration task, the SRIFT algorithm fully considers the influence of NRD in the feature extraction, as well as the feature description, so that it can extract a large number of stable and evenly distributed feature points. The SRIFT algorithm can also resist image scale and rotation distortion, and the registration image has a high coincidence degree with the reference image.

Quantitative comparison: Figure 11 quantitatively reflects the registration effect and accuracy of the eight algorithms on the nine sets of data with the three measurement indices introduced in Section 4.2. Figure 11a is the line chart of NCM, where the higher the value of NCM, the more keypoints are correctly matched, which reflects the ability of the different algorithms in the feature matching stage. Figure 11b is the line chart of the precision, where the higher the value, the higher the proportion of correct matching points in all the matching points, reflecting the ability of the different algorithms in the feature description stage. Figure 11c is the line chart of the RMSE, where the lower the value, the higher the registration accuracy, reflecting the overall registration ability of the different algorithms.

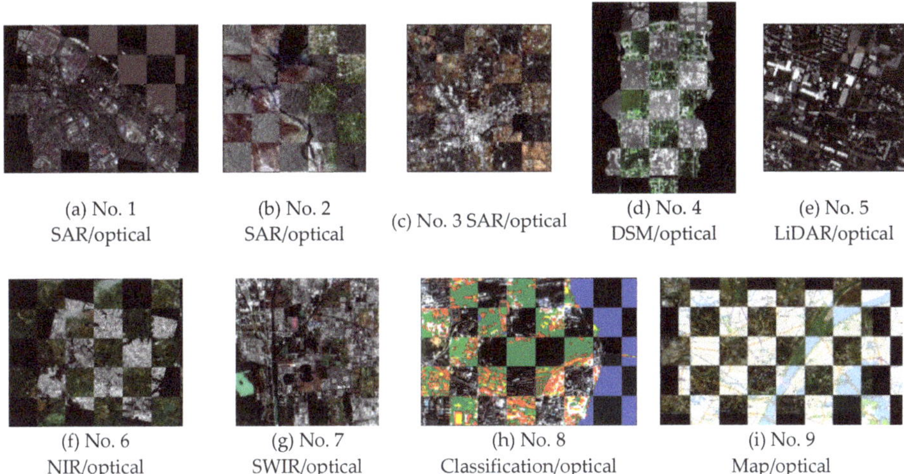

Figure 10. Multimodal remote sensing image registration display.

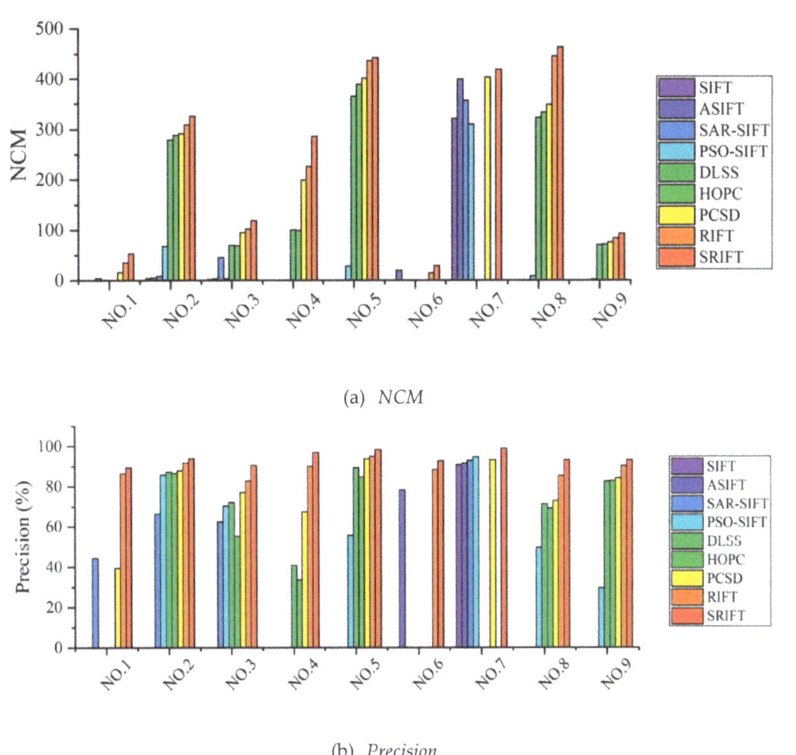

Figure 11. *Cont.*

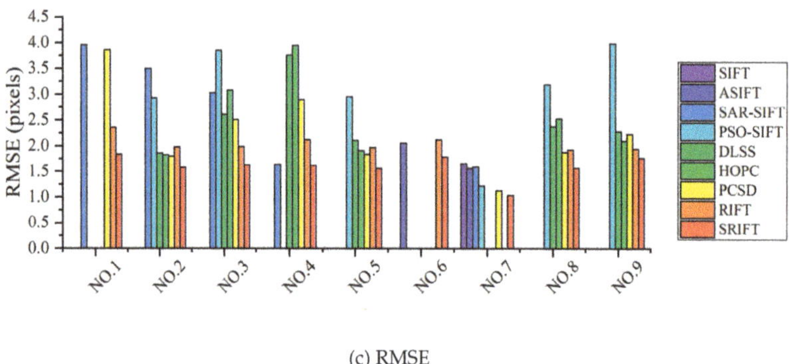

(c) RMSE

Figure 11. Performance comparison of the different descriptors on the input images.

As can be seen in Figure 11, SRIFT achieves the best precision. RIFT ranks second and PCSD ranks third. The basic idea of the HOPC and DLSS algorithms is to divide the images into blocks by counting the local information of each image block, and to then integrate the blocks into the overall information. Their performance generally lies in the middle level among the eight compared methods. The SIFT and ASIFT algorithms are not designed for multimodal data, so that they perform the worst of all. A detailed analysis of each method is presented in the following.

Due to the SIFT algorithm detecting feature points directly based on the intensity, and using gradient information for the feature description, which is sensitive to NRD, SIFT only obtains good registration results for the SWIR/optical case. In the SWIR/optical case, the sensors are on the same satellite, and the two sources have little difference in radiation mechanism, so that the image registration is relatively easy.

The results of the ASIFT algorithm are slightly better than those of SIFT. The registration results cannot be obtained for most of the images, but the registration results are better than those of SIFT in the complex set of geometric transforms in the NIR/optical case, because ASIFT is specially designed for affine transformation. ASIFT simulates the scale and the camera direction and normalizes the rotation and the translation.

The SAR-SIFT method was specially designed for SAR imagery, and it relies on a new gradient computation method adapted to SAR images. Therefore, the image registration results for the first three SAR datasets are satisfactory. However, the redefined gradient probability has difficulty in dealing with complex radiation distortion, and the multi-scale Harris detector has insufficient resistance to NRD.

PSO-SIFT achieves the best registration effect among the SIFT-related algorithms, which is due to the fact that PSO-SIFT applies multiple constraints, e.g., the feature distance, and hence results in a better registration.

The DLSS algorithm is an improved version of the LSS algorithm, which divides the template window into spatial regions called "cells." Each cell contains n × n pixels, and has an overlapping region of half a cell width with the neighboring cell. This method of division is essentially a template matching method, rather than a feature matching method. As a result, DLSS cannot resist the complex geometric distortion, and the registration performance in datasets 1, 4, 6, 7, and 9 is poor.

HOPC uses the Harris detector to detect the feature points. However, the Harris detector is very sensitive to NRD, and it is not universally suitable for all the different types of multimodal images. Therefore, overall, its registration effect is slightly worse than that of DLSS. Moreover, HOPC is similar to DLSS in the blocking strategy, so that HOPC also has a poor effect in images that cannot be registered by the DLSS algorithm.

The RIFT algorithm does not have scale invariance, so its registration effect in the SWIR/optical case is inferior. In order to achieve rotation invariance, RIFT transforms the initial layer to reconstruct a set of convolution sequences with different initial layers, and then calculates a maximum index map (MIM) from each convolution sequence to obtain a set of MIMs. This statistical method only establishes the relationship between the neighborhood pixels of the keypoints and the keypoints themselves, and it destroys the structural relationship between the neighborhood pixels of the keypoints. Therefore, for images with both scale and rotation transformation, the registration effect is weak.

In the feature description stage, the PCSD algorithm adopts a method similar to SIFT, which requires estimation of the main direction, and error in the primary direction estimation causes the extracted corresponding points to be deleted by mistake. In the feature extraction stage of PCSD, the points with the closest cosine similarity to the keypoints in the reference image are matched. However, this method does not take advantage of a phase congruency algorithm in the feature extraction, and its stable alignment points are insufficient.

The analysis of the experimental results confirms the powerful registration ability of the SRIFT algorithm. If the structural information of the images to be registered is similar, the SRIFT algorithm can register the images without considering the distortion of the image intensity.

4.2.4. Computational Cost

An experiment addressing the computational cost of the proposed method and that of some of the comparison methods (SIFT, SAR-SIFT, PCSD, HOPC, RIFT, SRIFT) was undertaken. The average run times of the different methods on the real datasets are shown in Table 4. The experiment is carried out on a computer with an Intel i7-6200U CPU @ 2.30GHz and 8 GB of RAM. All the methods were implemented in MATLAB.

Table 4. Computational cost comparisons (seconds).

Method	SIFT	SAR-SIFT	PCSD	HOPC	RIFT	SRIFT
Computational cost	2.04	15.25	46.82	57.63	38.65	64.70

According to Table 4, the SIFT algorithm has the highest computational efficiency, followed by the SAR-SIFT algorithm, because neither algorithm performs phase congruency calculation. Among the four methods that adopt phase congruency (PCSD, HOPC, RIFT, and SRIFT), RIFT shows the highest computational efficiency because it replaces phase congruency with maximum index graph. PCSD and HOPC are in the middle level, but the calculation time of the proposed method is relatively high. Due to the use of the multi-scale space, phase congruency, and RIC system to deal with the NRD, as well as the complex scale and rotation variation, the proposed method is relatively time-consuming. Algorithm optimization and efficiency improvement will be carried out in the future.

5. Conclusions

In this paper, we have presented a modality-free multimodal remote sensing image registration method named SRIFT, which has the advantage of scale, radiation, and rotation invariance, making it suitable for use with different multimodal remote sensing images. The building of a robust NDS space, the definition of a new concept called LOSPC, and the development of a new RIC system are the three main contributions of the proposed SRIFT method. The NDS space is constructed to resist the scale distortion of image pairs with a large difference in gradient distribution. LOSPC is computed in the NDS space of the images, in which the same kind of ground object will present similar structural distributions, and thus the features of these images are mapped into the same space. The idea of the RIC system is that the points in the neighborhood are statistically calculated through a continually changing local coordinate system, which is more suitable for the feature matching task than a global coordinate system, and realizes rotation invariance, without the need for the estimated orientation to be assigned. In the experimental analysis, two simulated datasets and nine sets of real data were used to qualitatively and quantitatively compare the registration performance of SIFT, ASIFT,

SAR-SIFT, PSO-SIFT, DLSS, HOPC, PCSD, RIFT, and the proposed SRIFT method. The registration performance of the SRIFT algorithm on the multimodal images with NRD was superior to that of the other state-of-the-art image registration methods. Our future study will focus on research into a correction model and error elimination for multimodal image registration.

Author Contributions: S.C. established the motivation, designed the method, developed the code, performed the experiments, and wrote the manuscript; M.X. and Y.Z. provided funding; A.M. and Y.Z. reviewed and improved the manuscript. All authors have read and agreed to the published version of the manuscript.

Funding: This research was supported by the National Key Research and Development Program of China under grant nos. 2018YFB0504801 and 2017YFB0504202, and in part by the Fundamental Research Funds for the Central Universities under grand no. 2042020kf0014, and in part by the National Natural Science Foundation of China under grant nos. 41622107 and 41801267.

Acknowledgments: The authors would like to thank Yuanxin Ye for sharing the code of the HOPC method and some of the experimental data, and Wenping Ma for sharing the code of the PSO-SIFT method.

Conflicts of Interest: The authors declare no conflicts of interest.

References

1. Zitová, B.; Flusser, J. Image registration methods: A survey. *Image Vis. Comput.* **2003**, *21*, 977–1000. [CrossRef]
2. Wu, Y.; Fan, J.; Li, S.; Wang, F.; Liang, W.; Wu, Y.; Fan, J.; Li, S.; Wang, F.; Liang, W. Fusion of synthetic aperture radar and visible images based on variational multiscale image decomposition. *J. Appl. Remote Sens.* **2017**, *11*, 025006. [CrossRef]
3. Hirschmuller, H. Stereo processing by semiglobal matching and mutual information. *IEEE Trans. Pattern Anal. Mach. Intell.* **2007**, *30*, 328–341. [CrossRef]
4. Jia, L.; Li, M.; Zhang, P.; Wu, Y. Sar image change detection based on correlation kernel and multistage extreme learning machine. *IEEE Trans. Geosci. Remote Sens.* **2016**, *54*, 5993–6006. [CrossRef]
5. Pratt, W. Correlation techniques of image registration. *IEEE Tans. Aerosp. Electron. Syst.* **1974**, *10*, 353–358. [CrossRef]
6. Mahmood, A.; Khan, S. Correlation-coefficient-based fast template matching through partial elimination. *IEEE Trans. Image Process.* **2012**, *21*, 2099–2108. [CrossRef] [PubMed]
7. Yang, L.; Tian, Z.; Zhao, W.; Yan, W.; Wen, J. Description of salient features combined with local self-similarity for sar image registration. *J. Indian Soc. Remote Sens.* **2017**, *45*, 131–138. [CrossRef]
8. Viola, P.; Wells, W.M., III. Alignment by maximization of mutual information. *Int. J. Comput. Vis.* **1997**, *24*, 137–154. [CrossRef]
9. Oliveira, F.P.M.; Tavares, J.M.R.S. Medical image registration: A review. *Comput. Methods Biomech. Biomed. Eng.* **2014**, *17*, 73–93. [CrossRef] [PubMed]
10. Wu, Y.; Miao, Q.; Ma, W.; Gong, M.; Wang, S. Psosac: Particle swarm optimization sample consensus algorithm for remote sensing image registration. *IEEE Geosci. Remote Sens. Lett.* **2018**, *15*, 242–246. [CrossRef]
11. Zhang, J.; Zareapoor, M.; He, X.; Shen, D.; Feng, D.; Yang, J. Mutual information based multi-modal remote sensing image registration using adaptive feature weight. *Remote Sens. Lett.* **2018**, *9*, 646–655. [CrossRef]
12. Hu, H.; Pun, C.-M.; Liu, Y.; Lai, X.; Yang, Z.; Gao, H. An artificial bee algorithm with a leading group and its application into image registration. *Multimed. Tools Appl.* **2019**, *79*, 14643–14669. [CrossRef]
13. Chen, S.; Li, X.; Zhao, L.; Yang, H. Medium-low resolution multisource remote sensing image registration based on sift and robust regional mutual information. *Int. J. Remote Sens.* **2018**, *39*, 3215–3242. [CrossRef]
14. De, C.E.; Morandi, C. Registration of translated and rotated images using finite fourier transforms. *IEEE Trans. Pattern Anal. Mach. Intell.* **1987**, *9*, 700–703.
15. Tong, X.; Ye, Z.; Xu, Y.; Liu, S.; Li, L.; Xie, H.; Li, T. A novel subpixel phase correlation method using singular value decomposition and unified random sample consensus. *IEEE Trans. Geosci. Remote Sens.* **2015**, *53*, 4143–4156. [CrossRef]
16. Brown, L.G. A survey of image registration techniques. *ACM Comput. Surv.* **1992**, *24*, 325–376. [CrossRef]
17. Zeng, Q.; Adu, J.; Liu, J.; Yang, J.; Gong, M. Real-time adaptive visible and infrared image registration based on morphological gradient and c_sift. *J. Real Time Image Process.* **2019**, *17*, 1103–1115. [CrossRef]
18. Rister, B.; Horowitz, M.A.; Rubin, D.L. Volumetric image registration from invariant keypoints. *IEEE Trans. Image Process* **2017**, *26*, 4900–4910. [CrossRef]

19. Hou, Y.; Zhou, S. Robust point correspondence with gabor scale-invariant feature transform for optical satellite image registration. *J. Indian Soc. Remote Sens.* **2017**, *46*, 1–12. [CrossRef]
20. Yan, L.; Wang, Z.; Liu, Y.; Ye, Z. Generic and automatic markov random field-based registration for multimodal remote sensing image using grayscale and gradient information. *Remote Sens.* **2018**, *10*, 1228. [CrossRef]
21. Xiang, Y.; Wang, F.; You, H. An automatic and novel sar image registration algorithm: A case study of the chinese gf-3 satellite. *Sensors* **2018**, *18*, 672. [CrossRef]
22. Xiang, Y.; Feng, W.; Ling, W.; You, H. An advanced rotation invariant descriptor for sar image registration. *Remote Sens.* **2017**, *9*, 686. [CrossRef]
23. Lowe, D.G. Distinctive image features from scale-invariant keypoints. *Int. J. Comput. Vis.* **2004**, *60*, 91–110. [CrossRef]
24. Morel, J.M.; Yu, G. Asift: A new framework for fully affine invariant image comparison. *Siam J. Imaging Sci.* **2009**, *2*, 438–469. [CrossRef]
25. Bay, H.; Ess, A.; Tuytelaars, T.; Van Gool, L. Speeded-up robust features (surf). *Comput. Vis. Image Und.* **2008**, *110*, 346–359. [CrossRef]
26. LeCun, Y.; Bengio, Y.; Hinton, G. Deep learning. *Nature* **2015**, *521*, 436–444. [CrossRef]
27. Reichstein, M.; Camps-Valls, G.; Stevens, B.; Jung, M.; Denzler, J.; Carvalhais, N.; Prabhat. Deep learning and process understanding for data-driven earth system science. *Nature* **2019**, *566*, 195–204. [CrossRef]
28. Niethammer, M.; Kwitt, R.; Vialard, F.-X. Metric learning for image registration. In Proceedings of the IEEE Conference on Computer Vision and Pattern Recognition, Nashville, TN, USA, 19–25 June 2019; pp. 8463–8472.
29. Shen, Z.; Han, X.; Xu, Z.; Niethammer, M. Networks for joint affine and non-parametric image registration. In Proceedings of the IEEE Conference on Computer Vision and Pattern Recognition, Long Beach, CA, USA, 29–31 October 2019; pp. 4224–4233.
30. Zhang, H.; Ni, W.; Yan, W.; Xiang, D.; Wu, J.; Yang, X.; Bian, H. Registration of multimodal remote sensing image based on deep fully convolutional neural network. *IEEE J. Sel. Top. Appl. Earth Obs. Remote Sens.* **2019**, *18*, 1–15. [CrossRef]
31. Ma, W.; Zhang, J.; Wu, Y.; Jiao, L.; Zhu, H.; Zhao, W. A novel two-step registration method for remote sensing images based on deep and local features. *IEEE Trans. Geosci. Remote Sens.* **2019**, *57*, 4834–4843. [CrossRef]
32. Merkle, N.; Luo, W.; Auer, S.; Müller, R.; Urtasun, R. Exploiting deep matching and sar data for the geo-localization accuracy improvement of optical satellite images. *Remote Sens.* **2017**, *9*, 586. [CrossRef]
33. Merkle, N.; Auer, S.; Müller, R.; Reinartz, P. Exploring the potential of conditional adversarial networks for optical and sar image matching. *IEEE J. Sel. Top. Appl. Earth Obs. Remote Sens.* **2018**, *11*, 1811–1820. [CrossRef]
34. Sedaghat, A.; Mohammadi, N. High-resolution image registration based on improved surf detector and localized gtm. *Int. J. Remote Sens.* **2018**, *40*, 2576–2601. [CrossRef]
35. Wang, S.; Quan, D.; Liang, X.; Ning, M.; Guo, Y.; Jiao, L. A deep learning framework for remote sensing image registration. *ISPRS J. Photogramm. Remote Sens.* **2018**, *145*, 148–164. [CrossRef]
36. Xu, C.; Sui, H.G.; Li, D.R.; Sun, K.M.; Liu, J.Y. An automatic optical and sar image registration method using iterative multi-level and refinement model. *Int. Arch. Photogramm. Remote Sens. Spat. Inf. Sci.* **2016**, *7*, 593–600. [CrossRef]
37. Aguilera, C.; Barrera, F.; Lumbreras, F.; Sappa, A.D.; Toledo, R. Multispectral image feature points. *Sensors* **2012**, *12*, 12661–12672. [CrossRef]
38. Chen, J.; Tian, J.; Lee, N.; Zheng, J.; Smith, R.T.; Laine, A.F. A partial intensity invariant feature descriptor for multimodal retinal image registration. *IEEE Trans. Biomed. Eng.* **2010**, *57*, 1707–1718. [CrossRef]
39. Shechtman, E.; Irani, M. Matching local self-similarities across images and videos. In Proceedings of the IEEE Computer Society Conference on Computer Vision and Pattern Recognition, Minneapolis, MN, USA, 17–22 June 2007; pp. 1–8.
40. Ye, Y.; Shen, L.; Hao, M.; Wang, J.; Xu, Z. Robust optical-to-sar image matching based on shape properties. *IEEE Geosci. Remote Sens. Lett.* **2017**, *14*, 564–568. [CrossRef]
41. Ye, Y.; Shen, L. Hopc: A novel similarity metric based on geometric structural properties for multi-modal remote sensing image matching. *ISPRS Ann. Photogramm. Remote Sens. Spat. Inf. Sci.* **2016**, *3*, 9–16. [CrossRef]
42. Li, J.; Hu, Q.; Ai, M. Rift: Multi-modal image matching based on radiation-variation insensitive feature transform. *IEEE Trans. Image Process.* **2020**, *29*, 3296–3310. [CrossRef]

43. Fan, J.; Wu, Y.; Li, M.; Liang, W.; Cao, Y. Sar and optical image registration using nonlinear diffusion and phase congruency structural descriptor. *IEEE Trans. Geosci. Remote Sens.* **2018**, *56*, 1–12. [CrossRef]
44. Perona, P.; Malik, J. Scale-space and edge detection using anisotropic diffusion. *IEEE Trans. Pattern Anal. Mach. Intell.* **1990**, *12*, 629–639. [CrossRef]
45. Kovesi, P. Phase congruency: A low-level image invariant. *Psychol. Res.* **2000**, *64*, 136–148. [CrossRef]
46. Kovesi, P. Phase is an important low-level image invariant. In *Pacific Rim Conference on Advances in Image and Video Technology*; Springer: Berlin/Heidelberg, Germany, 2007.
47. Kovesi, P. Image features from phase congruency. *J. Comput. Vis. Res.* **1999**, *1*, 115–116.
48. Fan, B.; Wu, F.; Hu, Z. Aggregating gradient distributions into intensity orders: A novel local image descriptor. In Proceedings of the IEEE Computer Society Conference on Computer Vision and Pattern Recognition, Providence, RI, USA, 20–25 June 2011; pp. 2377–2384.
49. Dellinger, F.; Delon, J.; Gousseau, Y.; Michel, J.; Tupin, F. Sar-sift: A sift-like algorithm for sar images. *IEEE Trans. Geosci. Remote Sens.* **2015**, *53*, 453–466. [CrossRef]
50. Ma, W.; Wen, Z.; Wu, Y.; Jiao, L.; Gong, M.; Zheng, Y.; Liu, L. Remote sensing image registration with modified sift and enhanced feature matching. *IEEE Geosci. Remote Sens. Lett.* **2017**, *14*, 3–7. [CrossRef]
51. Li, J.; Hu, Q.; Ai, M. Rift: Multi-modal image matching based on radiation-invariant feature transform. *arXiv* **2018**, arXiv:1804.09493.

Article

PWNet: An Adaptive Weight Network for the Fusion of Panchromatic and Multispectral Images

Junmin Liu *, Yunqiao Feng, Changsheng Zhou and Chunxia Zhang

School of Mathematics and Statistics, Xi'an Jiaotong University, Xi'an 710049, China; fyq9719@stu.xjtu.edu.cn (Y.F.); zhouchangsheng3@stu.xjtu.edu.cn (C.Z.); cxzhang@mail.xjtu.edu.cn (C.Z.)
* Correspondence: junminliu@mail.xjtu.edu.cn; Tel.: +86-029-82663155

Received: 31 July 2020; Accepted: 24 August 2020; Published: 29 August 2020

Abstract: Pansharpening is a typical image fusion problem, which aims to produce a *high resolution multispectral* (HRMS) image by integrating a high spatial resolution *panchromatic* (PAN) image with a low spatial resolution *multispectral* (MS) image. Prior arts have used either *component substitution* (CS)-based methods or *multiresolution analysis* (MRA)-based methods for this propose. Although they are simple and easy to implement, they usually suffer from spatial or spectral distortions and could not fully exploit the spatial and/or spectral information existed in PAN and MS images. By considering their complementary performances and with the goal of combining their advantages, we propose a *pansharpening weight network* (PWNet) to adaptively average the fusion results obtained by different methods. The proposed PWNet works by learning adaptive weight maps for different CS-based and MRA-based methods through an end-to-end trainable *neural network* (NN). As a result, the proposed PWN inherits the data adaptability or flexibility of NN, while maintaining the advantages of traditional methods. Extensive experiments on data sets acquired by three different kinds of satellites demonstrate the superiority of the proposed PWNet and its competitiveness with the state-of-the-art methods.

Keywords: pansharpening; component substitution; multiresolution analysis; neural networks; adaptive weight

1. Introduction

Due to technical limitations [1], current satellites, such as QuickBird, IKONOS, WorldView-2, GeoEye-1, can not obtain the high spatial resolution *multispectral* (MS) images, but only acquire an image pair with complementary features, i.e., a high spatial resolution *panchromatic* (PAN) image and a low spatial resolution MS image with rich spectral information. To get high-quality products, pansharpening is proposed with the goal of fusing MS and PAN images to generate *high resolution multispectral* (HRMS) image with the same *spatial* resolution of the PAN image and the *spectral* resolution of the MS image [2,3]. It can be cast as a typical kind of *image fusion* [4] or *super-resolution* [5] problems and has a wide range of real-world applications, such as enhancing the visual interpretation, monitoring the land cover change [6], object recognition [7], and so on.

Over decades of studies, a large number of pansharpening methods have been proposed in the literature of remote sensing [3]. Most of them can be categorized into the following two main classes [2,3,8]: (1) *component substitution* (CS)-based methods and (2) *multiresolution analysis* (MRA)-based methods. The CS class first transforms the original MS image into a new space and then substitutes one component of the transformed MS image by the histogram matched PAN image. The representative methods of the CS class are *Intensity-Hue-Saturation* (IHS) [9], *generalized IHS* (GIHS) [10], *principal component analysis* (PCA) [11], Brovey [12], among many others [13–17]. The MRA-based class is also known as the class of spatial methods, which extracts the high spatial frequencies of the high resolution PAN image through multiresolution analysis tools (e.g., wavelets or Laplacian pyramids) to enhance the spatial information of MS image. The representative methods belonging to the MRA-based class are *high-pass filtering* (HPF) [18], *smoothing filter-based intensity modulation* (SFIM) [19], the *generalized Laplacian pyramid* (GLP) [20,21], among many others [22,23]. The two class methods are fast and easy to implement. However, for the CS-based methods, local dissimilarities between PAN and MS images can not be eliminated, resulting in spectral distortion, and for the MRA-based methods, they have a relatively less spectral distortion but with limited spatial enhancement. From the above, the CS-based and the MRA-based methods usually have complementary performances in improving the spatial quality of MS images while maintaining the corresponding spectral information.

To balance the trade-off performances of the CS-based and MRA-based methods, the hybrid methods by combining both of these two classes have been proposed in recent years. For example, the *additive wavelet luminance proportional* (AWLP) [24] method is proposed by Otazu et al. via implementing the "à trous" wavelet transform in the IHS space. Shah et al. [25] proposed a method by combining an adaptive PCA method with the discrete contourlet transform. Liao et al. [26] proposed an framework, called *guided filter PCA* (GFPCA), which performs a *guided filter* in the PCA domain. Although the hybrid methods have an enhanced performance to the CS-based or MRA-based methods, these improvements are limited due to their hand-crafted design.

Recently, significant progress on improving the spatial and spectral qualities of the fused images for the classical methods has been achieved by *variational optimization* (VO)-based methods [27–31] and learning-based methods, among which *convolution neural network* (CNN)-based methods are the most popular, due to their powerful capability and the end-to-end learning strategy. For instance, Masi et al. introduced a CNN architecture with three layers in [32] for the pansharpening problem. Another novel CNN-based model, which is focused on preserving spatial and spectral information, is designed by Yang et al. in [33]. Inspired by these work, Liu et al. [34] proposed a two-stream CNN architecture with ℓ_1-norm loss function to further improve the spatial quality. Zheng et al. [35] proposed a CNN-based method by using deep hyperspectral prior and dual-attention residual network to deal with the problem of that the discriminative ability of CNNs is sometimes hindered. Though having great ability of automatically extracting features and the state-of-the-art performances, CNN-based methods usually require intensive computational resources [36]. In addition, unlike the CS-based and MRA-based methods, CNN-based methods are lack of interpretability and are more like a black-box game. A detailed summary and relevant works for the VO-based methods can be found in [2]. We do not discuss the VO class for more since this paper focuses on a combination of the other three classes.

In this paper, we propose a *pansharpening weight network* (PWNet) to bridge the classical methods (i.e., CS-based and MRA-based methods) and the learning-based methods (typically the CNN-based methods). On one hand, similar to the hybrid methods, PWNet can combine the merits of the CS-based and the MRA-based methods. On the other hand, similar to learning-based methods, PWNet is data-driven and is very effective and efficient. To achieve this, PWNet uses the CS-based and MRA-based methods as inference modules and utilizes CNN to learn adaptive weight maps for weighting the results of the classical methods. Unlike the above hybrid methods with hand-crafted design, the PWNet can be seen as an automatic and data-driven hybrid method for pansharpening. In addition, the structure of PWNet is very simple to ease training and save computational time.

The main contributions of this work are as follows:

- A model average network, called *pansharpening weight network* (PWNet), is proposed. The PWNet can be trained and is the first attempt to combining the classical methods via an end-to-end trainable network.
- PWNet integrates the complementary characteristics of the CS-based and MRA-based methods and the flexibility of the learning-based (typically the CNN-based) methods, providing an avenue to bridge the gap between them.

- PWNet is data-driven, and can automatically weight the contributions of different CS-based and MRA-based methods on different data sets. By visualizing the weight maps, we prove that the PWNet is adaptive and robust to different data sets.
- Extensive experiments on three kinds of data sets have been conducted and shown that the fusion results obtained by PWNet achieve state-of-the-art performance compared with the CS-based, MRA-based methods and other CNN-based methods.

The paper is organized as follows. In Section 2, we briefly introduce the background of the CS-based, MRA-based and learning-based methods. Section 3 introduces the motivation, network architecture, and other details of PWNet. In Section 4, we conduct the experiments, analyze the parameter setting and time complexity and present the comparisons with the-state-of-art methods at the reduced and full scales. Finally, we draw the conclusion in Section 5.

2. Related Work

Notations. We denote the *low resolution multispectral* (LRMS) image by $MS \in \mathbb{R}^{H\times W\times N}$, where H, W, and N are the width, the height, and the number of spectral bands of the LRMS image, respectively. We denote the high resolution PAN image by $P \in \mathbb{R}^{rH\times rW}$, where r is the spatial resolution ratio between MS and PAN, denote by $\widehat{MS} \in \mathbb{R}^{rH\times rW\times N}$ the reconstructed HRMS image. We let MS_k to represent the kth band of the LRMS image, where $k = 1, \ldots, N$, and let $\widetilde{MS}_k \in \mathbb{R}^{rH\times rW}$ to represent the upsampled version of MS_k by ratio r. For notational simplicity, we also denote by P the histogram matched PAN image. Based on these symbols, we next briefly introduce the main idea of the CS-based, MRA-based and learning-based methods.

2.1. The CS-Based Methods

The CS-based methods are based on the assumption that the spatial and spectral information of LRMS image can be separated by a projection or transformation of the original LRMS image [3,37]. The CS class usually has four steps: (1) upsample the LRMS image to the size of the PAN image; (2) use a linear transformation to project the upsampled LRMS image into another space; (3) replace the component containing the spatial information with the PAN image; (4) perform an inverse transformation to bring the transformed MS data back to their original space and then get the pansharpened MS image (i.e., the estimated HRMS). Due to the changes in low spatial frequencies of the MS image, the substitution procedure usually suffers from spectral distortion. Thus, spectral matching procedure (i.e., histogram matching) is often applied before the substitution.

Mathematically, above fusion process can be simplified without the calculation of the forward and backward transformation as shown in Figure 1, which leads the CS class to have the following equivalent form as

$$\widehat{MS}_k = \widetilde{MS}_k + g_k(P - I_L), \quad k = 1, \ldots, N, \tag{1}$$

where $g_1, \ldots, g_N$ are the injection gains, and I_L is a linear combination of the upsampled LRMS image bands and often called *intensity component*, defined as

$$I_L = \sum_{k=1}^{N} w_k \widetilde{MS}_k, \tag{2}$$

where $w_1, \ldots, w_N$ usually correspond to the first row of the forward transformation matrix, which is used to measure the degrees of spectral overlap between the MS and PAN channels.

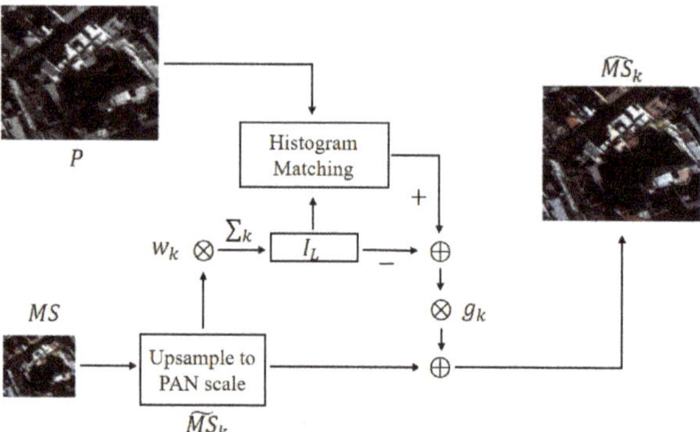

Figure 1. Flowchart of the CS-based methods for pansharpening.

Numerous CS-based methods have been proposed to sharpen the LRMS images according to Equation (1) and flowchart in Figure 1. The CS class includes IHS [9] which exploits the transformation into the IHS color space and its generalized version GIHS [10], PCA [11] based on the statistical irrelevance of each principal component, Brovey [12] based on a multiplicative injection scheme, *Gram-Schmidt* (GS) [13] which conducts the Gram-Schmidt orthogonalization procedure and by a weighted average of the MS bands minimizing the *mean square error* (MSE) with respect to a low-pass filtered version PAN image in the *adaptive* GS (GSA) [15], *band-dependent spatial detail* (BDSD) [14] and its enhanced version (i.e., *BDSD with physical constraints*: BDSD-PC) [16], *partial replacement adaptive component substitute* (PRACS) [17] based on the concept of *partial replacement* of the intensity component and so on. Each method differs from the others by the different projections of the MS images used in the process and by the different designs of injection gains. Although they show extreme performances in improving the spatial qualities of LRMS images, they usually suffer from heavily spectral distortions in some scenarios due to local dissimilarity or the not well-separated spatial structure with the spectral information. Refer to [3] for more detailed discussions about this.

2.2. The MRA-Based Methods

Unlike the CS-based methods, the MRA class is based on the operator of multi-scale decomposition or low-pass filter (equal to a single scale of decomposition) over the PAN image [3,37]. They first extract the spatial details over a wide range of scales from the high resolution PAN image or from the difference between the PAN image and its low-pass filtered version P_L, and then inject the extracted spatial details into each band of upsampled LRMS image. Figure 2 shows the general flowchart of the MRA-based methods.

Generally, for each band $k = 1, 2, \cdots, N$, the MRA-based methods can be formulated as

$$\widehat{MS}_k = \widetilde{MS}_k + g_k(P - P_L). \tag{3}$$

As we can see from above Equation (3), different MRA-based methods can be distinguished by the way of obtaining P_L and by the design of injection gains $g_1, g_2, \cdots, g_N$. Several methods belonging to this class have been proposed, such as HPF [18] using the box mask and additive injection, the SFIM [19], *decimated Wavelet transform using additive injection model* (Indusion) [23], the AWLP [24], GLP with *modulation transfer function* (MTF)-matched filter (denoted by MTF-GLP) [21], its HPM injection version (MTF-GLP-HPM) [22] and context-based decision version (MTF-GLP-CBD) [38], *a trous wavelet transform using the model* 3 (ATWT-TM3) [39], and so on.

The MRA-based methods highlight the extraction of multi-scale and local details from the PAN image, well in reducing the spectral distortion but compromising the spatial enhancement. To make up this problem, many approches have been proposed by the utilization of different decomposition schemes (e.g., *morphological filters* [40]) and the optimization of the injection gains.

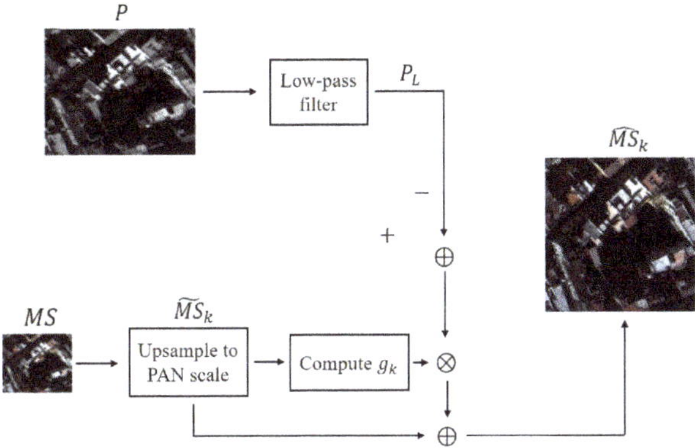

Figure 2. Flowchart of the MRA-based methods for pansharpening.

2.3. *The Learning-Based Method*

Apart from the traditional CS-based and MRA-based methods, the learning-based methods have been proposed or applied to the pansharpening, among which the CNN-based methods are the most popular [41]. The CNN-based methods are very flexible, and one can design a CNN with different architectures. Due to the end-to-end and data-driven properties, they achieve the state-of-the-art performances in some studies [32–35]. After a network architecture design, training image pairs with *low resolution* MS (LRMS) images as network input and *high resolution* MS (HRMS) images as network output, are needed to learn the network parameters θ. The learning procedure is based on the choice of loss function and optimization method, and the effect of learning is different from each other according to these choices of loss function and optimization strategy. However, these ideal image pairs are unavailable, and usually simulated based on scale invariant assumption by properly downsampling both PAN and the original MS images to a reduced resolution. Then, the resolution reduced MS images and the original MS can be used as an input-output pair.

Given the input-output MS pairs of $\widetilde{MS}^i$ with low resolution and Y^i with high resolution, and aided by the auxiliary PAN image P^i, $i = 1, 2, \ldots, n$, the CNN-based methods optimize the parameter by minimizing the following cost function

$$\mathcal{L}(\theta) = \sum_{i=1}^{n} ||f(P^i, \widetilde{MS}^i; \theta) - Y^i||_F^2 \tag{4}$$

where $f(P^i, \widetilde{MS}^i; \theta)$ denotes a neural network which takes θ as parameters, and $|| \cdot ||_F$ is the Frobenius norm, which is defined as the square root of the sum of the absolute squares of the elements.

To further improve the performances of CNN-based methods, recent work mainly resorts to the deep residual architecture [42] or to increase the depth of the model to extract multi-level abstract features [43]. However, these will require large number of network parameters and burden computation [36]. Unlike the above CNN-based methods that aim at generating the HRMS images or their residual images, we here to reduce the number of parameters and reduce the requirements on the computation capacity of the computer by learning weight maps for the CS-based and the MRA-based methods. Refer to the following section for more detailed discussions about this.

3. The Proposed PWNet Method

3.1. Motivation and Main Idea

According to the above analysis, the CS-based and MRA-based methods are simple and usually have complementary performances, i.e., the CS-based methods are good at spatial rendering but sometimes suffer from severe spectral distortions, while the MRA-based methods performance well in keeping the spectral information of the MS images but may have limited spatial enhancements. And the performances of the CS-based and MRA-based methods show data uncertainty, i.e., they have different fusion performances on different scenarios. The learning-based methods, especially the CNN-based methods, perform well in reducing spatial and spectral distortions due to their powerful feature extraction capabilities and data-driven training scheme. However, they usually need an extremely large data set to train the model parameters and are difficult to be interpretable.

Is there a way to make full use of the complementary performances of the CS-based and MRA-based methods at the same time reducing their data uncertainty? A straightforward idea is to firstly generate multiple fusion results by multiple methods (i.e., the CS-based and MRA-based methods), and then automatically combine them with weights based on performances within different scenarios to boost the fusion result. This may be realized by using a trainable CNN since it is data-driven and has strong abilities in the field of image processing.

Motivated by the above, we propose a novel model average method, referred to as *pansharpening weight netowrk* (PWNet), for the pansharpening. Specifically, rather than generating only one estimated HRMS image at a time, we use multiple inference modules to generate distinct estimated HRMS images at a time. Each inference module produces a distinct estimate of HRMS with bias, and multiple estimates have the positive and negative deviations. And then the biases can be complemented by averaging the multiple results, thus leading to the distortions of average are smaller than that of a single estimate. In order to make use of the simplicity and complementary characteristics of the CS-based and MRA-based methods, we choose them as inference modules, i.e., use each CS-based or MRA-based method as an inference module, and then design an end-to-end trainable network with the original MS and PAN images as input to simultaneously obtain weight maps for all fusion results obtained by the CS and MRA inference modules. Based on the powerful capability and data-driven training scheme of nerual network, the output weight maps are context and method dependent. Finally, we get an estimated HRMS image through adaptively averaging all the fusion results obtained by the CS-based and MRA-based methods. Figure 3 depicts the main procedures of the proposed PWNet for pansharpening.

3.2. Network Architecture

Pansharpening Weight Network (PWNet): The key of model average is to assign proper weights to different models depending on their performances. Unlike traditional model average methods that are based on hand-crafted design to assign weights, we resort the neural network to adaptively generate weights. The proposed PWNet is composed of two subnet: the *CS weight network* and the *MRA weight network*. For each subnet, the original LRMS and/or PAN images are took as input. Similar to [33], high-pass filter operation is first implemented in order to preserve edges and details. For the CS weight subnet, the high-pass filtered MS image is upsampled though a *transpose convolution* and concatenated with the high-pass filtered PAN image, and then fed into four *residual blocks* [44]. Different from this, the MRA weight subnet only passes the high-pass filtered PAN image into the subsequent residual blocks. Each of the residual blocks consists of three convolutional layers with each having a learnable filter of size 3×3 and a *rectified-linear unite* (ReLU) activation function [45]. To generate weight maps, the activation function of the output layer is set to be a *softmax* function. By considering computation efficiency and also for keeping the proportions between each pair of the MS bands unchanged, the PWNet only outputs one weigh map for each CS-based or MRA-based method to achieve a *pixel-wise* aggregation, which will be discussed below. The detailed architecture of PWNet for generating the weight maps is shown in Figure 4.

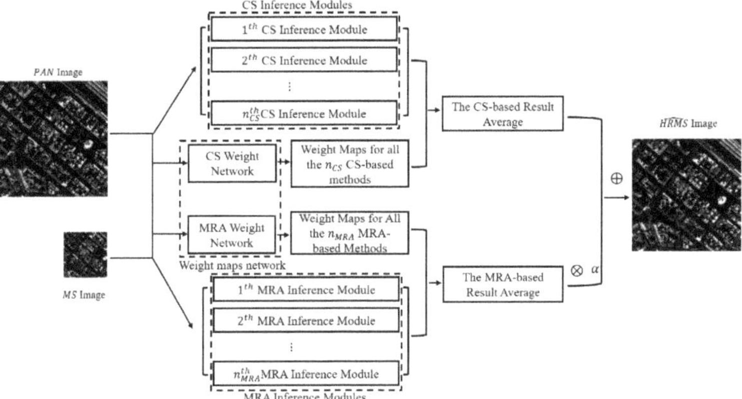

Figure 3. Blockdiagram of the proposed PWNet for pansharpening. The proposed method can be divided into three independent part, i.e., the weight maps network, the CS inference modules and the MRA inference modules. The weight maps network takes the reduced MS and PAN image as inputs, and outputs weight maps for each of the CS-based and MRA-based methods. At the same time, the CS inference modules and the MRA inference modules generate HRMS images according to specific pansharpening methods. Finally, the estimated HRMS image is obtained by averaging all HRMS images estimated by the selected CS-based and MRA-based methods with the weight maps generated by the weight maps network.

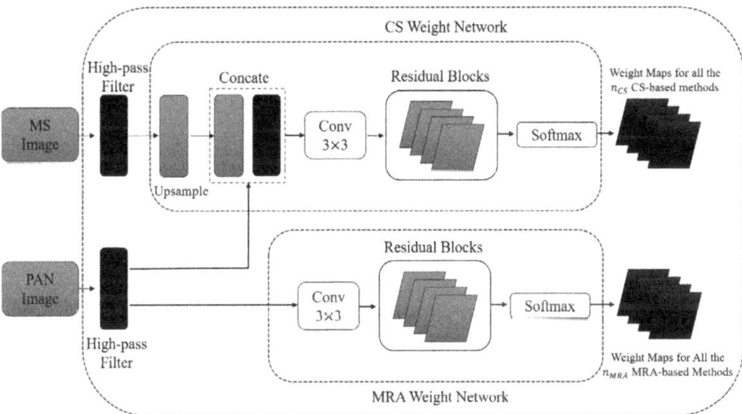

Figure 4. The architecture of proposed PWNet for generating the weight maps.

The CS-based Result Average: Suppose we have n_{CS} kinds of CS-based methods for our PWNet, and then, for the ith CS-based method, we have its estimated HRMS image as

$$\left[\widehat{MS}_{CS_{i1}}, \widehat{MS}_{CS_{i2}}, \cdots, \widehat{MS}_{CS_{iN}}\right] = CS_i(P, \widetilde{MS}) \tag{5}$$

$$i = 1, 2, \cdots, n_{CS} \tag{6}$$

where $\widehat{MS}_{CS_{ik}}, k = 1, 2, \cdots, N$, is the kth MS band and $CS_i(\cdot)$ denotes the ith CS-based method. According to the CS weight network, we get adaptive weight maps for the used CS-based methods as:

$$\left[W_{CS_1}, W_{CS_2}, \cdots, W_{CS_{n_{CS}}}\right] = f(G(P), G(MS); \theta_{CS}) \tag{7}$$

where W_{CS_i} is the ith output weight map generated by the CS weight network, $f(G(P), G(MS); \theta_{CS})$ denotes the CS weight network, θ_{CS} is the parameter set, the function $G(\cdot)$ is the high-pass filter. At last, we conduct pixel-wise multiplication for each estimated HRMS image and its corresponding weight map W_{CS_i} and sum the multiplication results to get the CS-based result average, i.e., for each band, we have the averaged result as

$$\widehat{MS}_{CS_k} = \sum_{i=1}^{n_{CS}} \widehat{MS}_{CS_{ik}} \odot W_{CS_i}, \tag{8}$$

$$k = 1, ..., N, \tag{9}$$

where $\widehat{MS}_{CS_k}$ denotes the result of the kth band of CS-based method module and $\odot$ denotes the point-wise multiplication.

The MRA-based Result Average: Consistent with the procedures of the CS-based method module, the averaged results for the n_{MRA} MRA-based methods can be given as

$$\widehat{MS}_{MRA_k} = \sum_{i=1}^{n_{MRA}} \widehat{MS}_{MRA_{ik}} \odot W_{MRA_i} \tag{10}$$

$$k = 1, ..., N, \tag{11}$$

where $\widehat{MS}_{MRA_{ik}}$ and W_{MRA_i} are the kth HRMS band obtained by the ith MRA-based method and the ith output weight map generated by the MRA weight network, which respectively are given as

$$\left[W_{MRA_1}, W_{MRA_2}, \cdots, W_{MRA_{n_{MRA}}}\right] = g(G(P); \theta_{MRA}) \tag{12}$$

and

$$\left[\widehat{MS}_{MRA_{i1}}, \widehat{MS}_{MRA_{i2}}, \cdots, \widehat{MS}_{MRA_{iN}}\right] = MRA_i(P, \widetilde{MS}), \quad i = 1, 2, \cdots, n_{MRA} \tag{13}$$

where $g(G(P); \theta_{MRA})$ denotes the MRA weight network, θ_{MRA} is the parameter set, and $MRA_i(\cdot)$ is the ith MRA-based method. Note that, different from the CS-based weight network, we only take the PAN image as the input of the MRA-based weight network since the MRA-base methods extract the spatial details only rely on the PAN image.

The Final Result Aggregation: After we have obtained the averaged results of the CS-based method module, $\widehat{MS}_{CS} = [\widehat{MS}_{CS_1}, \cdots, \widehat{MS}_{CS_N}]$, and the averaged results of the MRA-based method module, $\widehat{MS}_{MRA} = [\widehat{MS}_{MRA_1}, \cdots, \widehat{MS}_{MRA_N}]$, then we can aggregate them for compensating their spatial and/or spectral distortions. The final estimated HRMS image $\widehat{MS}$ can be given as

$$\widehat{MS} = \widehat{MS}_{CS} + \alpha \widehat{MS}_{MRA} \tag{14}$$

where α is a factor for balancing the contributions between the CS-based methods and the MRA-based methods.

The Loss function: To learn the model parameters $\theta = \{\theta_{CS}, \theta_{MRA}\}$, we would like to minimize the reconstruction errors between the estimated HRMS image, $\widehat{MS}$, and its corresponding ideal one, Y, i.e.,

$$\min_{\theta} \frac{1}{n} \sum_{i=1}^{n} ||\widehat{MS}^i - Y^i||_F^2 \tag{15}$$

where n is the number of training samples.

4. Experiments

4.1. Data Sets and Implementation Details

We conduct several experiments using three data sets respectively collected by the GeoEye-1, WorldView-2, and QuickBird satellites. Each data set is split into two nonoverlapping subset: a training data set and a testing data set. Each sample in the data set consists of an MS and PAN image pairs with the PAN image of size 64×64 and the MS image of size 16×16. In order to verify the generalization of the proposed method, we also perform pansharpening on a scene taken on other days for QuickBird satellite at the full resolution experiment. More detailed information of the three data sets is reported in Table 1.

Table 1. Information of the three data set.

Satellite	Resolution of PAN	Resolution of MS	Number of Training Data Set	Number of Test Data Set
GeoEye-1	0.5 m	2.0 m	182	21
WordView-2	0.5 m	2.0 m	224	32
QuickBird	0.6 m	2.4 m	529	60

As for the training of our proposed PWNet, due to the unavailable of the ideal HRMS images, similar to the other CNN-based pansharpening methods [32–34], we first follow the Wald's protocol [46] to generate the training input MS and PAN pairs by downsampling both the original PAN and MS images with scale factor $r = 4$ (i.e., the resolution of the MS and PAN images is reduced by applying the MTF-matched low-pass filters [21]), and then the the original MS images are treated as target outputs.

The PWNet is implemented in Tensorflow and trained on a Intel(R) Core(TM) i5-4210U CPU. We use the Adam algorithm [47] with an initial learning rate of 0.001 to optimize the network parameters. And we set the maximum number of epoch to 1000 and mini-batch sample size to 32. It takes about 1 h to train our network.

We first evaluate the methods at a reduced resolution, in addition to *visual analysis* on the experimental results, the proposed PWNet and other compared methods are also evaluated by five widely used quantitative metrics, namely, universal image quality index [48] averaged over the bands (Q_avg) and its four band extension, Q4 [3], *spectral angle mapper* (SAM) [49], *Erreur Relative Globale Adimensionnelle de Synthése* (ERAGS) [50] and the *spatial correlation coefficient* (SCC) [46]. The closer to one the Q_avg, Q4, and SCC, the better the quality of fused results, while the lower the SAM and ERGAS, the better the fusion quality.

We also evaluate the methods at a full resolution. In this case, the *quality with no-reference index* (QNR) [51] and its spatial index (D_S) and spectral index (D_λ) are employed for the quantitative assessment. It should be pointed out that the quantitative assessment at full resolution is challenging since these indexes (i.e., QNR, D_S and D_λ) are not computed with unattainable ground truth, but rely heavily on the original MS and PAN images [43]. This tends to quantify the similarity of certain components in the fused images to the low-resolution observations, which will lead biases in these indexes estimation. Due to this reason, some methods can generate images with high QNR values but poor image qualities [52].

In the following, we have carried out five sets of experiments to perform comprehensive analysis on the proposed PWNet, typically the effect of the hyperparameter α, the number of CS-based and MRA-based methods, the weight maps channels, and the quantitative, visual and running time comparisons with the CS-based, MRA-based and learning-based methods at reduced resolution and full resolution.

4.2. Analysis to the Hyper-Parameters α

There is a hyper-parameter α in our proposed PWNet, which is to balance the contributions of the CS-based and MRA-based methods. In this experiment, we will analyze this parameter to optimize the performance of PWNet. We fix the number of CS-based and MRA-based methods to six and change α from 0.1 to 1 with interval 0.1. The results obtained are shown in Table 2. As we can see from it, the PWNet attains constantly good performances when α varies from 0.7 to 0.9. Specially, the best results can be obtained for $\alpha = 0.7$. It is worth noting that when α goes to 1, the quantitative indexes seem to become worse. Thus, $\alpha = 0.7$ can be a relatively good choice in the following experiments.

Table 2. Quantiative results obtained by PWNet with different α.

α	Q_avg	SAM	ERGAS	SCC	Q4
0.1	0.9807	3.0250	2.5026	0.9831	0.9836
0.2	0.9825	2.9946	2.3140	0.9855	0.9854
0.3	0.9824	3.0216	2.2994	0.9858	0.9851
0.4	0.9831	**2.9901**	2.2947	0.9861	0.9856
0.5	0.9811	3.2168	2.3072	0.9854	0.9852
0.6	0.9807	3.3581	2.3935	0.9843	0.9854
0.7	**0.9834**	3.0276	**2.2191**	0.9866	**0.9863**
0.8	0.9831	3.0429	2.2504	**0.9867**	0.9859
0.9	0.9832	3.0543	2.2536	0.9862	**0.9863**
1	0.9493	3.2763	3.9663	0.9582	0.9546

4.3. Impact of the Number of the CS-Based and MRA-Based Methods

This experiment shows what influences would be produced by the different number of the CS-based and MRA-based methods under the condition of $n_{CS} = n_{MRA}$ for simplicity and also for keeping the balance between the CS class and the MRA class methods. The number of methods to be averaged is very important to balance the spectral and spatial information from the LRMS and high resolution PAN images. Too few methods might extract features incompletely and result in a poor performance, while too many would suffer from computational burden during testing. We would like to find a trade-off value according to the performance and running time. Therefore, we limit the range of n_{CS} (or n_{MRA}) to between 5 and 8.

Table 3 gives the quantitative results and the running time (in second) when the number of averaged methods varies from 5 to 8. Note that, n_{MRA} and n_{CS} equal 5, which means that the selected CS-based are PCA [11], GIHS [10], Brovey [12], GS [13], GSA [15] and the selected MRA-based methods are HPF [18], SFIM [19], AWLP [24], MTF-GLP-HPM [22], MTF-GLP-CBD [38], respectively. When both n_{CS} and n_{MRA} are equal to 6, the PRACS [17] and the ATWT-M3 [39] methods are added into the CS-based and MRA-based methods, respectively. When n_{MRA} and n_{CS} are equal to 7, we add the BDSD [14] into the CS weight network and the Indusion [23] into the MRA weight network, and the BDSD-PC [16] and the MTF-GLP [21] are added into the CS weight network and the MRA weight network as method modules when n_{MRA} and n_{CS} are equal to 8. As reported in Table 3, the performance of the proposed PWNet is improved when the number of averaged methods increases from 5 to 7, while it decays when n_{CS} and n_{MRA} are equal to 8. This reveals that the use of less number of averaged methods will reduce the performances of the PWNet and increasing the number of averaged methods will not continuously bring improvements but will suffer from more computation. In principle, our proposed PWNet is data-driven and thus can automatically weight different kinds of CS-based and/or MRA-based methods. In practice, we suggest two criteria for the selection of CS-based and MRA-based methods for the proposed PWNet. First, the number of CS-based methods and the number of MRA-based methods should be equal in order to keep the balance contribution of the CS class and MRA class. And then, to further improve the performances and robustness of the

PWNet, we suggest selecting the CS-based and MRA-based methods according to their performances reported in [3].

Table 3. Quantitative results and running time with different number of the averaged methods.

n_{CS} and n_{MRA}	Q_avg	SAM	ERGAS	SCC	Q4	Time
5	0.8781	5.9214	3.5179	0.8609	0.8761	**0.3800**
6	0.9079	5.7174	3.0616	0.9073	0.9062	0.5900
7	**0.9209**	**5.1654**	**2.8986**	**0.9131**	**0.9165**	0.9200
8	0.8781	5.4549	3.6942	0.8633	0.8615	1.6200

4.4. Impact on the Number of Weight Map Channels

In order to reduce the number of parameters and the computational cost, we set the weight map for each CS-based or MRA-based method to be one channel. That is, each MS band of a HRMS image obtained by a CS-based or MRA-based method share the same weight map. In general, the model capacity will be increased with the number of model parameters. Thus, we conduct the experiments based on different output weight map channels to verify whether the capacity of our model has suffered from the reduction of channels.

The results of PWNet with one shared weight map channel and with four different weight map channels for each CS-based or MRA-based method are reported in Table 4. As we can see from it, the PWNet with one shared weight map channel attains constantly good performances in terms of the five commonly used metrics on three different kinds of satellite data sets, while the PWNet with four different weight map channels for each method has a relatively poor performances with the same training conditions. This may due to that an under-fitting phenomenon caused by excessive parameters has happened in the PWNet with four weight map channels. It further verifies the advantages of the PWNet with one shared weight map channel, which can lower the training difficulty.

Table 4. Results of the proposed PWNet with one shared weight map channel or four different weight map channels for each CS-based or MRA-based method. The best results are highlighted in bold.

Satellite	Number of Channels	Index				
		Q_avg	SAM	ERGAS	SCC	Q4
WordView-2	1 channel	**0.9874**	**2.7582**	**2.3830**	**0.9813**	**0.9805**
	4 channels	0.3390	14.7379	14.2755	0.8378	0.8007
GeoEye-1	1 channel	**0.9207**	**4.2120**	**2.6868**	**0.9384**	**0.9117**
	4 channels	0.3585	19.9451	14.5300	0.8141	0.5585
QuickBird	1 channel	**0.8941**	**2.8252**	**1.8102**	**0.9480**	**0.8227**
	4 channels	0.3442	19.6107	14.6745	0.8867	0.1462

4.5. Comparison with the CS-Based and MRA-Based Methods

One key issue of the proposed PWNet method is whether its fusion result is better than that of each participating method. Only when the answer to this question is yes, we can claim that the proposed PWNet can produce appropriate weight maps for each CS-based or MRA-based method, so that the methods involved in model average can complement each other and the result can be improved. Here we set n_{MRA} and n_{CS} to 7 as we have proved in above that this setting has a good balance for performances and running times. The compared pansharpening methods include seven methods belonging to the CS class, namely, PCA [11], GIHS [10], Brovey [12], BDSD [14], GS [13], GSA [15], PRACS [17], and seven methods belonging to the MRA class such as HPF [18], SFIM [19], Indusion [23], AWLP [24], ATWT-M3 [39], MTF-GLP-HPM [22], MTF-GLP-CBD [38]. All methods follow the experimental settings recommended by the authors.

We first inspect the visual quality of the pansharpening results. Figures 5–7 present the pansharpened images on all the three data sets, obtained by our proposed PWNet and the other fourteen methods. As we can see from theses figures, the CS-based methods produce a relatively sharper spatial features in Figure 5a–g, but they suffer from spectral distortions, as highlighted in the small window, where the trees around buildings in Figure 5 and the bare soil in Figure 6 are a little darker than that of ground truth. In contrast, less spectral distortions are appeared in the results of the MRA-based methods, however, they show poor spatial rendering as they present a little blurring in Figure 6h–n, especially for the results of AWLP and ATWT-M3. Compared with the CS-based and MRA-based methods, the proposed PWNet can achieve more similar results to the ground truth. From the enlarged area in the upper left corner of the Figure 7o, we can see that PWNet has the best performance in both improving the spatial details and keeping spectral fidelity of the roads and trees. In summary of the visual analysis, the proposed PWNet method can debias the spectral and spatial distortions in the CS-based and the MRA-based methods, and can effectively combine the advantages of these two types of methods, thus shows better visual performances.

Besides visual inspection, we apply numeric metrics to assess the quality of pansharpened images. Tables 5–7 report the comparison results of the CS-based methods, MRA-based methods, and the proposed PWNet method on the three data sets. As we can find from these tables, for the WordView-2 and GeoEye-1 data sets, the BDSD method shows the best performances among the fourteen traditional methods, the AWLP achieves the best performance among the CS-based and MRA-based methods for the QuickBird data set. None of the CS-based and MRA-based methods systematically obtain the best performances for all the three data sets. The proposed PWNet yields results with the best spatial and spectral accuracy over the CS-based and MRA-based methods on all the three data sets. This proves once again that the proposed method can combine the advantages of the two types of methods to produce an optimal result.

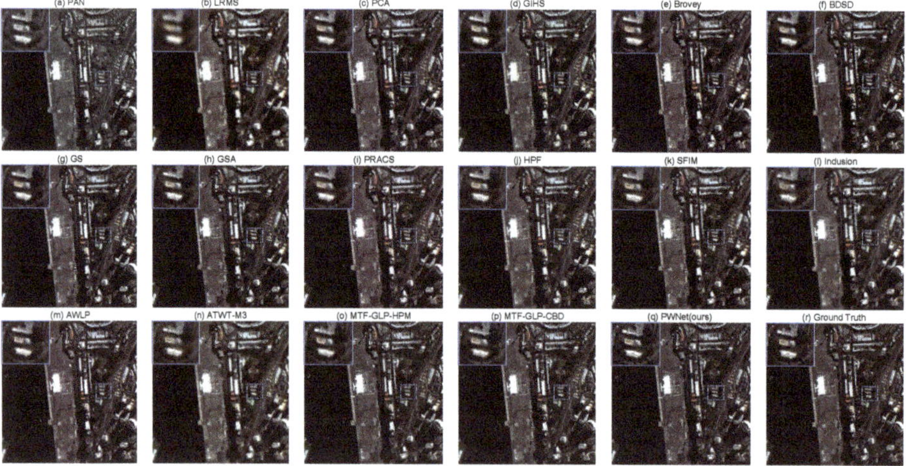

Figure 5. Visual comparison of the CS-based and MRA-based methods and the proposed PWNet method on the WorldView-2 images, (**a**) PAN; (**b**) LRMS; (**c**) PCA; (**d**) GIHS; (**e**) Brovey; (**f**) BDSD; (**g**) GS; (**h**) GSA; (**i**) PRACS; (**j**) HPF; (**k**) SFIM; (**l**) Indusion; (**m**) AWLP; (**n**) ATWT-M3; (**o**) MTF-GLP-HPM; (**p**) MTF-GLP-CBD; (**q**) PWNet (ours); (**r**) Ground Truth.

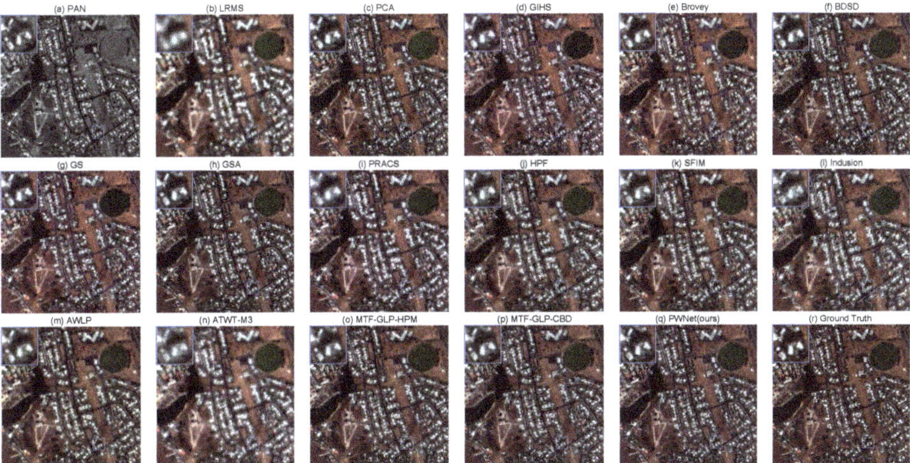

Figure 6. Visual comparison of the CS-based and MRA-based methods and the proposed PWNet method on the GeoEye-1 images, (**a**) PAN; (**b**) LRMS; (**c**) PCA; (**d**) GIHS; (**e**) Brovey; (**f**) BDSD; (**g**) GS; (**h**) GSA; (**i**) PRACS; (**j**) HPF; (**k**) SFIM; (**l**) Indusion; (**m**) AWLP; (**n**) ATWT-M3; (**o**) MTF-GLP-HPM; (**p**) MTF-GLP-CBD; (**q**) PWNet (ours); (**r**) Ground Truth.

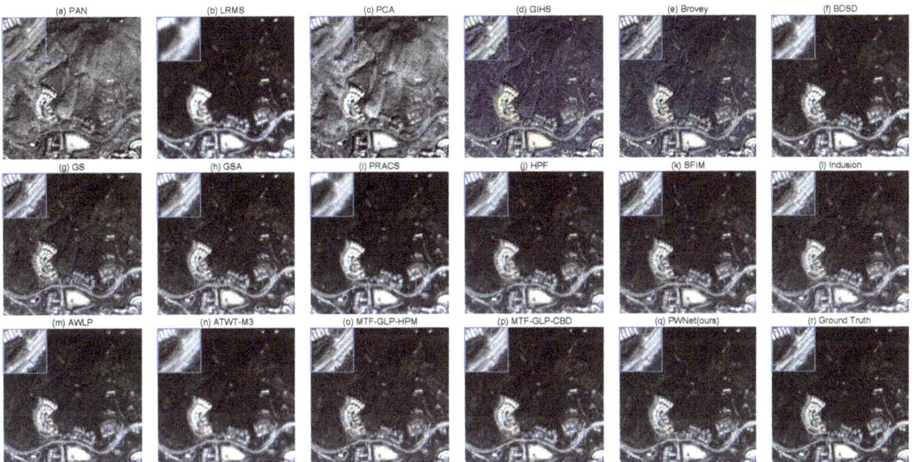

Figure 7. Visual comparison of the CS-based and MRA-based methods and the proposed PWNet method on the QuickBird images, (**a**) PAN; (**b**) LRMS; (**c**) PCA; (**d**) GIHS; (**e**) Brovey; (**f**) BDSD; (**g**) GS; (**h**) GSA; (**i**) PRACS; (**j**) HPF; (**k**) SFIM; (**l**) Indusion; (**m**) AWLP; (**n**) ATWT-M3; (**o**) MTF-GLP-HPM; (**p**) MTF-GLP-CBD; (**q**) PWNet (ours); (**r**) Ground Truth.

Table 5. Quantitative comparison of the CS-based and MRA-based methods and the proposed PWNet method on the WorldView-2 images. The best and second best results are highlighted in bold.

	Method	Q_avg	SAM	ERGAS	SCC	Q4
CS	PCA	0.8885	5.3974	5.8650	0.9419	0.8705
	GIHS	0.8858	5.6820	5.9937	0.9289	0.8777
	Brovey	0.8852	5.3037	5.9038	0.9357	0.8748
	BDSD	**0.9709**	4.4274	**3.2701**	**0.9641**	**0.9628**
	GS	0.8924	5.2195	5.7536	0.9428	0.8817
	GSA	0.9675	4.0382	3.6693	0.9542	0.9581
	PRACS	0.9224	4.5298	5.2480	0.9248	0.9143
MRA	HPF	0.9454	4.1524	4.4125	0.9521	0.9372
	SFIM	0.9498	4.2046	4.2015	0.9578	0.9426
	Indusion	0.8651	5.3162	6.4696	0.9025	0.8502
	AWLP	0.9617	**3.7299**	3.6508	0.9593	0.9529
	ATWT-M3	0.8723	6.1898	6.5592	0.9067	0.8696
	MTF-GLP-HPM	0.9676	3.9731	3.5227	0.9607	0.9589
	MTF-GLP-CBD	0.9620	4.0397	3.4418	0.9575	0.9620
	PWNet	**0.9836**	**3.5489**	**2.4584**	**0.9801**	**0.9723**

Table 6. Quantitative comparison of the CS-based and MRA-based methods and the proposed PWNet method on the GeoEye-1 images. The best and second best results are highlighted in bold.

	Method	Q_avg	SAM	ERGAS	SCC	Q4
CS	PCA	0.8110	6.3414	4.8720	0.8544	0.8175
	GIHS	0.8125	6.3189	4.8534	0.8539	0.8197
	Brovey	0.8063	6.5103	4.9593	0.8496	0.8120
	BDSD	**0.9211**	**5.3528**	**3.8133**	**0.8721**	**0.9309**
	GS	0.8119	6.3199	4.8650	0.8544	0.8188
	GSA	0.8959	6.3157	4.0425	0.8598	0.9079
	PRACS	0.8274	6.1486	4.7254	0.8483	0.8342
MRA	HPF	0.8562	6.3038	4.4650	0.8596	0.8638
	SFIM	0.8596	6.3688	4.4152	0.8599	0.8655
	Indusion	0.7518	7.1286	5.8135	0.7849	0.7579
	AWLP	0.8723	6.6096	4.3918	0.8548	0.8835
	ATWT-M3	0.7193	7.9694	5.7994	0.8118	0.7218
	MTF-GLP-HPM	0.9016	6.3340	3.9931	0.8639	0.9108
	MTF-GLP-CBD	0.9198	6.3207	3.9122	0.8638	0.9198
	PWNet	**0.9543**	**5.3245**	**2.8405**	**0.9380**	**0.9593**

To be interpretable, we also visualize the some weight maps of the selected traditional methods used in the PWNet, as shown in Figures 8–10. It can be seen from these figures that, the edges of the road and the buildings are extracted by the weight maps of both the CS-based and MRA-based methods. Typically, we can see that, for the WorldView-2 and GeoEye-1 data sets, the BDSD method plays an important role as its weight map is clearer than any others, as showin in Figures 8b and 9b, while the AWLP and MTF-GLP-CBD methods show a little greater contribution to the averaged results of the PWNet for the QuickBird data set, as can be see from Figure 10d,e. As for the PCA method, the weight maps are all black, which means that PCA method almost makes no contribution to the final result on all the three tested data sets. This conclusion is consistent with the previous visual inspection in Figures 5–7 and quantitative results reported in Tables 5–7. This proves the adaptive characteristic of our PWNet as it considers different performance of the selected CS-based and MRA-based methods.

From these experimental results, we can conclude that the proposed PWNet are adaptive and robust to different data sets.

Table 7. Quantitative comparison of the the CS-based and MRA-based methods and the proposed PWNet method on the QuickBird images. The best and second best results are highlighted in bold.

	Method	Q_avg	SAM	ERGAS	SCC	Q4
CS	PCA	0.5754	7.4697	4.5931	0.7911	0.7165
	GIHS	0.7965	4.2120	2.9488	0.9067	0.7906
	Brovey	0.8169	3.9059	2.7937	0.9187	0.8044
	BDSD	0.8977	4.2954	2.6979	0.9398	0.8906
	GS	0.7856	4.7382	3.1700	0.8874	0.7838
	GSA	0.8860	4.0225	2.5359	0.9252	0.8814
	PRACS	0.8510	3.8935	2.6258	0.9076	0.8319
MRA	HPF	0.8856	3.8858	2.4156	0.9278	0.8778
	SFIM	0.8902	3.8065	2.3617	0.9316	0.8818
	Indusion	0.7131	4.3943	3.5650	0.8664	0.7004
	AWLP	**0.9068**	**3.5775**	**2.2182**	**0.9425**	**0.8982**
	ATWT-M3	0.8059	4.5073	2.9060	0.8982	0.7912
	MTF-GLP-HPM	0.9056	3.7543	2.3366	0.9378	0.8985
	MTF-GLP-CBD	0.8877	4.0365	2.5360	0.9261	0.8877
	PWNet	**0.9196**	**3.5275**	**2.1635**	**0.9431**	**0.9109**

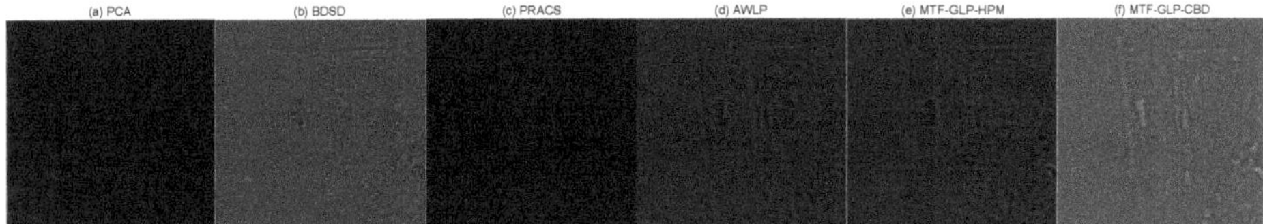

Figure 8. Visualization of weight maps on the WorldView-2 images, (**a**) PCA; (**b**) BDSD; (**c**) PRACS; (**d**) AWLP; (**e**) MTF-GLP-HPM; (**f**) MTF-GLP-CBD.

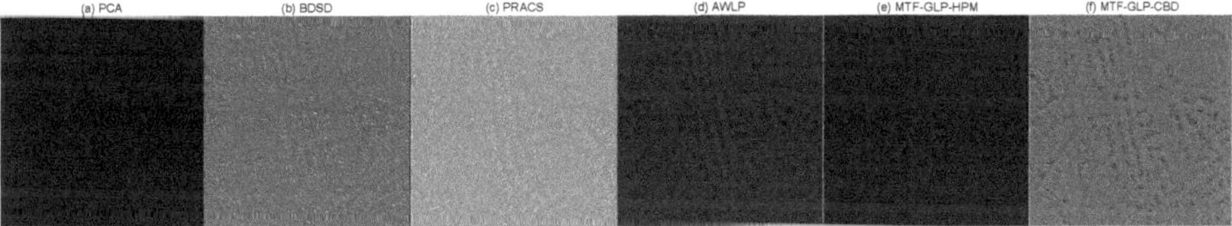

Figure 9. Visualization of weight maps on the GeoEye-1 images, (**a**) PCA; (**b**) BDSD; (**c**) PRACS; (**d**) AWLP; (**e**) MTF-GLP-HPM; (**f**) MTF-GLP-CBD.

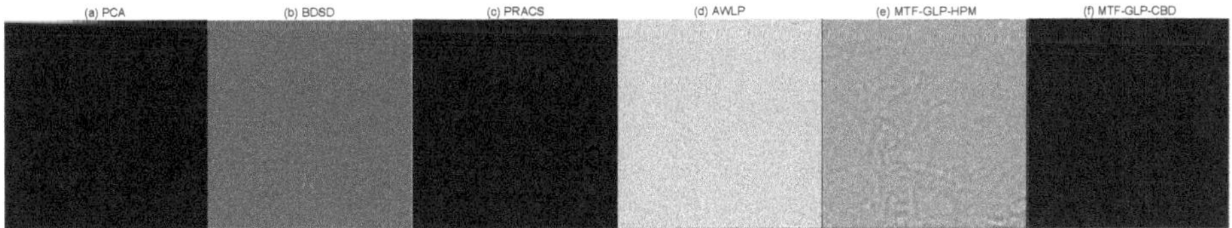

Figure 10. Visualization of weight maps on the QuickBird images, (**a**) PCA; (**b**) BDSD; (**c**) PRACS; (**d**) AWLP; (**e**) MTF-GLP-HPM; (**f**) MTF-GLP-CBD.

4.6. Comparison with the CNN-Based Methods

Currently, the proposed PWNet has shown its priority over the selected traditional CS-based and MRA-based methods. In this subsection, we are going to compare it with the CNN-based methods to verify its effectiveness. The other four *stat-of-the-art* (SOTA) methods including *pansharpening by convolutional neural networks* (PNN) [32], *deep residual pan-sharpening neural network* (DRPNN) [42], *multiscale and multidepth convolutional neural network* (MSDCNN) [43], are used as alternative methods for comparison. All the compared methods follow the experimental setting of their original papers. Note that the source codes of PNN is provided by the original authors and the codes of DRPNN, MSDCNN are available at https://github.com/Decri.

Figures 11–13 show some example regions selected from the pansharpened images on the three test data sets. In Figure 11, by magnifying the selected area in the image three times, it can be obviously seen that the other CNN-based methods have a little blurring to the ground truth, while the edges produced by our proposed PWNet method are more clear and natural as shown in zoomed areas. Although MSDCNN, DRPNN and PNN can produce better results with less spatial distortions, they sometimes suffer from a little spectral distortions, as shown in Figure 12b–d, where the bare soil is a little darker than the reference. This can also be seen in Figure 13b where the buildings in this scene is dark yellow while they are white in the ground truth as shown in Figure 13f. Compared with other CNN-based methods, the proposed PWNet shows a good balance between the injected spatial details and the maintain of original spectral information, this is clearly visible on the vegetable areas and textures (e.g., edges of the roof and road), as shown in Figures 11e–13e.

In addition, Table 8 shows the quantitative results for the three tested data sets obtained by the compared CNN-based methods and our proposed PWNet. It should be pointed out that, for each test experiment, we would choose one test sample randomly from the test data set rather than a cherry-picked sample, thus the results listed in Table 8 and Tables 5–7 are based on different PAN and MS image pairs and have different quantitative results. For better comparison, the best results among the four methods are highlighted in boldface. According to this table, one can see that performances of the proposed PWNet is better than the other three CNN-based methods in terms of the five indexes.

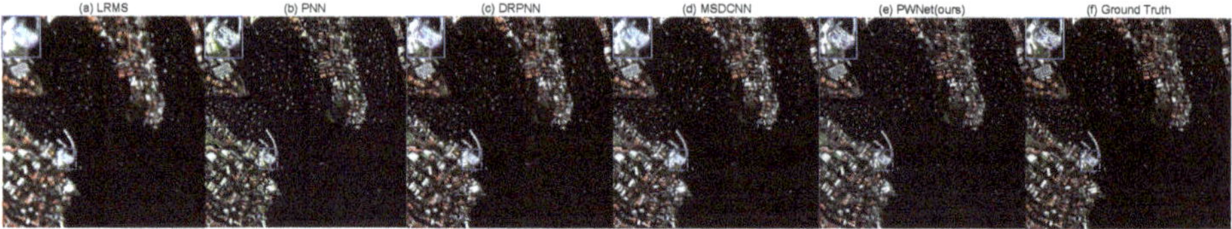

Figure 11. Visual comparison of the CNN-based methods on the WorldVie-2 data set, (**a**) LRMS; (**b**) PNN; (**c**) DRPNN; (**d**) MSDCNN; (**e**) PWNet (ours); (**f**) Ground Truth.

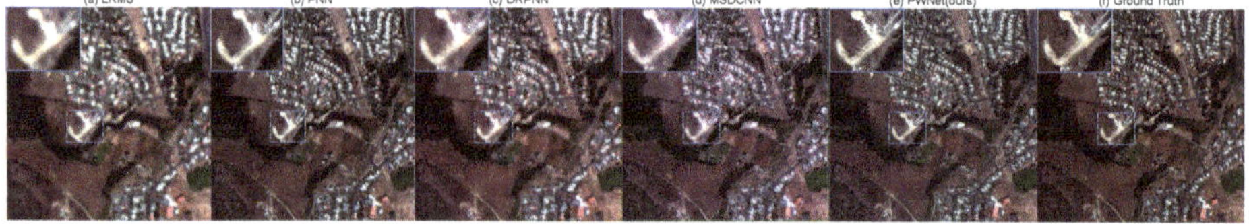

Figure 12. Visual comparison of the CNN-based methods on the GeoEye-1 data set, (**a**) LRMS; (**b**) PNN; (**c**) DRPNN; (**d**) MSDCNN; (**e**) PWNet (ours); (**f**) Ground Truth.

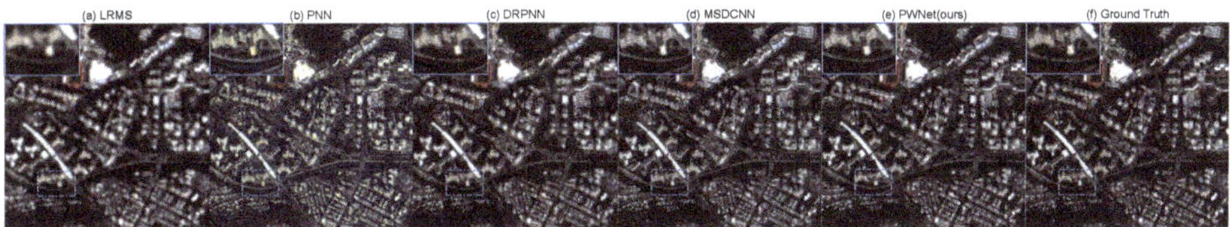

Figure 13. Visual comparison of the CNN-based methods on the QuickBird data set, (**a**) LRMS; (**b**) PNN; (**c**) DRPNN; (**d**) MSDCNN; (**e**) PWNet(ours); (**f**) Ground Truth.

Table 8. Quantitative comparison of the CNN-based methods on three test data sets. The best results are highlighted in bold.

	Method	Q_avg	SAM	ERGAS	SCC	Q4
WorldView-2	PNN	0.9479	5.2180	5.1040	0.8975	0.9256
	DRPNN	0.9292	4.4765	5.3072	0.8887	0.9030
	MSDCNN	0.9443	4.4135	6.8478	0.9407	0.8512
	PWNet	**0.9874**	**2.7582**	**2.3830**	**0.9813**	**0.9805**
GeoEye-1	PNN	0.8639	4.8190	3.5224	0.8888	0.8435
	DRPNN	0.7934	4.8685	3.7746	0.8931	0.7733
	MSDCNN	0.8462	4.9707	3.3115	0.9093	0.8518
	PWNet	**0.9207**	**4.2120**	**2.6868**	**0.9384**	**0.9117**
QuickBird	PNN	0.7047	3.3113	2.6980	0.9258	0.5950
	DRPNN	0.7648	2.4289	1.7544	0.9473	0.7750
	MSDCNN	0.7727	2.7034	1.8429	0.9450	0.7845
	PWNet	**0.8258**	**2.2163**	**1.5525**	**0.9644**	**0.8234**

4.7. Comparison at Full Resolution

The comparison results on three tested images at full resolution are shown in Figures 14–16 and Table 9. As we can see from the table that, for the WordView-2 and GeoEye-1 data sets, the DRPNN and PNN method respectively show the best performances, while the proposed PWNet holds the second best position for all the three data sets. On a whole, the CNN-based methods perform better than the traditional methods (i.e., the CS-based and MRA-based methods). By a visual inspection, the PNN, DRPNN, and MSDCNN methods tend to produce blurring results while the proposed PWNet is able to enhancing the spatial quality and shows clearly sharper fusion results, as shown in Figures 14–16r. As a summary, compared to the other methods at the full resolution, the proposed PWNet could consistently reconstruct sharper HRMS image with less spectral and spatial distortion.

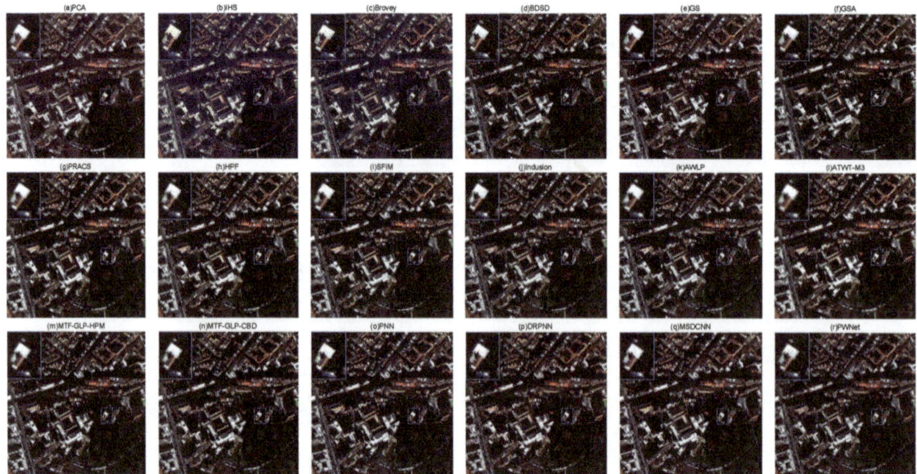

Figure 14. Visual comparison of different methods on the WorldVie-2 data set at full resolution, (**a**) PCA; (**b**) GIHS; (**c**) Brovey; (**d**) BDSD; (**e**) GS; (**f**) GSA; (**g**) PRACS; (**h**) HPF; (**i**) SFIM; (**j**) Indusion; (**k**) AWLP; (**l**) ATWT-M3; (**m**) MTF-GLP-HPM; (**n**) MTF-GLP-CBD; (**o**) PNN; (**p**) DRPNN; (**q**) MSDCNN; (**r**) PWNet (ours).

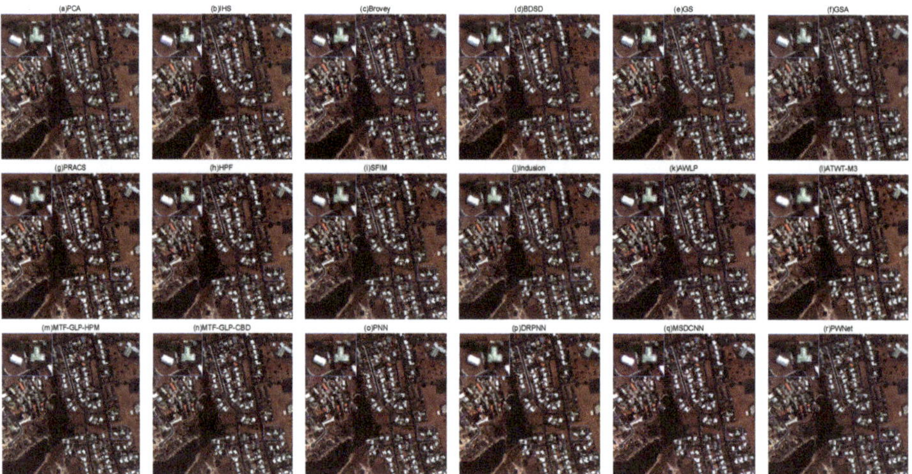

Figure 15. Visual comparison of different methods on the GeoEye-1 data set at full resolution, (**a**) PCA; (**b**) GIHS; (**c**) Brovey; (**d**) BDSD; (**e**) GS; (**f**) GSA; (**g**) PRACS; (**h**) HPF; (**i**) SFIM; (**j**) Indusion; (**k**) AWLP; (**l**) ATWT-M3; (**m**) MTF-GLP-HPM; (**n**) MTF-GLP-CBD; (**o**) PNN; (**p**) DRPNN; (**q**) MSDCNN; (**r**) PWNet (ours).

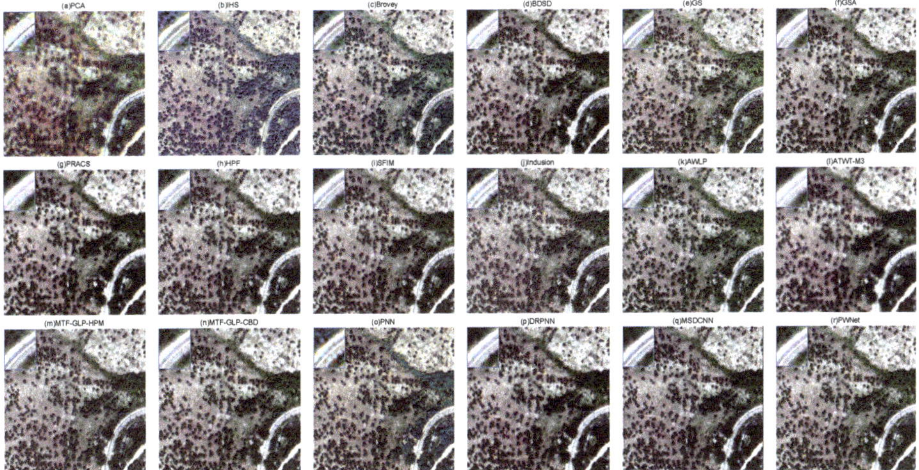

Figure 16. Visual comparison of different methods on the QuickBird data set at full resolution, (**a**) PCA; (**b**) GIHS; (**c**) Brovey; (**d**) BDSD; (**e**) GS; (**f**) GSA; (**g**) PRACS; (**h**) HPF; (**i**) SFIM; (**j**) Indusion; (**k**) AWLP; (**l**) ATWT-M3; (**m**) MTF-GLP-HPM; (**n**) MTF-GLP-CBD; (**o**) PNN; (**p**) DRPNN; (**q**) MSDCNN; (**r**) PWNet (ours).

Table 9. Performance comparison on three test data sets at full resolution. The best and second best results are highlighted in bold and underlined, respectively.

	GeoEye-1			WorldView-2			QuickBird		
Methed	D_λ	D_S	**QNR**	D_λ	D_S	**QNR**	D_λ	D_S	**QNR**
PCA	0.1020	0.1542	0.7595	0.0265	0.2143	0.7649	0.0527	0.1036	0.8491
GIHS	0.0347	0.1812	0.7903	0.0767	0.0845	0.8453	0.0819	0.0815	0.8432
Brovey	0.0337	0.1769	0.7954	0.0377	0.1008	0.8653	0.0545	0.0577	0.8910
BDSD	0.2103	0.3887	0.4827	0.1016	0.2569	0.6676	0.2748	0.3250	0.4895
GS	0.0379	0.1816	0.7873	**0.0167**	0.1049	0.8802	0.0443	0.0613	0.8971
GSA	0.2925	0.1226	0.6207	0.1079	0.1444	0.7633	0.0940	0.1126	0.8040
PRACS	0.0587	0.1533	0.7970	0.0231	0.0908	0.8882	0.0492	0.0845	0.8704
HPF	0.1500	0.2023	0.6780	0.0533	0.1028	0.8494	0.0603	0.0515	0.8913
SFIM	0.1469	0.1971	0.6849	0.0532	0.0974	0.8546	0.0587	0.0517	0.8926
Indusion	0.0867	0.1126	0.8105	0.0312	0.0544	**0.9161**	0.0691	**0.0444**	0.8895
AWLP	0.1621	0.1914	0.6776	0.0437	0.0908	0.8695	0.0575	0.0583	0.8875
ATWT-M3	0.1198	0.1782	0.7234	0.0617	0.0818	0.8615	0.0588	0.0806	0.8653
MTF-GLP-HPM	0.1810	0.1986	0.6563	0.0604	0.1096	0.8366	0.0827	0.0675	0.8554
MTF GLP CBD	0.2199	0.2009	0.6234	0.0752	0.1282	0.8062	0.0800	0.0673	0.8580
PNN	0.1042	**0.0404**	0.8596	0.0636	**0.0322**	0.9062	0.0306	0.0490	**0.9218**
DRPNN	**0.0268**	0.0764	**0.8988**	0.0359	0.0636	0.9027	**0.0199**	0.0954	0.8866
MSDCNN	0.0699	0.0795	0.8562	0.0282	0.0734	0.9004	0.0946	0.1068	0.8087
PWNet	0.0289	0.1127	0.8616	0.0172	0.0700	0.9139	0.0428	0.0559	0.9037

4.8. Running Time Analysis

In this subsection, we compare the running time of the proposed method with the others on a 64×64 LRMS and 256×256 PAN image pair. The experiments are performed by MATLAB R2016b on the same platform with Core i5-4210U/1.7 GHz/4G. The running times of different methods are listed in Table 10, in which the time is measured in second. From this table, it can be found that DRPNN is the most time-consuming method, because the number of hidden layers within DRPNN is more than the other CNN-based methods. In addition, the MSDCNN needs a little more time to obtain the fusion result than that of the proposed PWNet. In a word, the proposed PWNet method is more efficient than the CNN-based methods due to less hidden layers and that it only outputs weight maps rather than directly producing an estimated HRMS image.

Table 10. Running time comparison of different methods (in second).

Method	Time	Methed	Time
PCA	0.0900	Indusion	0.0600
GIHS	0.0100	AWLP	0.0400
Brovey	0.0010	ATWTM3	0.1100
BDSD	0.2700	MTF_GLP_HPM	0.0400
GS	0.0300	MTF_GLP_CBD	0.0500
GSA	0.1000	PNN	0.5900
PRACS	0.1000	DRPNN	6.5100
HPF	0.0100	MSDCNN	1.9900
SFIM	0.0100	PWNet	1.0200

5. Conclusions

In this paper, we presented a novel model average network for pansharpening, and is referred to as PWNet. The proposed PWNet attempts to integrate the complementary characteristics of the CS-based and MRA-based methods through an end-to-end trainable neural networks, and thus it is data-driven and able to adaptively weight the results of the classical methods depending on their performances. Experiments on several data sets collected by three kinds of satellites demonstrate that the pansharpened HRMS images by the proposed PWNet can not only enhance the spatial qualities but also can keep the spectral information of the original MS images. In addition, the proposed PWNet has some distribution structures. Thus, we will extend the proposed model to a distribution version by using the techniques of distributed processing [53] to further reduce the running time while maintaining the quality of the results.

Author Contributions: J.L. and C.Z. (Chunxia Zhang) structured and drafted the manuscript with assistance from C.Z. (Changsheng Zhou); Y.F. conceived the experiments and generated the graphics with assistance from C.Z. (Changsheng Zhou). All authors have read and agreed to the published version of the manuscript.

Funding: This research was funded by the National Key Research and Development Program of China under Grant 2018AAA0102201, and by the National Natural Science Foundation of China grant number 61877049.

Acknowledgments: The authors would like to thank Gemine Vivone for sharing the pansharpening Matlab toolbox [3], to Giuseppe Scarpa for providing their codes of PNN in [32], to Wei and Yuan in [42,43] for sharing their CNN-based codes for pansharpening.

Conflicts of Interest: The authors declare no conflict of interest.

References

1. Shaw, G.; Burke, H.K. Spectral imaging for remote sensing. *Linc. Lab. J.* **2003**, *14*, 3–28.
2. Meng, X.; Shen, H.; Li, H.; Zhang, L.; Fu, R. Review of the pansharpening methods for remote sensing images based on the idea of meta-analysis: Practical discussion and challenges. *Inf. Fusion* **2019**, *46*, 102–113. [CrossRef]
3. Vivone, G.; Alparone, L.; Chanussot, J.; Dalla Mura, M.; Garzelli, A.; Licciardi, G.A.; Restaino, R.; Wald, L. A critical comparison among pansharpenig algorithms. *IEEE Trans. Geosci. Remote Sens.* **2015**, *53*, 2565–2586. [CrossRef]
4. Ehlers, M.; Klonus, S.; Astrand, P.; Rosso, P. Multi-sensor image fusion for pansharpening in remote sensing. *Int. J. Image Data Fusion* **2010**, *1*, 25–45. [CrossRef]
5. Yue, L.; Shen, H.; Li, J.; Yuan, Q.; Zhang, H.; Zhang, L. Image super-resolution: The techniques, applications, and future. *Signal Process.* **2016**, *128*, 389–408. [CrossRef]
6. Souza, C.; Firestone, L.; Silva, M.; Roberts, D. Mapping forest degradation in the Eastern Amazon from SPOT 4 through spectral mixture models. *Remote Sens. Environ.* **2003**, *87*, 494–506. [CrossRef]

7. Mohammadzadeh, A.; Tavakoli, A.; Valadan Zoej, M.J. Synthesis of multispectral images to high spatial resolution: Road extraction based on fuzzy logic and mathematical morphology from pansharpened IKONOS images. *Photogramm. Rec.* **2006**, *21*, 44–60. [CrossRef]
8. Loncan, L.; De Almeida, L.B.; Bioucas-Dias, J.M.; Briottet, X.; Chanussot, J.; Dobigeon, N.; Fabre, S.; Liao, W.; Licciardi, G.A.; Simoes, M.; et al. Hyperspectral pansharpening: A review. *IEEE Geosci. Remote Sens. Mag.* **2015**, *3*, 27–46. [CrossRef]
9. Carper, W.; Lillesand, T.; Kiefer, R. Synthesis of multispectral images to high spatial resolution: The use of intensity-huesaturation transformations for merging SPOT panchromatic and multispectral image data. *Photogramm. Eng. Remote Sens.* **1990**, *56*, 459–467.
10. Tu T.; Su, S.C.; Shyu, H.C.; Huang, P.S. A new look at IHS-like image fusion methods. *Inf. Fusion* **2001**, *2*, 177–186. [CrossRef]
11. Chavez, P.S.; Kwarteng, A.W. Extracting spectral contrast in Landsat thematic mapper image data using selective principal component analysis. *Photogramm. Eng. Remote Sens.* **1989**, *55*, 339–348.
12. Gillespie A.; Kahle, A.B.; Walker, R.E. Color enhancement of highly correlated images-II. Channel ration and "Chromaticity" Transform techniques. *Remote Sens. Environ.* **1987**, *22*, 343–365. [CrossRef]
13. Laben, C.A.; Brower, B.V. Process for Enhancing the Spatial Resolution of Multispectral Imagery Using Pan-Sharpening. U.S. Patent 6011875, 4 January 2000.
14. Garzelli, A.; Nencini, F.; Capobianco, L. Optimal MMSE pan sharpening of very high resolution multispectral images. *IEEE Trans. Geosci. Remote Sens.* **2008**, *46*, 228–236. [CrossRef]
15. Aiazzi, B.; Baronti, S.; Selva, M. Improving component substitution pansharpening through multivariate regression of MS+Pan data. *IEEE Trans. Geosci. Remote Sens.* **2007**, *45*, 3230–3239. [CrossRef]
16. Vivone, G. Robust band-dependent spatial-detail approaches for panchromatic sharpening. *IEEE Trans. Geosci. Remote Sens.* **2019**, *57*, 6421–6433. [CrossRef]
17. Choi, J.; Yu, K.; Kim, Y. Synthesis of multispectral images to high spatial resolution: A new adaptive component-substitution based satellite image fusion by using partial replacement. *IEEE Trans. Geosci. Remote Sens.* **2011**, *49*, 295–309. [CrossRef]
18. Chavez, P. S., Jr.; Sides, S.C.; Anderson, A. Comparison of three different methods to merge multiresolution and multispectral data: Landsat TM and SPOT panchromatic. *Photogramm. Eng. Remote Sens.* **1991**, *57*, 295–303.
19. Liu, J. Smoothing filter-based intensity modulation: A spectral preserve image fusion technique for improving spatial details. *Int. J. Remote Sens.* **2000**, *21*, 3461–3472. [CrossRef]
20. Aiazzi, B.; Alparone, L.; Baronti, S.; Garzelli, A.; Selva, M. An MTF based spectral distortion minimizing model for pan-sharpening of very high resolution multispectral images of urban areas. In Proceedings of the 2nd GRSS/ ISPRS Joint Workshop Remote Sensing and Data Fusion URBAN Areas, Berlin, Germany, 22–23 May 2003; pp. 90–94.
21. Aiazzi, B.; Alparone, L.; Baronti, S.; Garzelli, A.; Selva, M. MTF tailored multiscale fusion of high-resolution MS and Pan imagery. *Photogramm. Eng. Remote Sens.* **2006**, *72*, 591–596. [CrossRef]
22. Vivone, G.; Restaino, R.; Dalla Mura, M.; Licciardi, G.; Chanussot, J. Contrast and error-based fusion schemes for multispectral image pansharpening. *IEEE Trans. Geosci. Remote Sens.* **2014**, *11*, 930–934. [CrossRef]
23. Khan, M.M.; Chanussot, J.; Condat, L.; Montavert, A. Indusion: Fusion of multispectral and panchromatic images using the induction scaling technique. *IEEE Trans. Geosci. Remote Sens.* **2008**, *5*, 98–102. [CrossRef]
24. Otazu, X.; González-Audìcana, M.; Fors, O.; Nùñez, J. Introduction of sensor spectral response into image fusion methods. Application to wavelet-based methods. *IEEE Trans. Geosci. Remote Sens.* **2005**, *43*, 2376–2385. [CrossRef]
25. Shah, V.P.; Younan, N.H.; King, R.L. An efficient pan-sharpening method via a combined adaptive PCA approach and contourlets. *IEEE Trans. Geosci. Remote Sens.* **2008**, *46*, 1323–1335. [CrossRef]
26. Liao, W.; Huang, X.; Van Coillie, F.; Gautama, S.; Pižurica, A.; Philips, W.; Liu, H.; Zhu, T.; Shimoni, M.; Moser, G.; et al. Processing of Multiresolution Thermal Hyperspectral and Digital Color Data: Outcome of the 2014 IEEE GRSS Data Fusion Contest. *IEEE J. Sel. Top. Appl. Earth Observ. Remote Sens.* **2015**, *8*, 2984–2996. [CrossRef]
27. Ballester, C.; Vselles, V.; Igual, L.; Verdera, J.; Rouge, B. A variational model for P+XS image fusion. *Int. J. Comput. Vis.* **2006**, *69*, 43–58.

28. Moller, M.; Wittman, T.; Bertozzi, A.; Berger, M. A variational approach for sharpening high dimensional images. *SIAM J. Imaging Sci.* **2012**, *5*, 150–178. [CrossRef]
29. Zhu, X.; Bamler, R. A sparse image fusion algorithm with application to pansharpening. *IEEE Trans. Geosci. Remote Sens.* **2013**, *51*, 2827–2836. [CrossRef]
30. Deng, L.; Vivone, G.; Guo, W.; Mura, M.; Chanussot, J. A variational pansharpening approach based on reproducible kernel hilbert space and heaviside function. *IEEE Trans. Image Process.* **2018**, *27*, 4330–4344. [CrossRef]
31. Palsson, F.; Ulfarsson, M.; Sveinsson, J. Model-based reduced-rank pansharpening. *IEEE Geosci. Remote Sens. Lett.* **2020**, *17*, 656–660. [CrossRef]
32. Masi, G.; Cozzolino, D.; Verdoliva, L.; Scarpa, G. Pansharpening by convolutional neural networks. *Remote Sens.* **2016**, *8*, 594. [CrossRef]
33. Yang, J.; Fu, X.; Huang, Y.; Ding, X.; Paisley, J. PanNet: A deep network architecture for pan-sharpening. In Proceedings of the International Conference on Computer Vision, Venice, Italy, 22–29 October 2017; pp. 5449–5457.
34. Liu, X.; Wang, Y.; Liu, Q. Synthesis of multispectral images to high spatial resolution: Remote sensing image fusion based on two-stream fusion network. *IEEE Trans. Geosci. Remote Sens.* **2018**, *46*, 428–439.
35. Zheng, Y.; Li, J.; Li, Y.; Guo, J.; Wu, X.; Chanussot, J. Hyperspectral Pansharpening Using Deep Prior and Dual Attention Residual Network. *IEEE Trans. Geosci. Remote Sens.* **2020**, *10*, 1–18. [CrossRef]
36. Han, S.; Pool, J.; Tran, J.; Dally, W. Learning both weights and connections for efficient neural network. In Proceedings of the Advances in Neural Information Processing Systems 2015, Montreal, QC, Canada, 7–12 December 2015; pp. 1135–1143.
37. Ghamisi, P.; Rasti, B.; Yokoya, N.; Wang, Q.; Hofle, B.; Bruzzone, L.; Bovolo, F.; Chi, M.; Anders, K.; Gloaguen, R.; et al. Multisource and multitemporal data fusion in remote sensing: A comprehensive review of the state of the art. *IEEE Geosci. Remote Sens. Mag.* **2019**, *7*, 6–39. [CrossRef]
38. Alparone, L.; Wald, L.; Chanussot, J.; Thomas, C.; Gamba, P.; Bruce, L. Comparison of pansharpening algorithms: Outcome of the 2006 GRS-S DataFusion Contest. *IEEE Trans. Geosci. Remote Sens.* **2007**, *45*, 3012–3021. [CrossRef]
39. Ranchin, T.; Wald, L. Fusion of high spatial and spectral resolution images: The ARSIS concept and its implementation. *Photogramm. Eng. Remote Sens.* **2000**, *66*, 49–61.
40. Restaino, R.; Vivone, G.; Dalla Mura, M.; Chanussot, J. Fusion of multispectral and panchromatic images based on morphological operators. *IEEE Trans. Image Process.* **2016**, *25*, 2882–2895. [CrossRef]
41. Zhong, J.; Yang, B.; Huang, G. Remote sensing image fusion with convolutional neural network. *Sens. Imaging* **2016**, *17*, 10. [CrossRef]
42. Wei, Y.; Yuan, Q.; Shen, H.; Zhang, L. Boosting the Accuracy of Multispectral Image Pansharpening by Learning a Deep Residual Network. *IEEE Geosci. Remote Sens. Lett.* **2017**, *14*, 1795–1799. [CrossRef]
43. Yuan, Q.; Wei, Y.; Meng, X.; Shen, H.; Zhang, L. A Multiscale and Multidepth Convolutional Neural Network for Remote Sensing Imagery Pan-Sharpening. *IEEE J. Sel. Top. Appl. Earth Obs. Remote Sens.* **2018**, 3, 978–989. [CrossRef]
44. He, K.; Zhang, X.; Ren, S.; Sun, J. Deep residual learning for image recognition. In Proceedings of the IEEE Conference on Computer Vision and Pattern Recognition 2016, Las Vegas, NV, USA, 27–30 June 2016; pp. 770–778.
45. Nair, N.; Hinton, G.E. Rectified linear units improve restricted boltzmann machines. In Proceedings of the 27th International Conference on Machine Learning, Haifa, Israel, 21–24 June 2010; pp. 807–814.
46. Wald, L.; Ranchin, T.; Mangolini, M. Data fusion: Definitions and architectures: Fusion of satellite images of different spatial resolutions: Assessing the quality of resulting images. *Photogramm. Eng. Remote Sens.* **1997**, *63*, 691–699.
47. Kingma, D.P.; Ba, J. Adam: A Method for Stochastic Optimization. *arXiv* **2014**, arXiv:1412.6980.
48. Wang, Z.; Bovik, A.C. A universal image quality index. *IEEE Signal Process. Lett.* **2002**, *9*, 81–84. [CrossRef]
49. Yuhas, R.H.; Goetz, A.F.; Boardman, J.W. Discrimination among semi-arid landscape endmembers using the Spectral Angle Mapper (sam) algorithm. In Proceedings of the Summaries of the Third Annual JPL Airborne Geoscience Workshop, Pasadena, CA, USA, 1–5 June 1992; 147–149.
50. Zhou, J.; Civco, D.; Silander, J. Chanussot, A wavelet transform method to merge landsat tm and spot panchromatic data. *Int. J. Remote Sens.* **1998**, *19*, 743–757. [CrossRef]

51. Alparone, L.; Aiazzi, B.; Baronti, S.; Garzelli, A.; Nencini, F.; Selva, M. Multispectral and panchromatic data fusion assessment without reference. *Photogramm. Eng. Remote Sens.* **2008**, *74*, 193–200. [CrossRef]
52. Qu, Y.; Baghbaderani, R.K.; Qi, H.; Kwan, C. Unsupervised Pansharpening Based on Self-Attention Mechanism. *arXiv* **2020**, arXiv:2006.09303v1.
53. Silva, G.; Medeiros, R.; Jaimes, B.; Takahashi, C.C.; Vieira, D.; Braga, A. CUDA-Based Parallelization of Power Iteration Clustering for Large Datasets. *IEEE Access* **2017**, *5*, 27263–27271. [CrossRef]

MDPI
St. Alban-Anlage 66
4052 Basel
Switzerland
Tel. +41 61 683 77 34
Fax +41 61 302 89 18
www.mdpi.com

Remote Sensing Editorial Office
E-mail: remotesensing@mdpi.com
www.mdpi.com/journal/remotesensing

www.ingramcontent.com/pod-product-compliance
Lightning Source LLC
LaVergne TN
LVHW070136170726
843515LV00007B/1807

* 9 7 8 3 0 3 6 5 6 7 9 8 3 *